MATHEMATICAL AND PHYSICAL MODELLING OF MICROWAVE SCATTERING AND POLARIMETRIC REMOTE SENSING

Monitoring the Earth's Environment Using Polarimetric Radar: Formulation and Potential Applications

by

A.I. KOZLOV

Moscow State Technical University of Civil Aviation, Russia

L.P. LIGTHART

Delft University of Technology,
International Research Centre for Telecommunications-Transmission and Radar, The Netherlands

and

A.I. LOGVIN

Moscow State Technical University of Civil Aviation, Russia

Managing and Technical Editors:
I.M. BESIERIS
The Bradley Department of Electrical and Computer Engineering.
Virginia Polytechnic Institute and State University, Blacksburg, Virginia, U.S.A.
L.P. LIGTHART
Delft University of Technology,
International Research Centre for Telecommunications-Transmission and Radar, The Netherlands
E.G. PUSONE
Delft University of Technology,
International Research Centre for Telecommunications-Transmission and Radar, The Netherlands

KLUWER ACADEMIC PUBLISHERS
DORDRECHT / BOSTON / LONDON

A C.I.P. Catalogue record for this book is available from the Library of Congress.

ISBN 1-4020-0122-3

Published by Kluwer Academic Publishers,
P.O. Box 17, 3300 AA Dordrecht, The Netherlands.

Sold and distributed in North, Central and South America
by Kluwer Academic Publishers,
101 Philip Drive, Norwell, MA 02061, U.S.A.

In all other countries, sold and distributed
by Kluwer Academic Publishers,
P.O. Box 322, 3300 AH Dordrecht, The Netherlands.

Cover illustration: KLL-Sphere

Printed on acid-free paper

Printed in the Netherlands.

FOR RADAR KNOWLEDGE FUSION

TABLE OF CONTENTS

Chapter 3: Physical and Mathematical Modelling

Chapter 4: Summary of Available Scattering Methods

PART III – DIAGNOSTICS OF THE EARTH'S ENVIRONMENT USING POLARIMETRIC RADAR MONITORING: FORMULATION AND POTENTIAL APPLICATIONS

Chapter 5: Basic Mathematical Modelling for Random Environments

Chapter 6: Review of Vegetation Models

Chapter 7: Electrodynamic and Physical Characteristics of the Earth's Surfaces

Chapter 8: Reflection of Electromagnetic Waves from Non-Uniform Layered Structures

Chapter 12: Implementing Solutions to Inverse Scattering Problems: Signal Processing and Applications

PART IV: CONCLUDING REMARKS

Chapter 13: Review of Potential Applications of Radar Polarimetry

Chapter 14: Historical Development of Radar Polarimetry in Russia

PREFACE

An agreement involving a project entitled "Theoretical Modelling of Microwave Scattering" was signed by Prof. Dr. V.G. Vorobiev, Rector of Moscow State Technical University of Civil Aviation (MSTUCA), and Prof. Dr. L.P. Ligthart, Director of the International Research Centre for Telecommunications-transmission and Radar (IRCTR), Delft University of Technology (DUT), on September 21, 1994. Within the framework of that agreement scientists and experts of MSTUCA have conducted scientific research in collaboration with IRCTR.

The agreement came after a long period of essentially no formal exchange of information between Russia (former USSR) and the Western World. And yet significant technical developments were made by Russians during that period, as evidenced by their known success in space research and satellites. This monograph, based on developments in theoretical modelling of microwave scattering and applications to radar polarimetry during the past two decades, is intended to serve two goals: first, to establish a bridge for exchanging and documenting research experiences between Russia and the Western World; second, to provide a useful reference for scientists or engineers interested in radar polarimetry in the presence of diverse scattering environments.

Delft, October 2001

ACKNOWLEDGEMENTS

Radar remote sensing focuses on the analysis of electromagnetic scattered field vectors as can be measured with full-polarimetric radar. In combination with new polarization models, it leads to an improved classification and/or identification of radar objects.

Applying high-resolution polarization radar necessitates a full understanding of the mechanism of wave scattering from objects. In this respect, it can be stated that results with ground-based radar sensors are not yet fully understood and fundamental research is essential to prevent us from 'data graveyards'. The large-scale measurement campaigns provide large databanks, but it does not mean that the information contained in the data is useful.

External stimulation of fundamental research on the characteristics of electromagnetic scattering from water, land and atmosphere in the microwave domain has not yet been given sufficient attention. Funding this research is essential to progress in the understanding of the interaction between waves and matter. To give priority to this background research within the outlines of national programs emphasizing utility and commercialization is not the right approach.

Thanks to the support of the Netherlands Ministry of Education, Culture and Science, the Netherlands Science Foundation (NWO) and the Netherlands Technology Foundation (STW), an international research program on earth observation using polarization radar between the International Research Centre for Telecommunications-transmission and Radar (IRCTR) of Delft University of Technology and the Moscow State Technical University of Civil Aviation (MSTUCA) was initiated in 1994. The 2-year program was entitled "Theoretical Modelling of Microwave Scattering" with the following themes:

- Summarizing microwave remote sensing fundamentals, i.e., microwave scattering from objects, surfaces and volume distributed targets;
- Differentiation between surface and volume scattering on the basis of polarimetric analysis of mono-static reflections;
- Polarimetric scattering models consisting of two layers and describing different kinds of surfaces, including interfaces between the atmosphere and ground or sea;
- Polarimetric scattering models consisting of three layers and describing interfaces between atmosphere, vegetation and ground;
- Spatial and temporal statistics of polarimetric scattering from rough surfaces.

The basic idea for setting up this program was to summarize existing knowledge on radar remote sensing modelling in countries of the Former Soviet Union (FSU), which was not fully available to western research institutes and organizations until the 90's. This knowledge is of interest because polarization radar in FSU countries has a longer history than in western countries. Thanks to the conversion programs to apply radar techniques for civil applications, these important initiatives have been undertaken by the Netherlands and resulted into detailed knowledge and insights in IRCTR.

Applications, in which FSU knowledge is integrated in IRCTR programs, are being described in a new 4-year IRCTR-MSTUCA program entitled "Modelling and Verification of Earth-Based Radar Objects", which started in 1997.

The evaluation of the first program took place in 1997 and then the idea was born of writing this monograph based on the various reports delivered as part of the program agreements. The different chapters of the monograph were discussed at various meetings held in 1997 and 1998. In 1998, it became clear that summarizing the radar remote sensing modelling from only the FSU perspective was not sufficient. For that reason the authors looked into the possibility to include the general accepted theoretical modelling concepts on radar remote sensing developed in the Western World. Using additional funding from the Netherlands Ministry of Education, Culture and Science, we were able to appoint Dr. E. Pusone as an IRCTR scientist to prepare a draft of Part II of the monograph and asked Prof. Dr. I.M. Besieris from Virginia Polytechnic Institute & State University as an IRCTR guest scientist to review critically the manuscript.

The authors would like to particularly thank Prof. Dr. V.L. Kouznetsov for his important contributions to the research contained in Part III of the monograph, E.M.H.M. Ligthart-Versaevel for her help in editing the original Russian reports which form the basis of this monograph, Prof. Dr. I.M. Besieris and Dr. E. Pusone for their help in editing and reviewing the manuscript, Prof. A. Yarovoy of IRCTR for his review in PART IV, and Ms. G.T. Liem and Ms. W. Murtinu for their typing of the text and figures of the entire manuscript. Also, they would like to acknowledge contributions from the following scientists who participated in the project: Prof. Dr. A.V. Prochorov, Ass. Prof. Dr. V.L. Mendelson, Ass. Prof. Dr. A.J. Korabliev, Ass. Prof. G.N. Andreev, Ass. Prof. Dr. A.V. Starugh, Dr. L.A. Kozlova, Dr. O.A. Logvin and A. J. Ovsinsky.

The authors are also grateful to the following publishing companies for giving them the permission to reproduce certain figures in this monograph: NATO RTA-AGARD, Artech House, IEEE Press, IEEE Transactions, IOP Publishing Ltd, Mc Graw Hill Co,

IEE London, AGU-Radio Science, NATO Saclantcen, NATO SHAPE Technical Centre and to Kluwer Publishing Company for their unfailing cooperation.

PART I

INTRODUCTION

A. Scope of the subject

Monitoring the environment is most urgent and is receiving more attention from scientists worldwide. Various means can be used; a main role, however, is played by radar. Efficient use of radar for receiving authentic volumetric information of investigated objects is one of the major issues for radar experts, previously involved in the field of radio location. The scope of the subject in this monograph is to provide a review of available polarimetric radar techniques for solving practical inverse problems in remote sensing of various types of scatterers on the earth's surface (vegetation, ocean, terrain, etc.)

B. Description of the research program

The research program includes the following topics:

(a) Summarizing microwave remote sensing fundamentals, i.e., microwave scattering from objects, surfaces, and volume-distributed targets.
(b) Differentiation between surface and volume scattering on the basis of polarimetric analysis of mono-static reflections.
(c) Polarimetric scattering from models consisting of two layers and description of different kinds of surfaces, including interfaces between the atmosphere and ground, atmosphere and sea, etc.
(d) Polarimetric scattering from models involving three layers and description of interfaces between atmosphere, vegetation and ground.
(e) Spatial and temporal statistics of polarimetric scattering from rough surfaces.

The major research aspects are listed as follows:

1. All natural formations are represented from the point of view of electrodynamics. Their electrical characteristics depend on various physical parameters such as salinity, humidity, temperature, pressure, density, etc. These physical quantities influence the main electrodynamic characteristics, e.g., the complex dielectric permittivity, with a real part and an imaginary part, the latter characterizing the loss tangent. The solution of Maxwell's equations depends on the way the complex permittivity changes in the structures and their boundaries. The variety of

natural formations does not allow the construction of a uniform generalized electromagnetic scattering model.

2. A sequential approach is selected whereby assumptions of constant permittivities and flat boundaries are not needed. The problems are solved using deterministic and/or stochastic approximations.

3. The following models are considered:

   Isotropic model:
   - Layered models with a constant permittivity in each horizontal layer;
   - Layered models with an exponential change in the permittivity with depth;
   - Layered models with a polynomial change in the permittivity with depth;
   - Models with random changes in the permittivity with depth.

   Anisotropic models:
   - Layered models with geometrical inhomogeneities;
   - Models with a rough half-space;
   - Layered models with one and two rough boundaries;
   - Models of structures with internal ruptures;
   - Models using volumetric scattering.

4. In all scattering models, the polarization, frequency and angle of incidence effects are taken into account.

In microwave scattering, the electromagnetic fields and their characteristics are the sources of information that allow us to classify the sensed objects. Unwanted effects, resulting in loss of information during the registration of the scattered radiowave signals, may limit the classification potential.

Mathematical and physical modelling plays an important role in remote sensing processes. Mathematical modelling makes use of the statistical characteristics of parameters derived from the reflected signals. Physical modelling is related to radiowave scattering theory and provides the first step for finding appropriate descriptions of radar returns.

Different approaches for solving radiowave scattering problems are used in this monograph; specifically:
- The method of small perturbations and iterative updatings;
- The method of tangent planes;
- The method of volumetric scattering.

The potential of these methods and a summary of the main findings are the primary goals in this work.

C. Outline of the monograph

An introduction to mathematical and physical modelling of microwave scattering for remote sensing with use of polarimetry is provided in Part II of the Monograph. This part is composed of four chapters (1-4).

In Chapter 1, an introduction is given to the inverse problem and the advantages of using polarization information to detect and classify remote objects. Also, the effects of other parameters, such as the frequency of electromagnetic waves and the geometry configuration of scattering environments, are described.

Chapter 2 contains a description of polarimetric radar realizations and a general expression of measured voltage as a function of polarimetric parameters of the electromagnetic field and of the scattering coefficients (scattering cross section) characterizing the polarimetric response of a remotely sensed object.

Chapter 3 is dedicated to modelling of interactions (physical model) of electromagnetic waves with a scattering medium, and scattered signal statistics (mathematical model) characterizing the type of scatterers or the nature of scattering surfaces under investigation.

A summary of the available scattering methods used to predict the scattered power measured by a polarimetric radar is given in Chapter 4. The methods described are based on small perturbations models, Kirchhoff and transport theory of scattering in random media. The limits of applicability of these methods are also described, with particular emphasis on scattering from rough surfaces. Effects, such as depolarization of electromagnetic waves, are given attention with examples and results for multiple scattering geometries.

Part III of the monograph contains applications of polarimetric radar monitoring for various types of earth environments. This part is structured in six chapters (5-10).

Chapter 5 contains background mathematical modelling for randomly inhomogeneous media with applications to radar remote sensing of vegetation-covered ground surfaces.

A review of electrodynamic models for vegetation for a frequency band covering millimeter to meter waves is provided in Chapter 6. The biometrical characteristics of vegetation are described, and examples for various types of crops are provided. Also, modelling is described for electromagnetic reflection from deterministic and random vegetation layers.

Chapter 7 consists of a survey of the available literature on the dependence of the complex permittivity of earth surface materials on wave frequency and polarization and on physical parameters, such as temperature, moisture, salinity and medium density.

Chapter 8 is devoted to a study of the reflection of electromagnetic waves from layered structures under different polarization conditions. Various permittivity profiles in the layered medium are considered: linear, exponential and polynomial. The ensemble-averaged reflection coefficients from the reflecting layers are computed.

Chapter 9 illustrates specific examples of reflection of electromagnetic waves from special structures characterized by internal ruptures (e.g., ice ravines). The effect of polarization is studied for these examples in the approximation of geometrical optics.

The coherent scattering of horizontally polarized electromagnetic waves by a finite layer of vegetation covering the ground is examined in Chapter 10. The vegetation-atmosphere interface is modelled as a random rough surface.

Chapter 11 contains a new method for solving inverse problems for remote sensing using polarimetric information. This new method determines the complex permittivity from the measurements of the polarization ratio (for example vertical versus horizontal) for smooth as well as for random rough surfaces.

Chapter 12 is devoted to signal processing aspects of remote sensing. Examples are given of applications of synthetic aperture radar (SAR) with use of Doppler information. Other examples of remote sensing with use of polarimetry are illustrated based on radar imaging techniques, radar altimeter and atmospheric radar monitoring.

Part IV is the concluding part of the monograph. This part is composed of two chapters (13 and 14).

Chapter 13 is a review of applications and potentials of radar polarimetry. An historical overview of Russian research on radar polarimetry is provided in Chapter 14.

PART II

AN INTRODUCTION TO MATHEMATICAL AND PHYSICAL MODELLING OF MICROWAVE SCATTERING AND POLARIMETRIC REMOTE SENSING

CHAPTER 1

Introduction to Inverse Radar Scattering Problems

1.1 Theoretical aspects

Presently, ecological problems attract great attention. That is why the decision to institute ecological monitoring is of current interest and a search for the best methods of radar monitoring is being carried out in many countries. The main problem with ecological monitoring is the interpretation of radar measurements. The method of measurements consists of transmitting an electromagnetic signal of a given form through a medium under investigation and picking-up at the receiver the signal scattered after the interaction with the medium. The received signal is distorted by the scattering medium. The interpretation of the modifications on the received signal are made at the receiver by a signal processor.

The signal processing is concerned with both detection and estimation. Detection is defined as the determination of the existence or non-existence of a signal at the receiver, for example based on crossing or not crossing of a predefined threshold level. Estimation is the quantification of the parameters or descriptors of the signal, the medium, or the contents of the medium.
The question of estimation leads to a class of signal processing problems called "inverse problems". This is not a well-defined concept, but is best described in terms of its relation to the "forward problem". The forward problem can be stated as follows: If the source transmits electromagnetic energy of a given form through the medium, what does the receiver receive? Conversely, the "inverse problem" is concerned with the question: If the receiver receives a signal of a given form, what does this tell us about either the medium, its contents, or the source?

The inverse problem is concerned with imaging, surface profiling, target classification, tomography, etc. As an example, consider the inverse source problem associated with the notion of deconvolution; the latter can be represented mathematically by the integral equation

$$\vec{f}(\vec{r}) = \int G(\vec{r},\vec{r}')\vec{g}(\vec{r}')d\vec{r}' \tag{1.1}$$

where $G(\vec{r},\vec{r}')$ is the Green's function, $\vec{g}(\vec{r}')$ is a source distribution and $\vec{f}(\vec{r})$ is the received field at position $\vec{r}$. The goal in this case is to infer information about the source distribution from the received signal and the properties of the medium (channel) embodied in the Green's function.

A specific realization of the inverse source problem arises in connection with the inhomogeneous Helmholtz equation in an unbounded, homogeneous, isotropic medium, viz.,

$$\left(\nabla^2 + k^2\right)\vec{f}\left(\vec{r}\right) = \vec{g}\left(\vec{r}\right) \tag{1.2}$$

where k denotes the wavenumber. In this case the Green's function is given explicitly as follows:

$$G\left(\vec{r},\vec{r}'\right) = \frac{1}{4\pi}\frac{1}{\left|\vec{r}-\vec{r}'\right|}e^{jk\left|\vec{r}-\vec{r}'\right|} \tag{1.3}$$

In a typical remote sensing application, the problem is reduced to the definition of parameters depending on the physical characteristics of the scattering medium, or the remotely sensed object (scatterer) under investigation. For example, the parameter may be the electric permittivity of the scattering environment. The scattered electric field measured at a distance R from the scattering volume V (cf. Fig. 1) is given by

$$\vec{E}_s\left(\vec{k}\right) = \frac{k_s^2}{4\pi R}\int_V \vec{E}_{\text{int}}\left(\vec{r}\right)\Delta\varepsilon_r\left(\vec{r}\right)e^{-j\vec{k}\cdot\vec{r}}d\vec{r} \tag{1.4}$$

within the framework of the Born approximation and under far field conditions $R >> V^{1/3}$. Here, $\vec{k}_i$ is the wavevector associated with the incident field, $\vec{k}_s$ is the wave vector of the scattered field, $\vec{k} = \vec{k}_i - \vec{k}_s$, $k_s = \left|\vec{k}_s\right|$, $\vec{E}_{\text{int}}\left(\vec{r}\right)$ is the electric field seen within the scattering volume V and $\Delta\varepsilon_r\left(\vec{r}\right) = \left[\varepsilon\left(\vec{r}\right) - \varepsilon_0\right]/\varepsilon_0$, where $\varepsilon(\vec{r})$ is the permittivity within V and ε_0 is the vacuum permittivity.

It should be noted in Eq. (1.4) that if $\vec{E}_{\text{int}}$ is a constant vector within the scattering volume, the scattered field is proportional to the Fourier transform of the relative permittivity; specifically,

$$\vec{E}_s = \frac{k_s^2\vec{E}_{\text{int}}}{4\pi R}\int_V \Delta\varepsilon_r\left(\vec{r}\right)e^{-j\vec{k}\cdot\vec{r}}d\vec{r} \tag{1.5}$$

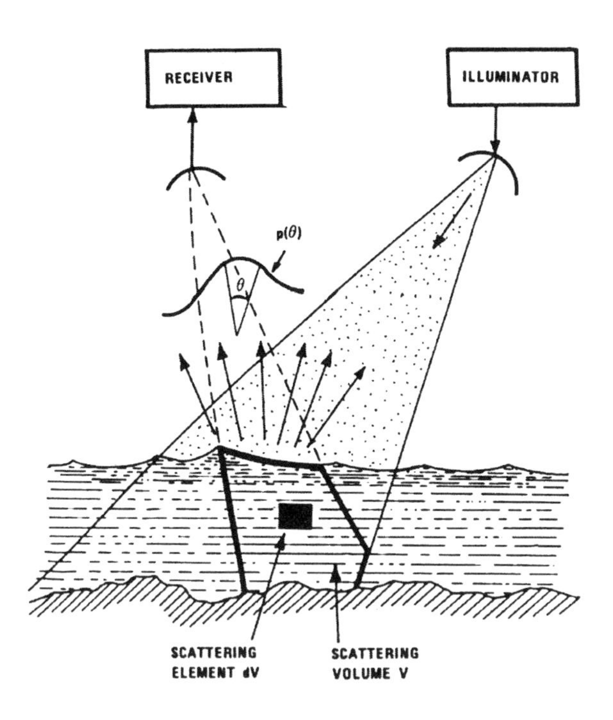

Fig. 1.1 Scattering from irregularities in the scattering volume (Reproduced with permission of NATO RTA, D.T. Gjessing, NATO AGARD-LS-93, Fig. 4.5 Part 12, 1978)

Thus, $\Delta\varepsilon_r(\vec{r})$ can formally be derived by an inverse Fourier transform of the scattered field. Suppose that $\Delta\varepsilon_r(\vec{r})$ is a random function of position and the statistical fluctuations are homogeneous and isotropic. It follows, then, from Eq. (1.5) that

$$\langle\left|\vec{E}_s\left(\vec{k}\right)\right|^2\rangle = \frac{k_s^4\left|\vec{E}_{\text{int}}\right|^2}{\left(4\pi R\right)^2}\int_V d\vec{r}_2\int_V d\vec{r}_1 R_{\varepsilon_r}\left(\left|\vec{r}_2-\vec{r}_1\right|\right)e^{-j\vec{k}\cdot(\vec{r}_2-\vec{r}_1)} \tag{1.6}$$

where R_{ε_r} is the correlation function of $\Delta\varepsilon_r(\vec{r})$. Introducing a center-of-mass and difference coordinate change of variables, viz.,

$$\vec{r}_s = \frac{\vec{r}_2 + \vec{r}_1}{2} \quad , \quad \vec{r}_d = \vec{r}_2 - \vec{r}_1 \tag{1.7}$$

we obtain

$$\langle \left|\vec{E}_s\left(\vec{k}\right)\right|^2 \rangle = \frac{k_s^4 \left|\vec{E}_{\text{int}}\right|^2}{\left(4\pi R\right)^2} \int_V d\vec{r}_s \int_V d\vec{r}_d R_{\varepsilon_r}\left(r_d\right) e^{-j\vec{k}\cdot\vec{r}_d} \tag{1.8}$$

where $r_d = |\vec{r}_d|$. With $\vec{k}\cdot\vec{r}_d = kr_d \cos\theta$, this expression reduces to the angular power spectrum (illustrated as $p(\theta)$ in Fig. 1) given by

$$\langle \left|\vec{E}_s\left(\vec{k}\right)\right|^2 = \frac{k_s^4 \left|\vec{E}_{\text{int}}\right|^2 V}{\left(4\pi R\right)^2} \int_V d\vec{r}_d R_{\varepsilon_r}\left(r_d\right) e^{-j\vec{k}\cdot\vec{r}_d} \tag{1.9}$$

Measurements of this quantity can be used to derive the autocorrelation of the relative permittivity.

It is also interesting to note that if $\Delta\varepsilon_r(\vec{r})$ in Eq. (1.4) is constant within the scattering volume, but the internal field $\vec{E}_{\text{int}}$ varies randomly, we have

$$\vec{E}_s\left(\vec{k}\right) = \frac{k^2 \Delta\varepsilon_r}{4\pi R} \int_V \vec{E}_{\text{int}}\left(\vec{r}\right) e^{-j\vec{k}\cdot\vec{r}} d\vec{r} \tag{1.10}$$

If the statistical fluctuations of $\vec{E}_{\text{int}}(\vec{r})$ are homogeneous and isotropic, we obtain

$$\langle \left|\vec{E}_s\left(\vec{k}\right)\right|^2 \rangle = \frac{k_s{}^4 \left(\Delta\varepsilon_r\right)^2 V}{\left(4\pi R\right)^2} \int_V R_{E_{\text{int}}}\left(r_d\right) e^{-j\vec{k}\cdot\vec{r}_d} d\vec{r}_d \tag{1.11}$$

where $R_{E_{\text{int}}}$ denotes the autocorrelation of $\vec{E}_{\text{int}}(\vec{r})$.

The transformation pairs between the scattering medium characteristics (e.g. permittivity, spatial distribution of scatterers, etc.) and the received scattered field indicate a "codification" of the scattered field as a function of the parameters of the medium. By "decoding" the information of the received signal (changes), we may obtain information on the parameters (source/encoder) of the scatterers. We can then classify the remote-sensed scatterers by evaluation of the parameters and pattern recognition.

1.2 Pattern recognition and evaluation parameters

The most substantial part of the pattern recognition problem is the decision process leading to a classification of remote-sensed objects under specified criteria. There are two main approaches for constructing classifiers: deterministic and statistical [*Tuchkov*, 1985]. The former can be expressed in terms of criteria based on a partition consisting of N mutually non-overlapping areas, each corresponding to specified classes. The latter can be expressed in terms of criteria based on statistical decision theory and the theory of hypothesis testing, e.g., the maximum likelihood Bayes test. Methods leading to the construction of decision-making deterministic procedures can be divided into those based on the concept of a decision function and methods based on the concept of classes and features in the so-called feature space.
The necessity of increasing the classification potential in radar remote sensing applications is based on the concept of a classification distance in the feature space. A feature can be, for example, the scattering cross section of an object. The feature space can then be defined in the polarization domain (3-D) as

VV: backscatter and incident fields vertically polarized.
HH: backscatter and incident fields horizontally polarized.
VH: backscatter field in vertical polarization, incident field in horizontal.
HV: backscatter field in horizontal polarization, incident field in vertical.

The definition of a class depends upon which physical characteristics we want to measure. If it is the roughness of the surface of a scatterer, the class is the "degree" of roughness. The degree can be specified statistically, for example by the standard deviation of surface height variations around the mean, and the correlation of these variations in space.

A good image of a rough terrain can be reconstructed if the radar is capable of distinguishing a large number of different classes (degree of roughness) of the surface scatterers, and at the same time avoiding miss-classification, meaning that rough scattering regions are not placed in the same class. The task of a classifier is to determine to which class most likely the scatterers belong to. One procedure is to compare the measured features of the scatterers (scattering cross-section) with "training" data obtained for known classes (degree of roughness) of scatterers. By signal processing, the contrast between two different types (classes) of scatterers can be enhanced to reduce confusion. With use of the minimum classification distance criteria the scattering object is ''placed'' in a class with values of the scattering cross section nearest to the "training'' data. This method of classification incorporates the "nearest neighbor" and the "L nearest neighbors" approach. This non-parametric approach is very efficient when solving problems in which the objects, belonging to one class, are characterized by a limited degree of variability. All methods using the concept of distance assume the availability of one or several standards per class.

In real situations of remote sensing, it is often necessary to apply statistical methods. *A priori* information is not available as a rule. A similar situation exists in the theory of recognition, that is the so-called *a priori* uncertainty. There are two kinds of *a priori* uncertainties: parametric and non-parametric: In the first case, it is assumed that a probability distribution of the feature per class is known with sufficient accuracy to limit the number of unknown parameters. In the second case, there is no *a priori* information about probability models for classes in the feature space. In order to overcome *a priori* uncertainty, special iterative procedures called training algorithms are applied. There are two kinds of training: with a supervisor, and without a supervisor (for possible applications of modern neural techniques see [*Brooks*, 1996]).

From the aforementioned methods, it can be concluded that the method of non-parametric uncertainty should be applied for remote sensing systems. However, the implementation of this method in real time is quite limited and may prove to be very laborious. That is why it is useful to assume specific models for the probability density distribution of feature vectors and for the probability of appearance of a specified object within the corresponding class. In this case, the problem of pattern recognition is simplified but the reliability of the received information is also reduced. Therefore, in real events, it is necessary to look for trade-off decisions.

It should be noted that the problem of parameter estimation is substantially less difficult than the problem of pattern recognition.

1.3 Conditions for implementing inverse scattering techniques

The main difficulty in solving the inverse problem is that it can be what mathematicians refer to "ill-posed." This problem is due to noisy data, not enough data, or a combination of the two. A problem is "well-posed" when a unique solution exists that is stable to changes on the data.

The inverse problem has a unique solution if the initial data (in particular, values of the complex reflection coefficient) are defined over an infinite range of frequencies. In practice, this condition cannot be met physically and the values of the scattered field are limited by frequency range. This range is dependent on technical limitations of the instruments, the finite bandwidth of a remote sensing signal spectrum and the finite pass-band of the receiver. In practice, it is then possible that more than one object can be distinguished within these constraints, i.e., there is more than one solution of the inverse problem. The problem of non-uniqueness can be dealt with by introducing *a priori* information about a scattering object (electromagnetic modelling of the object).

A stability problem arises when, given the existence of a solution, the solution is extremely sensitive to small perturbation in the data. As an example, consider a measurement of the scattered field $f(\vec{r})$. Let this measurement have an error Δf and f_0 be the true value. Then, based on Eq. (1.1), we have

$$\vec{f}(\vec{r}) = \vec{f}_0(\vec{r}) + \Delta\vec{f}(\vec{r}) = \int G(\vec{r},\vec{r}')\left[\vec{g}(\vec{r}') + \Delta\vec{g}(\vec{r}')\right]d\vec{r}' \tag{1.12}$$

with

$$\vec{f}_0(\vec{r}) = \int G(\vec{r},\vec{r}')\vec{g}(\vec{r}')d\vec{r}' \tag{1.13}$$

From Eqs (1.10) and (1.11) we obtain

$$\Delta\vec{f}(\vec{r}) = \int G(\vec{r},\vec{r}')\Delta\vec{g}(\vec{r}')d\vec{r}' \tag{1.14}$$

It follows from Eq. (1.14) that Δf can be thought of as a "weighted average" of $\Delta g(\vec{r}')$, where $G(\vec{r},\vec{r}')$ is a weighting kernel. Thus, we are free to select a function $\Delta g(\vec{r}')$ whose weighted average is as close to zero as desired, but still can produce large errors in $g(\vec{r}')$.

An explanation of how the scattering problem can be ill-posed is to note that the scattered field, which is the source of the observed data, can depend rather weakly on very large changes in the scattering object. Thus, a small amount of noise, or simply the addition of new data, can represent a very large fictitious change in the evaluated parameters of the scattering object, which is the object of interest.

Much of the effort expended in the inverse problem is focused on particular formulations of electromagnetic modelling of the object and detection-estimation algorithms. The solution of these problems will be successful under a number of conditions: (a) High reliability of results received during the measurement process; (b) High information contents of the specified parameters; (c) Robustness of the experimental data. In this work attention is given to radar monitoring. We examine now in more detail the aforementioned conditions.

Reliability of experimental data first of all is determined by the accuracy of the measuring instrumentation and by the number of instruments. The larger the number of instruments simultaneously engaged in measurements, the more reliable the final result. Here, we should clarify the words "the number of instruments." Measurements can be carried out simultaneously at different frequencies, e.g., when multi-channel radar systems are used. It is obvious that an increase in the number of channels will lead to more accurate results. However, it complicates to a great extent the construction of the radar. That is why there is another way to increase, at least double, the number of channels. Namely, the application of a remote sensing signal at one frequency, but with two different polarizations. For example, the signal at one frequency can be emitted with horizontal and vertical polarization and with specified amplitude-phase characteristics, or with left-hand and right-hand circular polarization and under similar amplitude-phase conditions. Then, with four frequency channels available, we deal with eight independent measuring devices.

The second condition for a successful solution of the scattering problem is high information content of the specified parameters. This is associated with how much information concerning the remotely sensed object is obtained from the radar signal. It is obvious that the parameters of a remote-sensing signal vary when incident electromagnetic waves are affected by surface interactions. Moreover, some parameters of the remote sensing signal do not vary during the interactions at all, and so they do not contribute to the determination of the properties of a remotely sensed object. On the other hand, for certain surfaces some parameters vary to a great extent; as a consequence these parameters give much required information about the surfaces. The above-mentioned aspects deal with the information contents of the specified parameters. Studies show that one of the most informative characteristic properties of the remote sensing signal is the polarization of the corresponding wave.

The third condition concerns the reproducibility of the received results because the objects and their geometrical configuration may change differently in time. This means that the electromagnetic scattering depends on the season of the year, daytime, atmospheric pressure, temperature of the ambient environment, geographic coordinates of the object, meteorological conditions, etc. In order to receive robust and reproducible results, it is necessary to carry out repeated, and to some extent monotonous, measurements with several polarizations of the remote sensing wave. In other words, the object is analyzed in many different ways, which leads to an improvement in measurement robustness.

In order to increase the level of the three most significant conditions applied to a radar remote sensing system, it is necessary to use polarization characteristics of transmitted and received waves. That is why it is possible to speak about polarization diagnostics of the environment, and the underlying terrain in particular. The incorporation of specific devices (e.g. polarizers integrated in the radar channels) into the remote sensing radar system is one of the most advanced methods for meeting the aforementioned three conditions.

1.4 Polarimetric radar

1.4.1 Effects of polarization

A polarimetric radar measures the scattered field for each transmitting and receiving polarization combination. The ratio between the scattered and the transmitted field for each polarization combination is called the scattering coefficient (usually a complex number, with amplitude and phase). If, for example, the radar system is configured to measure all possible combinations available from the vertically polarized and horizontally polarized antennas, a "scattering matrix" can be determined composed of four complex scattering coefficients. The polarimetric radar has distinct advantages compared to the conventional fixed-polarization radar. The conventional radar measures a single scattering coefficient for a specific polarization combination (mostly co-polar). The result of this single channel approach is that only one component of the scattered wave (which is vector-valued) is measured, and any additional information contained in the polarization properties of the scattered field is lost. The polarimetric radar, instead, ensures that all the information of the scatterer is retained in the scattered wave vector. In our example, the information on the scatterer is obtained by measuring the four complex scattering coefficients.
These four complex coefficients characterize the scatter cross section of a scattering surface for any incident or scattered polarization (e.g., right circular, left circular, elliptical, etc.)

As an illustration of the effect of polarization, we consider an aggregate of scatterers modelled as short vertical dipoles. An horizontally polarized incident wave (perpendicular to the plane of incidence) does not interact with the scatterers. A vertically polarized wave (in the plane of incidence) interacts strongly with the scatterers. If the direction of polarization is rotated by an angle relative to the horizontal component, only the vertical component interacts with the scatterers.

1.4.2 Effects of frequency

The frequency is an important signal parameter in the interaction of the signal with a scattering media. The depth of penetration of the signal in the media, and the scattering process from a rough surface, are important factors.

For most media, the penetration depth increases with the radar wavelength.

Scattering from a rough surface is strongly dependent on the frequency. In the case of a constant roughness spectrum, the scattering cross section increases as the fourth power of the frequency. Even if the surface roughness spectrum decreases as the square or the cube of the spectral frequency (i.e., ~ k^{-2} or k^{-3}), the backscatter cross section increases as a function of frequency. Scattering from a rough surface is dependent also on the radio wavelength λ compared to the "scale of roughness L." The scattering cross section can have a maximum value when the wavelength and the scale of roughness have a comparable size ($L/\lambda \sim 1$).

1.4.3 Effects of angle of incidence

The scattered signal is strongly dependent on the angle of incidence (measured with respect to the normal vector on the averaged planar surface), and it can provide information on the slopes distribution of the scattering surface. Therefore, different scattering surfaces can be distinguished and classified based on the angular scattered signal spectrum.

CHAPTER 2

Description of Remote Sensing by Radar Polarimetry

Our goal in this chapter is to analyze the process of radar monitoring based on polarimetric data. This analysis is made in two steps. First, by looking into the physical process of how data can be generated by a scattering object (source/encoder) and extracted by a polarimetric radar (receiver/decoder) for the interpretation and classification of the remote-sensed object. Secondly, by investigating the procedures of implementing solutions to inverse scattering problems using polarimetric radar techniques within the framework of specific applications.

2.1 Physical process of encoding-decoding of polarimetric data

Information about scattering objects can be extracted by means of decoding of received polarimetric signals. The coding takes place during the interaction of the incident radiowaves with scattering objects. The transfer of information from an object to the radar can be modelled as a communications system composed of four stages:

a) Information encoding by the source (object).
b) Signal communication through the propagation channel. In this stage, signal distortions may occur due to random noise perturbations, or unwanted reflected signals and multi-path.
c) Signal reception and information decoding.
d) Interpretation of the polarimetric data and classification of the object.
   These stages are illustrated with the block diagram in Fig. 2.1:

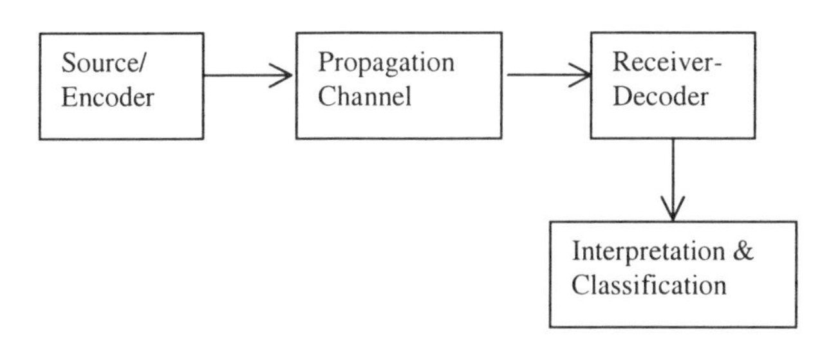

Fig. 2.1 Block diagram of a (polarimetric) radar data channel

The encoding-decoding process is based on the fact that the polarization of the back-scattered field depends on the target properties and, in general, differs from the polarization of the field incident on the remote-sensed object. For example, a thin straight wire is distinguished from a homogeneous sphere by observing the variation of the received signal amplitude as the linear polarization is rotated. The echo from the sphere is not modulated, whereas the received signal from the wire varies between a maximum and a minimum. In other words, the "coding" produced by the scatterer on the polarized incident field gives an unmodulated back-scatter signal for the case of a spherical object and a modulated (oscillating between a maximum and a minimum value) signal for the case of the wire.

To understand the coding process, we need detailed knowledge on scattering. However, the signal-scatterer interaction mechanism is very complex. Satisfactory solutions are only available in cases for which the parameters characterizing the scatterer size are very large or just very small compared to the radio wavelength. Approximate theories, such as Rayleigh scattering, Born and Van de Hulst approximations, can be applied [*Newton*, 1969].

The encoding-decoding process is complicated by a number of factors. The most significant are related to the complexity of the scatterer, which, in general, is inhomogeneous and random, and to the effects of the propagation, e.g. signal distortion due to noise or reflection from unwanted objects and multi-path.

2.1.1 Effects of propagation

The design of an optimal encoder-decoder for remote sensing becomes complex by the fact that changing the frequency, polarization, or the incidence angle of radiation, the conditions of propagation (and signal-scatterer interactions) change. This effect leads to a variation in the structure of the coding of the scattered field. The noise and multi-path in the propagation channel are the causes of variations in the code. These variations can be random.

The signal is distorted when the wave interacts (e.g. through multiple scattering) with volumetric objects or particles, as in clouds, precipitation and atmospheric turbulence. Many publications have been devoted to the investigation of the mechanisms underlying these processes of multiple scattering by random media [*Brussaard*, 1990; *Thurai*, 1992; *Kuznetsov*, 1994]. Multiple scattering is a limiting factor in radar remote sensing. For example, if vegetation is investigated, we have multiple scattering from vegetation and also from earth's surface (see a model of vegetation in Fig. 2.2).

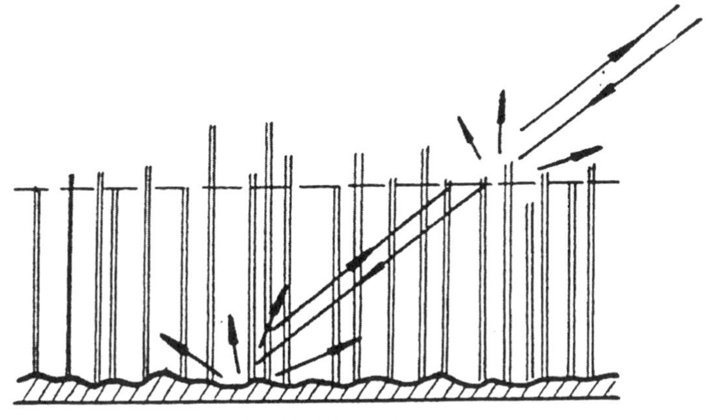

Fig. 2.2 Model of vegetation cover

The reflections from the earth's surface are the noisy part of the signal and must be separated from the vegetation reflections. This suppression of the unwanted reflections from earth can be made using polarimetry. In this example, the unwanted reflections can be suppressed by transmitting a rotated polarization vector (see Sec. 1.4) and assuming that the contribution of the vertical component of the received scattered field is primarily due to the vegetation canopy and the horizontal component to the soil.

Due to the generally complex geometry of the vegetation (canopy) and the randomness of the soil surface, it is not possible to separate completely the signal in two parts, each of which is the response to the interaction of one component only (either the vegetation or soil surface). The extraction of information on vegetation requires laborious signal processing. One method of processing used is to evaluate by means of the method of Lagrange multipliers [*Ulaby*, 1990] the optimum transmit-receive antenna polarization that maximizes the contrast between vegetation and soil. This is tantamount to finding the optimum rotation angle of the polarization vector that maximizes the signal (reflection from vegetation) to clutter (reflection from soil) ratio.

2.2 Physical realization of a polarimetric radar

In previous descriptions of polarimetry we have examined how it can be used to improve the detection and classification of objects, compared to the conventional single channel radar. An illustration of polarimetric radar principle used for remote sensing purposes is shown in Fig. 2.3.

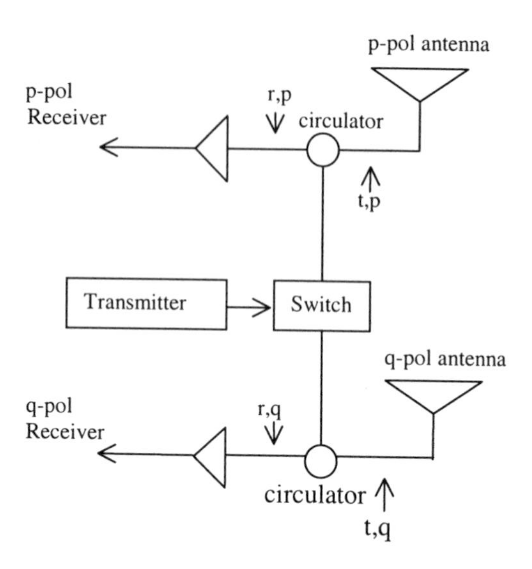

Fig. 2.3 Illustration of polarimetric radar principle

We have indicated in Fig. 2.3, the paths:
t,p: path of the p-polarized transmitted signal
t,q: path of the q-polarized transmitted signal
r,p: path of the p-polarized received (scattered) signal
r,q: path of the q-polarized received (scattered) signal

The radar measures the four scattering matrix elements by first transmitting a pulse through the p-polarized field of the antenna, and subsequently receiving signals simultaneously on both orthogonally polarized (p-q) feeds. The next transmission utilizes the q-polarized antenna feed for transmitting, again followed by simultaneous recording from both antenna ports.

2.2.1 Computation of the polarimetric radar received voltages

The power density incident upon an object at a distance r_1 from the transmit antenna is given by

$$\Pi_i = \frac{W_t}{4\pi r_1^2} G_t \tag{2.1}$$

where W_t is the transmit power and G_t is the transmit antenna gain. The electric field incident upon the object is therefore given by

$$\vec{E}^i = \sqrt{\Pi_i Z} e^{-j\vec{k}\cdot\vec{r}_1} \vec{P}^t \tag{2.2}$$

where $\vec{P}^t$ is the polarization vector, given by

$$\vec{P}^t = \frac{\vec{E}^t}{\left|\vec{E}^t\right|} \tag{2.2a}$$

Z is the impedance of the propagation medium, $\vec{k}$ is the wavevector and $\vec{E}^t$ is the transmit electric field vector. In case of free space, we have

$$Z = 120\pi \;\; \Omega \tag{2.2b}$$

From Eqs (2.1), (2.2), (2.2a) and (2.2b), we obtain the expression

$$\vec{E}^i = \sqrt{30 W_t G_t} \frac{e^{-j\vec{k}\cdot\vec{r}_1}}{r_1} \vec{P}^t \tag{2.2c}$$

The transmit electric field vector in Eq. (2.2a) is expressed as a linear combination of orthogonal complex components, viz.,

$$\vec{E}^t = E_p^t \vec{p} + E_q^t \vec{q} \tag{2.3a}$$

where $\vec{p}$ and $\vec{q}$ are unit vectors and

$$E_p^t = \left|\vec{E}_p{}^t\right| e^{j\delta_p^t} \tag{2.3b}$$

$$E_q^t = \left|\vec{E}_q^t\right| e^{j\delta_q^t} \tag{2.3c}$$

$$\left|E_p^t\right| = \left|\vec{E}^t\right| \cos\gamma_t \tag{2.3d}$$

$$\left|E_q^t\right| = \left|\vec{E}^t\right| \sin\gamma_t \tag{2.3e}$$

The angle γ_t is defined by the polarization ratio:

$$\tan\gamma_t = \left|E_q^t\right| / \left|E_p^t\right| \tag{2.3f}$$

and δ^t_p, δ^t_q are respectively the phase angles of the $\vec{p}$ and $\vec{q}$ components of the transmit electric field $\vec{E}^t$. From Eqs (2.3a), (2.3b), (2.3c), (2.3d) and (2.3e) is derived:

$$\vec{E}^t = e^{j\delta_p^t} \left|\vec{E}^t\right| \left\{\cos\gamma_t \vec{p} + \sin\gamma_t e^{j\delta_{pq}^t} \vec{q}\right\} \tag{2.4}$$

where

$$\delta_{pq}^t = \delta_q^t - \delta_p^t \tag{2.5}$$

The phase δ^t_p is taken as reference. The polarization vector is given by the ratio

$$\vec{P}^t = \frac{\vec{E}^t}{\left|\vec{E}^t\right|} = \cos\gamma_t \vec{p} + \sin\gamma_t e^{j\delta_{pq}^t} \vec{q} = \begin{pmatrix} \cos\gamma_t \\ \sin\gamma_t e^{j\delta_{pq}^t} \end{pmatrix} \tag{2.6}$$

The polarization of the (transmitted) wave is in general elliptical. The polarization ellipse of the scattered wave will in general be different from that of the transmitted field. The polarization ellipse parameters i.e. the angles β (orientation) and χ (ellipticity) are related to the wave parameters γ and δ by the expressions [*Boerner*, 1991]

$$\tan 2\beta = \frac{2\tan\gamma\cos\delta}{1-\tan^2\gamma} = (\tan 2\gamma)\cos\delta \tag{2.7a}$$

$$\sin 2\chi = \frac{2\tan\gamma\sin\delta}{1+\tan^2\gamma} = (\sin 2\gamma)\sin\delta \tag{2.7b}$$

The orientation β is given by the angle between the major axis of the ellipse and a reference direction [Fig. 2.4]. The ellipticity χ is defined by the ratio [Fig. 2.4]:

$$\tan\chi = \pm\frac{a_\eta}{a_\xi} \tag{2.7c}$$

where a_η , a_ξ are the minor and the major axes of the ellipse, respectively. The sense of rotation of the electric field $\vec{E}(\omega) = \mathrm{Re}\left[\vec{E}\ e^{j\omega t}\right]$ describing the ellipse in its polarization plane is determined by the sign (+ clockwise, - counterclockwise for a travelling wave in $\hat{k}$ direction) in the Eq. (2.7c). Another parameter of the ellipse is its amplitude given by:

$$A = \sqrt{a_\eta^{\ 2} + a_\xi^{\ 2}} \tag{2.7d}$$

The wave parameters δ, γ are given by the phase difference $\delta_q - \delta_p$ of the two orthogonal components E_q , E_p, and by the polarization ratio $|E_q|/|E_p|$, respectively.

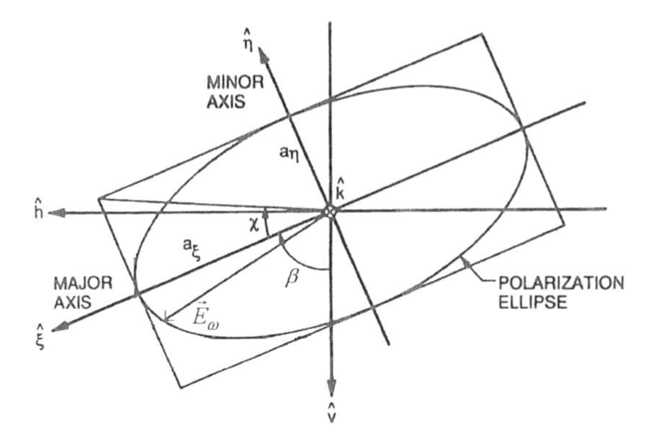

Fig. 2.4 Polarization ellipse
(In the q=v, p=h polarization plane for a travelling wave in $\hat{k}$ direction)
(Reproduced with permission of Artech House, F.T. Ulaby, "Radar Polarimetry for Geoscience Applications", Fig. 1.6, Ch. 1, 1990)

The polarization ratio is represented in the Fig. 2.5 below:

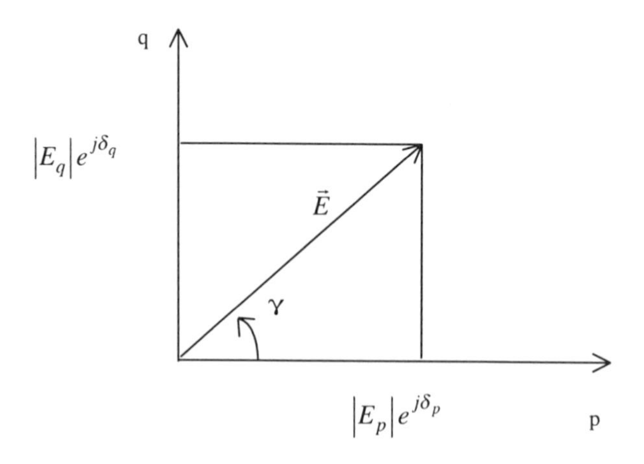

Fig. 2.5 Representation of the polarization ratio in the orthogonal (p, q) basis

The electric field $\vec{E}^s$ scattered from a scattering object is given by

$$\vec{E}^s = [S]\vec{E}^i \tag{2.8}$$

For an arbitrary polarization pair ($\vec{p},\vec{q}$), it is convenient to express the scattered field in the matrix form

$$\begin{pmatrix} E_p^s \\ E_q^s \end{pmatrix} = \begin{pmatrix} S_{pp} & S_{pq} \\ S_{qp} & S_{qq} \end{pmatrix} \begin{pmatrix} E_p^i \\ E_q^i \end{pmatrix} \tag{2.9}$$

In Eq. (2.8), [S] is the scattering matrix [cf. Eq. 2.9]:

$$[S] = \begin{pmatrix} S_{pp} & S_{pq} \\ S_{qp} & S_{qq} \end{pmatrix} \tag{2.10}$$

E_p^i , E_q^i are the p and q (orthogonal) complex components of the electric field incident upon the scatterer and E_p^s , E_q^s are the p and q components of the electric field scattered by the scattering object. In this expression, S_{pp} , S_{pq} , S_{qp} , S_{qq} are the scattering coefficients for the four polarization combinations.
The scattering coefficient S_{pq} is a complex number composed of a magnitude $s_{pq} = |S_{pq}|$ and a phase angle ψ_{pq}. Specifically,

$$S_{pq} = s_{pq} e^{j\psi_{pq}} \tag{2.11}$$

with

$$s_{pq} = \left| \frac{E_p^s}{E_q^i} \right| \tag{2.12}$$

and

$$\psi_{pq} = \psi_p^s - \psi_{\mathrm{q}}^{\mathrm{i}} \tag{2.12a}$$

The coefficients S_{pp}, S_{qq} are the co-polar terms (diagonal elements of the scattering matrix), and S_{pq}, S_{qp} are the cross - polar (non-diagonal elements of the scattering matrix). The cross-polar terms describe the effects of the wave depolarization (polarization changes from $\vec{p}$ to $\vec{q}$ and $\vec{q}$ to $\vec{p}$) when scattering from asymmetric objects or propagating through inhomogeneous and/or random media. This can be the case for the random vegetation canopy described in Sec. 2.1.1. If the amplitude and phase of the scattering matrix elements are known from measurements, we can obtain full information on the polarization properties of the medium or the scatterer (object) under investigation [*Huynen*, 1998].

The electric field measured by a receiving antenna of effective aperture A_e placed at distance r_2 from the scatterer is calculated by [cf. Eq. (2.8)]

$$\vec{E}^r = \sqrt{A_e} \frac{e^{-j\vec{k} \cdot \vec{r}_2}}{r_2} [S] \vec{E}^i \tag{2.13}$$

where $[S]$ is the scattering matrix defined in Eq. (2.10) and

$$\vec{E}^r = \begin{pmatrix} E_p^r \\ E_q^r \end{pmatrix}, \qquad \vec{E}^i = \begin{pmatrix} E_p^i \\ E_q^i \end{pmatrix} \tag{2.14}$$

The effective antenna aperture A_e is defined by [*Silver*, 1984]

$$A_e = \frac{G_r \lambda^2}{4\pi} \tag{2.15}$$

where G_r is the receive antenna gain and λ the radio wavelength. From Eqs (2.1), (2.2), (2.6), (2.11), (2.12) and (2.15), we calculate for the total path the received electric field:

$$\vec{E}^r = \frac{1}{4\pi} \sqrt{W_t G_t G_r Z} \frac{\lambda}{r_1 r_2} e^{-j\vec{k}\cdot(\vec{r}_1 + \vec{r}_2)} [S] \vec{P}^t \tag{2.16}$$

The polarization vector $\vec{P}^r$ of the receiving antenna, in analogy with the transmit antenna [cf. Eq. (2.6)] is given in matrix form as [*Skolnik*, 1970; *Kostinski*, 1986]

$$\vec{P}^r = \begin{pmatrix} \cos\gamma_r \\ \sin\gamma_r e^{j\delta_{pq}^r} \end{pmatrix} \tag{2.17}$$

where γ_r is the angle between $\vec{p}$ and $\vec{E}^r$ at the receiving antenna, and δ_{pq}^r is the phase difference between the two orthogonal ($\vec{p}, \vec{q}$) components of $\vec{E}^r$.

The voltage measured by the receiving antenna is proportional to the scalar product of the received electric field and the polarization vector [*Boerner*, 1991; *Huynen*, 1965]

$$V^r = h_l \vec{P}^r \cdot \vec{E}^r \tag{2.18}$$

where h_l is the constant of proportionality which depends on the size of the antenna over which the received electric field $\vec{E}^r$ acts and induces the voltage V^r [*Berkowitz*, 1965; *Collin*, 1985]. The constant of proportionality will be computed in the calibration procedure [*Sarabandi et al.*, 1990].

From Eqs (2.16), (2.17) and (2.18) we obtain

$$V^r = h_l C_0 \frac{e^{-\vec{k}\cdot(\vec{r}_1+\vec{r}_2)}}{r_1\, r_2}\left\{\vec{P}^r \cdot [S]\vec{P}^t\right\} \tag{2.19}$$

with the quantity C_0 given by

$$C_0 = \frac{\lambda}{4\pi}\sqrt{W_t G_t G_r Z} \tag{2.20}$$

From Eqs (2.6), (2.10), (2.11), (2.17) and (2.19) the voltages are calculated for each polarization pair, specifically,

$$V^r_{pp} = \frac{h_l C_0}{r_1 r_2} e^{-j\vec{k}\cdot(\vec{r}_1+\vec{r}_2)} s_{pp} \cos\gamma_r \cos\gamma_t e^{j\psi_{pp}} \tag{2.21}$$

$$V^r_{pq} = \frac{h_l C_0}{r_1 r_2} e^{-j\vec{k}\cdot(\vec{r}_1+\vec{r}_2)} s_{pq} \cos\gamma_r \sin\gamma_t e^{j\left(\psi_{pq}+\delta^t_{pq}\right)} \tag{2.22}$$

$$V^r_{qp} = \frac{h_l C_0}{r_1 r_2} e^{-j\vec{k}\cdot(\vec{r}_1+\vec{r}_2)} s_{qp} \sin\gamma_r \cos\gamma_t e^{j\left(\psi_{qp}+\delta^r_{pq}\right)} \tag{2.23}$$

$$V^r_{qq} = \frac{h_l C_0}{r_1 r_2} e^{-j\vec{k}\cdot(\vec{r}_1+\vec{r}_2)} s_{qq} \sin\gamma_r \sin\gamma_t e^{j\left(\psi_{qq}+\delta^t_{pq}+\delta^r_{pq}\right)} \tag{2.24}$$

with δ^t_{pq} and δ^r_{pq} being the phase difference between the complex field components ($\vec{p},\vec{q}$) at the transmitter and receiver, respectively.

After the calibration procedure [*Sarabandi et al.,* 1990], the scattering matrix elements with amplitudes s_{pp}, s_{pq}, s_{qp}, s_{qq} and phases $\psi_{pp},\psi_{pq},\psi_{qp},\psi_{qq}$ can be computed from the measured voltages [cf. Eqs (2.21)-(2.24)].
One can see from Eqs (2.21)-(2.24) how the amplitude and phase of the received polarized field are modified with respect to the transmit field by the amplitudes and phases of the scattering coefficients. This modification reflects into a change of the polarization state of the received (scattered) wave [see Eqs (2.7a) and (2.7b)]. This modification can be measured from the changes of the orientation and ellipticity

angles of the received wave with respect to the transmitted wave. The interpretation of this change is the essence of the inverse problem for a polarimetric radar.

2.3 Methods of measurements of polarimetric data

After interaction with a spatially inhomogeneous random medium, electromagnetic radiation becomes the input to a receiving antenna. The receiving antenna response to the incident field (superposition of waves coming from different directions, with different amplitudes) is given by the product of the incident field with a weighting matrix describing the receiving antenna pattern. We indicate the weighting matrix by $G_{\alpha\beta}$, where the first index α corresponds to the main polarization of the antenna system and the second index β can have values corresponding to co-polarization or cross-polarization. Furthermore, we assume that $\alpha = 1$ for the case of a horizontal orientation of the vector $\vec{E}$ and $\alpha = 2$ for the case of a vertical polarization. The index β corresponds to the polarization of the wave incident on the antenna. For $\alpha = \beta$, the antenna receives co-polarly. When α is different from β, $G_{\alpha\beta}$ describes the response to the cross-polar wave of the receiving system.

The full output response of an antenna receiving a field with arbitrary polarization α can be expressed in the following form:

$$U_\alpha = \sum_\beta \int \int dk_x\left(\bar{\theta},\bar{\varphi}\right) dk_y\left(\bar{\theta},\bar{\varphi}\right) G_{\alpha\beta}(r,\theta,\varphi,\bar{\theta},\bar{\varphi};k_0) E_\alpha \exp\left\{i\sqrt{k_0^2 - k_x^2 - k_y^2}\, z\right\} \tag{2.25}$$

Here, E_α is the received field of polarization α scattered from a remote-sensed object located at distance z. And:

$$k_x = k_x\left(\mathrm{k}_0,\bar{\theta},\bar{\varphi}\right) \tag{2.25a}$$

$$k_y = k_y\left(\mathrm{k}_0,\bar{\theta},\bar{\varphi}\right) \tag{2.25b}$$

$$x = x\left(r,\theta,\varphi\right) \tag{2.25c}$$

$$y = y\left(r,\theta,\varphi\right) \tag{2.25d}$$

It should be pointed out that in the paraxial approximation,

$$k / k_0 << 1 \tag{2.26}$$

where:

$$k = \sqrt{k_x^2 + k_y^2} \text{ and } k_0 = \omega / c \tag{2.26a}$$

The cross-polar component of the weighting matrix $G_{\alpha\beta}\left(r,\theta,\varphi,\bar{\theta},\bar{\varphi};k_0\right)$ is small, i.e.,

$$G_{\alpha\beta}\left(r,\theta,\varphi,\bar{\theta},\bar{\varphi};k_0\right) / G_{\alpha\alpha}\left(r,\theta,\varphi,\bar{\theta},\bar{\varphi};k_0\right) << 1 \tag{2.27}$$

For remote sensing of geophysical objects, condition (2.26) may not be met. This may result in an inaccuracy of the receiving antenna output response calculation. Depending on the application, the error can become significant.

Methods of observation of the scattered field can be based on coherent or incoherent polarimetric radars. A coherent polarimetric radar relies on the phase measurements of the signal in all polarization (channel) combinations. A very accurate phase calibration of the polarization channel is required. Errors, or loss of information of the phase relationship between the polarization channels, result in an incorrect estimation of the elements of the scattering matrix, and may lead to false conclusions regarding the scatterer (object). By preserving the phase information, the coherent polarimetric radar ensures complete characterization of the scattering matrix. This leads to a complete description of the scatterer under investigation.

With incoherent radar reception, on the other hand, we measure only the amplitude (intensity) of the signal, without preserving the phase. In this case, we obtain information on the magnitude of the elements of the scattering matrix, and not on the phase. This leads to an “incomplete” description of the scatterer.

2.4 Radar techniques for polarimetric remote sensing

Various radar techniques are used for polarimetric remote sensing:

- Monostatic and multistatic radars.
- Radars with dual or multi-antenna systems for measuring partial field coherence and cross correlation distance.

- Radars with multi-frequency systems for measuring correlation properties of the field in the frequency domain.
- Doppler polarimetry radars.
- Synthetic Aperture Radar (SAR).
- Radars for atmospheric monitoring.
- Radar with polarization control.

2.4.1 Monostatic and multistatic radars

A monostatic radar has co-located transmitting and receiving antennas. It can be realized with a single antenna using a microwave switch (circulator). In this configuration, the radar measures the field backscattered from the area illuminated by the antenna (see Fig. 2.6). The scattering cross sections (proportional to the modulus square of the scattering matrix elements) are calculated from backscatter field measurements. For coherent measurements, the phase reference is given by a single clock at the TX-RX site.

A multistatic radar can be realized with one transmitter and multiple receivers placed at different distant locations from the transmitter (see Fig. 2.7). In this configuration, the scattering cross section measured by the radar is based on forward scatter propagation. The coherence of the field measurements can be maintained if all phase relationships between the receivers (clocks) and the transmitter (clock) are known. The multistatic radar can provide distinct advantages (except for the complexity of coherence); among them are the following:

- We can probe, at higher resolution, different remote areas and measure field (space) correlations.
- We can measure the variation of the scattered field with changes of the scattering angle.
- The clutter area is reduced by the narrower RX beam.
- We can combine all receivers (each in polarization diversity V, H) weighting the amplitude and phase of the received signals in order to realize a highly coherent series of measurements. This can result in an improved mapping of the illuminated area.

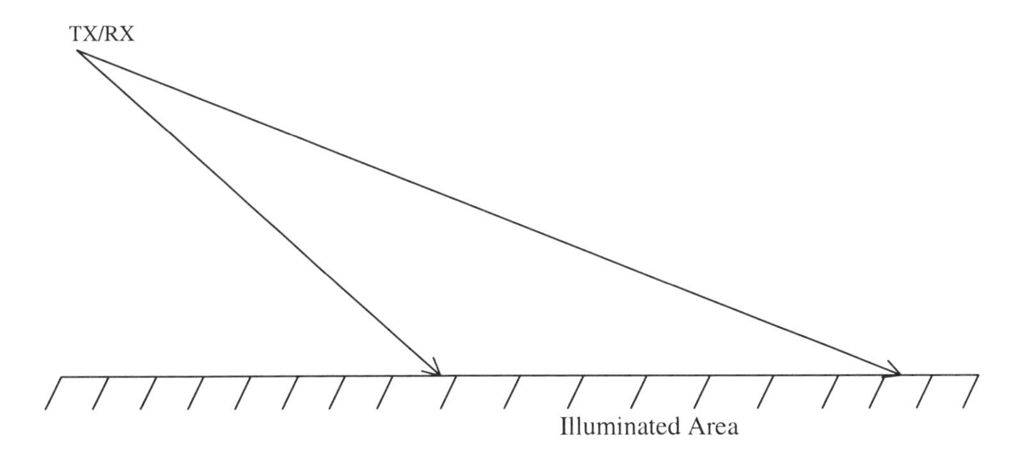

Fig. 2.6 Monostatic radar

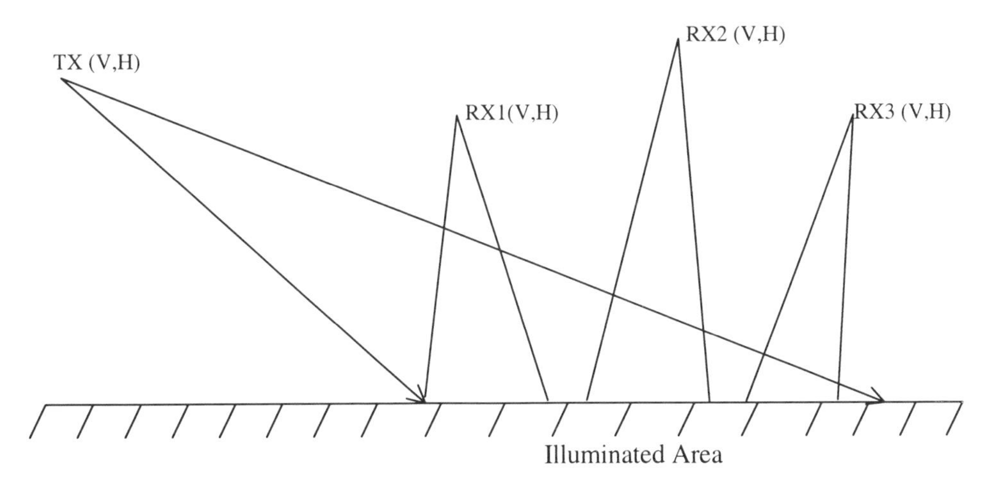

Fig. 2.7 Multi-static radar

2.4.2 *Multi-antenna radar system for measuring field space coherence and correlation distance*

We can obtain information on field space correlation from the knowledge of the angular spectrum of the scattered field [*Gjessing*, 1978; *Kochin*, 1990, 1992]. In Sec. 1.1 we derived Eq. (1.9) which learns that the angular spectrum ($p(\theta)$ in Fig. 1) emitted from a scattering area is the Fourier transform of the spatial autocorrelation function of the field over this area. Applying an inverse Fourier transformation, we can thus derive the spatial field correlation. This can be realized with an antenna array by measuring the angular spectrum $p(\theta)$ of the scattered field. If, for example, this spectrum is a $\sin(x)/x$ function, then the Fourier transform is a rectangular function. If the width of this function (correlation distance of the field) is "L", then we have a relationship between the 3-dB beamwidth of this spatial field pattern and the correlation distance given by

$$\theta_{3-dB} = 0.88\lambda / L \tag{2.28}$$

where λ is the wavelength of the electromagnetic wave. From equation (2.28), the correlation distance is derived from a measurement of θ_{3-dB}, that is

$$L = 0.88\lambda / \theta_{3-dB} \tag{2.29}$$

It is interesting to note that with this method of measuring field correlation it is possible to obtain information on the (transverse) distribution of the scatterers in the remote sensed area. In the geometrical configuration shown in Fig. 2.8, we have

$$x = R\sin\theta \tag{2.29a}$$

In the small angle (far-field) approximation,

$$\theta \sim \frac{x}{R} \tag{2.29b}$$

The angular spectrum $p(\theta)$ at a distance R from the scatterer is proportional to

$$p(\theta) \sim \sigma(x/R) \tag{2.30}$$

where $\sigma(x)$ is the distribution of the scattering cross section of the object (see Fig. 2.8).

If, for example, the object gives constant scattering between limits Δx apart, the angular spectrum $p(x/R)$ is rectangular. The width is $\Delta x/R$, Δx being the width of the object. The Fourier transform of this rectangular angle of arrival spectrum is a $\sin(x)/x$ function, the width of which is given by [*Achmanov*, 1981; *Gjessing*, 1978; *Rytov*, 1978]:

$$L \sim \lambda R/\Delta x \tag{2.31}$$

L being the field correlation distance. From (2.31), the "size Δx" of the object can be derived. The basic principle underlying the measurement of (transverse) distribution of scatterers by means of spatial correlation measurements of the field is illustrated in Fig. 2.8. The distribution of the scatterers forming the object is obtained by measuring the amplitude and phase of the wave backscattered from the object at a number of points in which the elements of the antenna (array) are placed. The analysis of the wavefront using an antenna array is illustrated in Fig. 2.9. An example of theoretical results on the spatial correlation function is given in Fig. 2.10 [*Gjessing*, 1978].

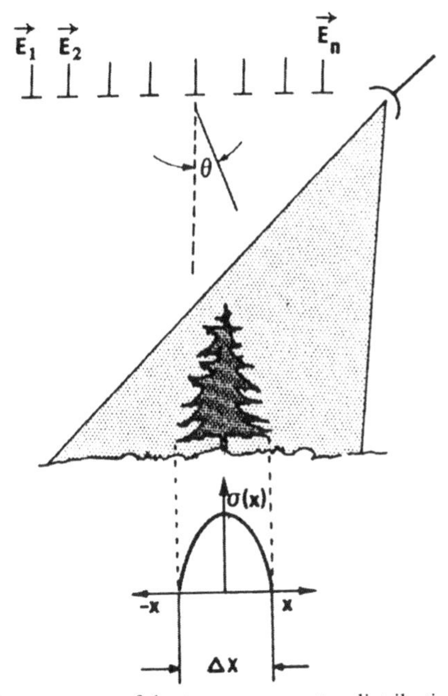

Fig. 2.8 Measurements of the transverse scatter distribution from a correlation measurement of the field strength.
(Reproduced with permission of NATO RTA, D.T. Gjessing, NATO AGARD-LS-93, Fig. 4.7, Part 12, 1978)

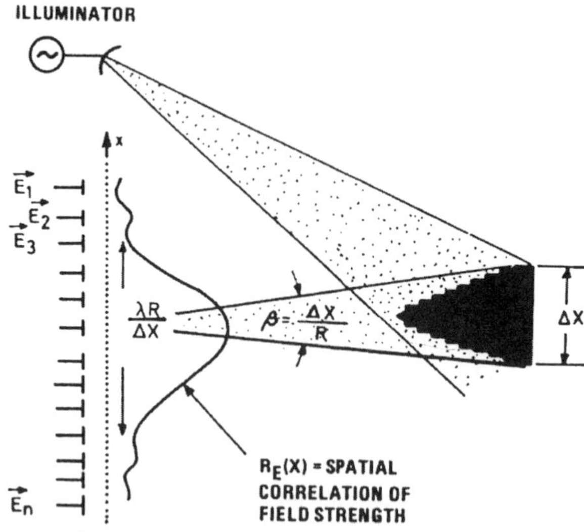

Fig. 2.9 Wavefront analysis using an array matrix (Reproduced with permission of NATO RTA, D.T. Gjessing, NATO AGARD-LS-93, Fig. 6.10, Part 12, 1978)

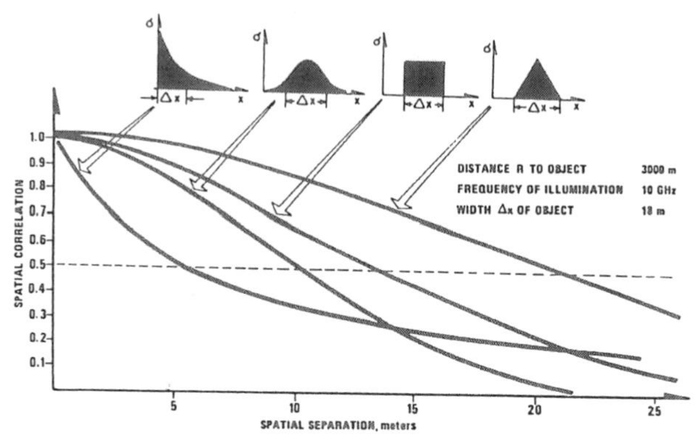

Fig. 2.10 Theoretical results of a spatial correlation function (Reproduced with permission of NATO RTA, D.T. Gjessing, NATO AGARD-LS-93, Fig. 6.11, Part 12, 1978)

2.4.3 Multi-frequency radar system for measuring field correlations in the frequency domain and the frequency correlation bandwidth

We can obtain information on the field frequency correlation from knowledge of the delay spectrum of the scattered field. The delay spectrum is the distribution of the scattered field as a function of the delays produced by the (random) scatterers in the medium under investigation. Under the hypothesis of stationarity and uncorrelated scatterers, the frequency correlation properties of the scattered field can be measured from the Fourier transformation of the delay spectrum [*Schwartz*, 1966]. In Sec. 1.1 we derived Eq. (1.9) which expresses the scattering angular spectrum (proportional to the scattering cross section) as the Fourier transform of the autocorrelation of permittivity fluctuations in space. Thus, frequency correlation analysis of the field can be used to obtain information on the random medium (see block diagram in Fig. 2.11):

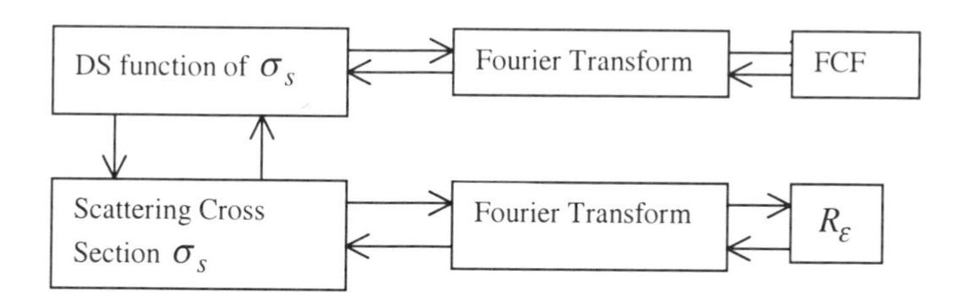

Fig. 2.11 Method of frequency correlation analysis

Here, the following notation is used:

DS = Delay Spectrum
σ_s = Scattering cross section
FCF = Frequency correlation function
R_ε = Autocorrelation of permittivity fluctuations

This method can be realized by measuring the correlation properties of the field (in the frequency domain) from a multi-frequency illumination of the surface under investigation. The correlation as a function of frequency spacing is computed for pairs of CW signals spaced by a variable interval of frequency Δf .

As an example of interest to remote sensing problems, we examine this method for the case of scattering process from vegetation [*Gjessing*, 1978]. The geometry of the scattering process is described in Fig. 2.12. We assume that the ground surface being illuminated consists of coniferous trees having needles distributed in depth in such a way that the shadowing effect becomes progressively more dominant as the wave progresses. We assume an exponential shadowing effect so that the illuminated scattering facets are distributed in an exponential manner in depth. This leads to a set of waves at the receiver, which interfere and the result is a limited correlation bandwidth of the scattering surface. The calculation of this bandwidth is reported as given by [*Gjessing*, 1978]:

The delay function of the reflected waves becomes (see Fig. 2.12)

$$P(\tau) = e^{-\alpha\tau} \tag{2.32}$$

because the distribution in depth of the scattering cross section is assumed to be exponential. The $1/e$ width of this delay function is given by

$$\tau_0 = \frac{1}{\alpha} \tag{2.33}$$

The Fourier transform of the delay function assumes the form

$$F(\omega) = \frac{1}{\alpha + j\omega} \tag{2.34}$$

We shall now calculate the autocorrelation function $R(\Delta\omega)$ in the frequency domain. The voltage V_1 of the signal scattered backwards at frequency ω is given by

$$V_1 = F(\omega) = (\alpha + j\omega)^{-1} \tag{2.35}$$

Similarly, the voltage V_2, at frequency $\omega + \Delta\omega$, becomes

$$V_2 = F(\omega + \Delta\omega) = \left[\alpha + j(\omega + \Delta\omega)\right]^{-1} \tag{2.36}$$

The normalized complex autocorrelation of these two voltages is then given by

$$R(\Delta\omega) = \frac{\int_{-\infty}^{+\infty} (\alpha + j\omega)^{-1} \left[\alpha - j(\omega + \Delta\omega)\right]^{-1} d\omega}{\int_{-\infty}^{+\infty} \left(\alpha^2 + \omega^2\right)^{-1} d\omega} \tag{2.37}$$

Carrying out the integration, we arrive at the following expression for the modulus of the autocorrelation function:

$$\left|R(\Delta\omega)\right| = \left[1 + \left(\frac{\Delta\omega}{2\alpha}\right)^2\right]^{1/2} \tag{2.38}$$

Defining now the correlation bandwidth in the frequency domain to be the half-width of the autocorrelation function, it is found [*Gjessing*, 1978]:

$$\Delta\omega_{3-dB} = 2\alpha\sqrt{3} \tag{2.39}$$

or

$$\Delta F = \frac{\sqrt{3}}{\pi}\frac{c}{\Delta z} \quad Hz \tag{2.40}$$

where $\Delta z = c \cdot \tau_0$ is the penetration depth of the electromagnetic wave, c is the speed of light and τ_0 is defined by Eq. (2.33). It should be noted from Eq. (2.40) that a higher penetration Δz causes a smaller correlation bandwidth. This means that the effect of a finite depth is a decorrelation on the frequency properties of the scattered waves.

In conclusion, measuring the correlation properties of the scattered signal in the frequency domain allows one to obtain information on the distribution in depth σ_z of the contributing scatterers (see Fig. 2.12). Specific analytical results based on the frequency correlation function are given in Fig. 2.13.

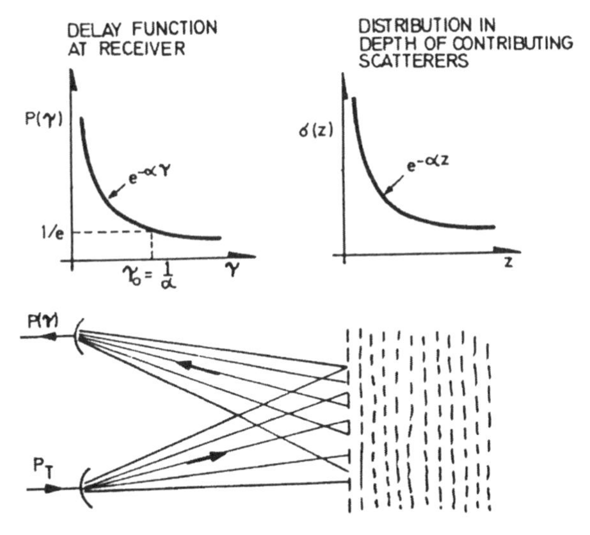

Fig. 2.12 Geometry of the scattering process
(Reproduced with permission of NATO RTA, D.T. Gjessing, NATO AGARD-LS-93, Fig. 4.9 Part 12, 1978)

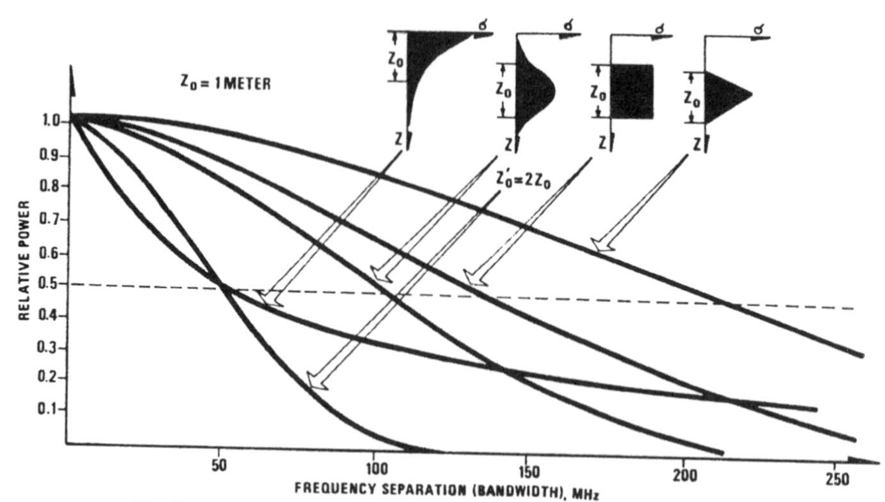

Fig. 2.13 Theoretical results based on the frequency correlation method (Reproduced with permission of NATO RTA, D.T. Gjessing, NATO AGARD-LS-93, Fig. 6.6, Part 12, 1978)

2.4.4 Doppler-polarimetric radar

By means of successive measurements using a coherent radar, we can get information about the dynamic processes of the objects under investigation. Doppler- polarimetric methods in radar remote sensing are based on this principle [*Doviak*, 1988].

Doppler-polarimetry is a methodology for the determination of both Doppler velocity (radial component) and the polarization dependence of a moving scatterer. When the scatterer is moving, the phase of the received scattered signal is determined by the polarization-dependent properties of the scatterer and by the radial velocity of the scatterer. Thus, we cannot distinguish simultaneously in the phase measurements between phase changes due to polarization-dependent properties of the scatterer and changes due to Doppler velocity [*Niemeijer*, 1996]. To resolve the ambiguity in phase, the design of the polarimetric radar waveform is based on the following considerations:

(a) When the Doppler velocity is known, the Doppler frequency-induced phase change in the received signal can be compensated for.

(b) Only those received signals with equal polarizations are considered for Doppler analysis.

The polarimetric radar waveform can be realized by using the polarimetric radar described in Sec. 2.2 and by repeating periodically the sequence composed of the orthogonal polarization states p,q: Transmit p=V, Receive p=V and q=H; Transmit q=H, Receive p=V and q=H. The timing sequence diagram is illustrated in Fig. 2.14:

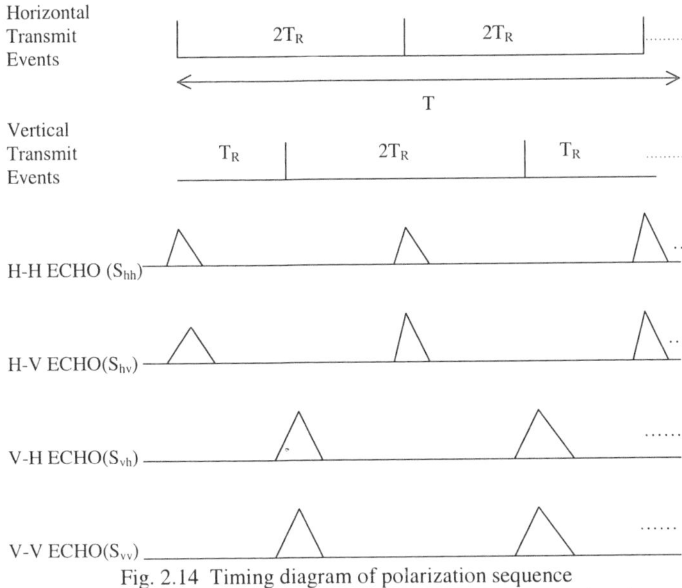

Fig. 2.14 Timing diagram of polarization sequence
T_R: Time interval between consecutive pulses; T: Total integration time;
$S_{hh}, S_{hv}, S_{vh}, S_{vv}$: Scattering elements for the four polarization pairs

This results in the measurement of four received voltages in two pairs V_{VV} and V_{HV}, V_{VH} and V_{HH} that are used to compute the scattering matrix elements $S_{hh}, S_{hv}, S_{vh}, S_{vv}$ [cf. Eqs (2.21)-(2.24)] and the Doppler spectrum of the sensed area.

For each of the four sequences of received voltages, we then obtain the Doppler spectrum by FFT. From analysis of the Doppler spectrum, we can compute the parameters Doppler shift and Doppler spread for each polarization channel. From the results of this Doppler analysis, we may characterize the dynamic behavior of the sensed object.

The four scattering matrix elements are calculated from the four received voltages. From the analysis of the Doppler-compensated scattering elements, information on the type of target can be obtained. The amplitude (radar cross section) measures backscatter strength (size of the target), and the phase (delays) can provide information on the geometrical structure (shape) of the target.

We provide below an example of calculation of the scattering matrix and Doppler:

1. Compute the scattering matrix from the two pairs of voltages V_{HH}, V_{HV} and V_{VH}, V_{VV} spaced by the time interval T_R.
2. Select a time sequence of scattered signals measured in the same polarization channel; for example, the H-echo signals (see Fig. 2.12). We have two H-echo sequences. One for the H polarization transmission, the other for V transmission.
3. Fourier transform each H-echo receive sequence.
4. Calculate the Doppler spectrum for a series of Doppler resolution cells.
5. Extract amplitude and phase information of the Doppler signal for each resolution cell.
6. Repeat the same procedure from step 2 to 5 for the two V-echo time sequences.
7. Compute the scattering matrix from the two pairs V_{HH}, V_{HV} and V_{VV}, V_{VH} for each Doppler bin.

We examine, next, the parameters that are important for the polarimetric radar design.

(a) Correlation time

It must be noted that the design of this polarization radar relies on the time correlation properties of the scattered signals. Many scattering elements on the surface can contribute at any given time. The result is that a set of waves reaches the receiver.
These will have different Doppler and different phases and amplitudes. This situation may lead to scintillations, e.g. fading, that change (randomly) in time. Therefore, the signals will decorrelate. No significant decorrelation occurs if the correlation time of the signals is much longer than the interval between two successive measurements (echoes). In the time interval T_R between two polarization states (see Fig. 2.14), one scattering matrix is acquired at the radar receiver. If τ is the correlation time of the propagation channel, the condition for coherent measurement is

$$\tau >> T_R \tag{2.41}$$

We also note that the correlation time can be considered in terms of Doppler. The correlation time τ is inversely proportional to the width of the Doppler spectrum (Doppler spread σ_F) causing scintillations, viz.,

$$\tau = 1/\sigma_F \tag{2.42}$$

where σ_F is proportional to the distribution of the velocity of the scatterers. It is important to remark that a small Doppler spread gives long correlation times of the signal according to Eq. (2.42).

Consider the following illustrative numerical example:

$$\sigma_F \sim 2F_0\sigma_v / c \tag{2.43}$$

where

$$\begin{aligned} F_0 &= 10 \text{ GHz} \\ c &= 3\cdot 10^8 \text{ m/s} \quad \text{(speed of light)} \\ \sigma_v &= 1 \text{ m/s} \\ \sigma_F &\sim 2.\ 10^2/3 \ \ \text{s}^{-1} \end{aligned} \tag{2.44}$$

From Eq. (2.42), the correlation time τ is given by

$$\tau = 1/\sigma_F = 15ms \tag{2.45}$$

To satisfy the condition (Eq. 2.41) of coherent measurements, T_R should take values much smaller than 15ms. If the velocity distribution of the scatterers is increased to $\sigma_v = 10m/s$, we have a correlation time $\tau = 1.5ms$, which is 10 times smaller than the previous one. In this case, the interpulse period T_R should be reduced to values smaller than 1ms.

(b) Ambiguity

It is also interesting to note that combining Eqs (2.41) and (2.42), we obtain

$$\sigma_F << 1/T_R \tag{2.46}$$

That is, the Doppler spread σ_F must be much smaller than the separation $1/T_R$ between two spectral lines of the signal.

It is worth mentioning that $1/T_R$ defines also the maximum unambiguous Doppler velocity $V_{\max}$ for a sequence of pulses with time repetition T_R. Increasing the interval T_R, the maximum unambiguous Doppler velocity is reduced. So, we need to make T_R very short. Combining Eqs (2.46) and (2.43), we obtain for the Doppler velocity spread

$$\sigma_v << c/2F_0T_R \tag{2.47}$$

Let F_0 = 10 GHz, for example. We have then,

$$\sigma_v << 1.5 \cdot 10^{-2}/T_R \quad \text{m/s} \tag{2.48}$$

For $T_R = 1ms$, we obtain

$$\sigma_v << 15m/s \tag{2.49}$$

For an increase (by a factor of 10) of the time interval, e.g. $T_R = 10ms$, the Doppler velocity spread requirement (2.49) should be decreased to 1.5m/s. The maximum unambiguous Doppler velocity is then reduced by a factor of 10.

(c) Resolution

The frequency increment (resolution) Δf in FFT, is given by:

$$\Delta f = 1/T \tag{2.50}$$

where T is the total integration time of the FFT.

If we need for example a minimum resolution Δv in Doppler speed of the order of 1m/s, we have a Doppler frequency resolution for the radial component at X-band (λ=0.033m) given by

$$\Delta f_D = 2\,\Delta v/\lambda \sim 60\text{Hz} \tag{2.51}$$

For a frequency resolution of 60Hz, we require [cf. Eq. (2.50)] a total FFT integration time T ~ 16ms

(d) Effects of relative motion radar-sensed object

We examine the effects on coherence of measurement in the presence of an object moving in the sensed area. The fluctuations in the random motion of an object cause a phase error in the radar signal; specifically,

$$\sigma_\varphi = \frac{2\pi}{\lambda}\sigma_y \tag{2.52}$$

where σ_φ is the standard deviation of the signal phase, and σ_y is the standard deviation of the fluctuations of the motion around a y-direction. The standard deviation on the (Doppler) frequency is given by

$$\sigma_f = \frac{2}{\lambda}\sigma_{\dot{y}} = \frac{2}{\lambda}\sigma_{\Delta v} \tag{2.53}$$

where $\dot{y}$ is the time variation of the motion along the y direction. This variation depends on the speed of the object, on the deviation angles from a straight line motion and on the "aspect" angle of the object with respect to the radar. If we assume, as a simple example, a target speed of 20m/s and the variation of speed equals $\Delta v =$ 0.2m/s, we may find from Eq. (2.53) at X-band

$$\sigma_f \sim 13Hz \tag{2.54}$$

that gives, using Eq. (2.42), a correlation time

$$\tau \sim 66ms \tag{2.55}$$

We now examine the effects of motion composed of a translation + rotation of the object. The object is considered to be composed of two point scatterers, the center of gravity O moving at constant velocity, and a point P rotating around O at constant rotation speed. The model of the object and its motion is given in Fig. 2.15:

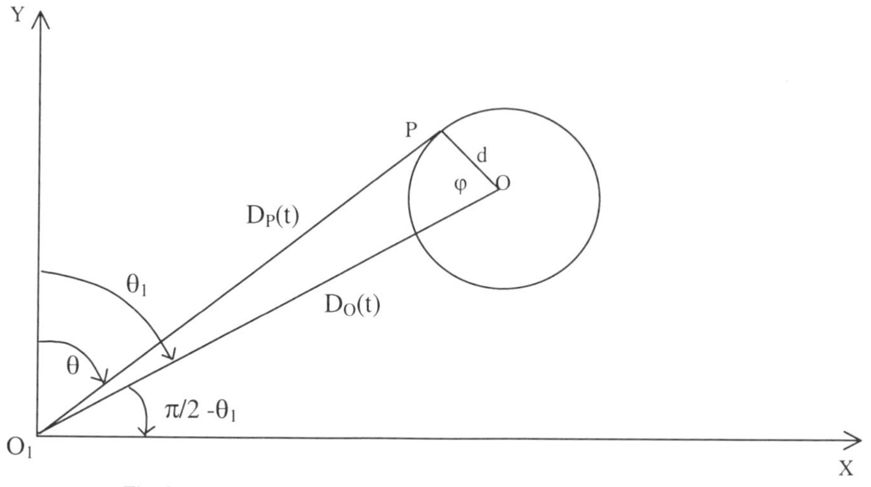

Fig. 2.15 Object (2-scatterer model) motion analysis (translation + rotation)

The total signal received by scattering from the two points, point O (center of gravity) and point P at distance d from point O, is given by

$$S_T = A_0 e^{\left[j\omega\left(t-\frac{2D_0(t)}{c}\right)\right]} + A_P e^{\left[j\omega\left(t-\frac{2D_P(t)}{c}\right)\right]} \tag{2.56}$$

where

$$D_0(t) = \mathrm{v}_0 t \text{ (uniform rectilinear motion of point O)} \tag{2.57}$$

$$D_P(t) = D_0(t) - d\cos\varphi(t) \text{ (rotation of P around O)} \tag{2.58}$$

and φ is the rotation angle of P around O.

By FM discrimination (time derivative of the received signal) we can derive the Doppler information from the amplitude variation of the received (Doppler) signal given by (see detailed computation in appendix A):

$$\sqrt{\dot{S}^2_{T_{Doppler}}} = \frac{4\omega \mathrm{v}_0}{c}\left[2(1+\cos\alpha)\left(1+\frac{\mathrm{v}_P}{\mathrm{v}_0}\sin\varphi\right)+\left(\frac{\mathrm{v}_P}{\mathrm{v}_0}\sin\varphi\right)^2\right]^{1/2} \tag{2.59}$$

where α is defined by

$$\alpha = \frac{2\omega d}{c}\cos\varphi \tag{2.60}$$

and v_P is the rotation speed (defined by $\dot{\varphi}/d$) of point P around O.

The maximum value (for α=0 and $\varphi = \pi/2$) of the Doppler signal is derived from Eq. (2.59) for $\mathrm{v}_P = 2\mathrm{v}_0$:

$$\sqrt{\dot{S}^2_{T_{Doppler}}}_{Max} = 8\left(\frac{2\omega \mathrm{v}_0}{c}\right) \tag{2.61}$$

The Doppler spread in this case is given by

$$\sigma_{f_{Doppler}} = 16\frac{\omega}{c}\sigma_{vo} \Big/ 2\pi = \frac{16}{\lambda}\sigma_{vo} \tag{2.62}$$

yielding the correlation time

$$\tau = \frac{1}{\sigma_{f_{Doppler}}} = \frac{2\pi\ c}{16\omega\sigma_{v_0}} \tag{2.63}$$

For a variation of target speed $\sigma_{vo} = 0.2 m/s$, we have in X band a Doppler spread of about 107Hz, and a correlation time of around 9ms. It is noticeable that this value of correlation computed for this example of rotation of the target is about 8 times shorter than the value computed in the first example without rotation [see Eqs (2.51)-(2.55)]. This result indicates that the type of motion has effects on correlation. In this example, the effect of rotation reduces the correlation. The time interval T_R between two polarization states must therefore be reduced to obtain coherent measurements [cf. Eq. 2.41] of the scattering matrix.

CHAPTER 3

Physical and Mathematical Modelling

The general physical problem in remote sensing is to measure the scattering of electromagnetic waves from (random) media. The design of remote sensing radars requires knowledge of radio-wave scattering processes for various types of media.

Intensive experimental investigations have been carried out in many countries in order to extract statistics of signals reflected from various geometries under different conditions. However, consideration of all types of surfaces and all types of geometric configurations is practically impossible. That is why modelling of the radio-wave scattering processes is necessary. The possibility of creating complicated scattering environments (e.g. surfaces with arbitrary degrees of roughness) and using signals at various frequencies, with various degrees of coherence and with different polarizations, are some of the advantages of using modelling of radio-wave scattering processes.

In remote sensing problems we distinguish two kinds of modelling, physical and mathematical:

- Physical modelling is based on the physics of the interaction between an electromagnetic field and the scattering medium. The goal for physical models is to obtain detailed insight during pattern recognition and to make optimal use of available information (e.g. surface or volume scattering).

- Mathematical modelling is based on the statistics of the scattered signal parameters. A mathematical model makes use of the echo-signal statistics and compares them with known statistical distribution functions.

3.1 Physical modelling

In the interaction of an electromagnetic field with the atoms in a dielectric medium, the atoms become small electromagnetic oscillators (electric oscillating dipoles) radiating waves in all directions. Atoms and (non-polar) molecules become polarized upon the action of an external electromagnetic field. For natural polar molecules, the effect of an electromagnetic field is to align all molecules in the same direction as the field. The permittivity is a function of the time response of the medium (relaxation time), thermodynamics conditions (pressure, temperature) and the frequency of the field. It can also depend on the direction of the field if the scattering medium is not isotropic. In general, the permittivity in a three-dimensional space is characterized by a polarization tensor. Each component of the tensor relative to vacuum is a complex number, given by

$$\varepsilon_r = \varepsilon' - j\varepsilon'' \tag{3.1}$$

A specific model for the relative permittivity ε_r based on the physical properties of the medium is given by Debye's theory [*Fieschi*, 1976]. The real part ε' is given by

$$\varepsilon' = \varepsilon_\infty + \frac{\varepsilon_s - \varepsilon_\infty}{1 + (\omega\tau_l)^2} \tag{3.2}$$

Where

ε_s = static relative permittivity at zero frequency
ε_∞ = infinite frequency relative permittivity
ω = angular frequency
τ_l = relaxation time of the medium

The imaginary part ε'' is equal to

$$\varepsilon'' = \frac{(\varepsilon_s - \varepsilon_\infty)\omega\tau_l}{1 + (\omega\tau_l)^2} \tag{3.3}$$

The relaxation time depends on the temperature [*Landau,* 1998]. For example, at room temperature we have a relaxation time for sea water of approximately $10^{-11}s$ [*Fieschi,* 1976; *Stratton,* 1941].

The effective conductivity σ_c of the medium can be derived from Eq. (3.3) by multiplying with $\omega\varepsilon_0$:

$$\sigma_c = \varepsilon_0 \frac{(\varepsilon_s - \varepsilon_\infty)\omega^2\tau_l}{1 + (\omega\tau_l)^2} \tag{3.4}$$

ε_0 being the permittivity of vacuum $\left(8.854 \cdot 10^{-12} F/m\right)$. For example, sea water at room temperature, with relaxation time $\tau_l = 10^{-11} s$, $\varepsilon_s = 81$, $\varepsilon_\infty = 1.8$ and for a frequency of $10GHz$, has conductivity equal to $20 mhos(S)/m$.

The permittivity of air is not frequency dependent in the microwave spectrum. It is related to the refractive index n, by the formula

$$n = \sqrt{\varepsilon_r} \tag{3.5}$$

Usually, the quantity N, known as the refractivity, is measured by radio-sonde. It is given by

$$N = (n-1)10^6 \tag{3.6}$$

From a model [*Levy,* 1989] of the refractivity N of the air widely used for atmospheric radar applications, an expression of the permittivity is derived as a function of the thermodynamic parameters of the atmosphere using Eqs (3.5) and (3.6); specifically,

$$\varepsilon_r = 1 + \frac{155.2 \cdot 10^{-6}}{T} p + \frac{0.746}{T^2} e_h \tag{3.7}$$

Where

T = temperature (Kelvin) of the air
p = air pressure (millibars)
e_h = vapor pressure (millibars)

The vapor pressure can be expressed with empirical formulas in terms of the relative humidity and saturated vapor pressure, the latter being a function of temperature [*Levy,* 1989; *Kireev,* 1968].

In remote sensing problems we are interested in radar waves interacting with media (oscillating dipoles) on the surfaces of discontinuities between different media (with different permittivity properties) and between a defined object (scatterer) and the medium in which the object is embedded; also, in a region where there are permittivity fluctuations.

3.1.1 Wave-surface scattering

To illustrate the effects of the discontinuity between two media with different permittivities on an incident plane electromagnetic wave, let us consider the case of an interface separating a vacuum half space (upper medium) and a dielectric half-space (lower medium) with permittivity ε_r. If the interface (surface) is perfectly smooth, the incident field excites the atomic oscillators in the propagation medium so that the scattered field consists of two waves components: one reflected at an angle equal to the angle of incidence in the upper medium and the other at an angle different from the incident angle (refracted wave) in the lower medium. If the surface is rough relative to wavelength, some of the energy of the incident wave is scattered in all directions. A qualitative relationship between surface roughness and scattering is illustrated in Fig. 3.1.

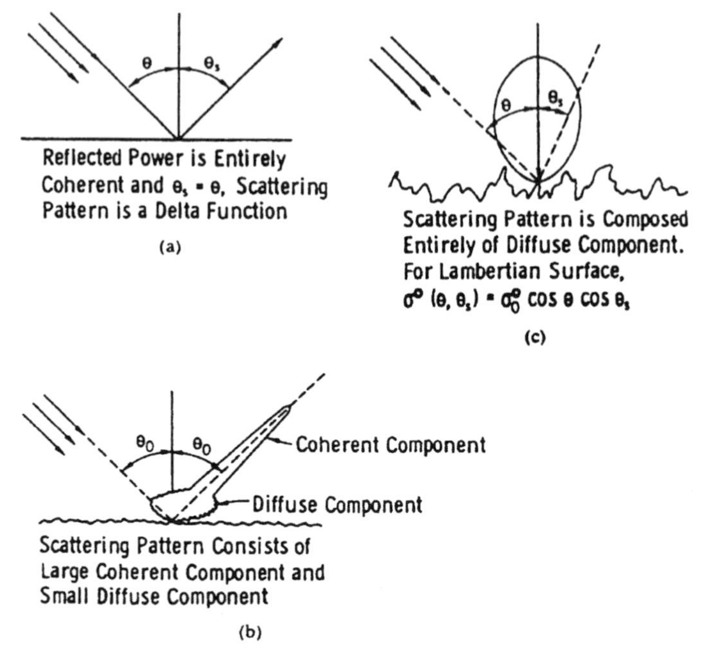

Fig. 3.1 Relative contributions of coherent and diffuse scattering components for different surface roughness conditions: (a) specular, (b) slightly rough and (c) very rough (Lambertian surface)
(Reproduced with permission of Artech House, F.T. Ulaby, "Microwave Remote Sensing", Vol. II, Fig. 11.4, Ch. 11, 1982)

We examine, next, a few physical models of surface scattering:

a. Point scatterers

Within the framework of this model, the surface behaves as a series of radiating points. We can make various assumptions depending on the physical characteristics of the radiating points (scatterers). The assumption of isotropic scatterers is valid for dimensions of the scatterers (or particles of the scattering medium) smaller than the wavelength. The assumption of uncorrelated scattering is valid if the separation between the scatterers is larger than the wavelength. The assumptions of isotropic and uncorrelated scatterers can be applicable, for example, for vegetation-covered surfaces having leaves with sizes small compared to wavelength and separated by many wavelengths.

In general, the total electric field E received from N_s scatterers is the resultant of the phasors of the individual scatterers:

$$E = \sum_{k=1}^{N_s} E_{0k} e^{-j\varphi_k} e^{-j\frac{2\omega R_k}{c}} \tag{3.8}$$

Here, E_{0k} and φ_k are respectively the amplitude and phase associated with the k-th scatterer (scattering properties of the scatterer) and determine the response (or radiation pattern) of the scatterer. Both E_{0k} and φ_k are in general determined by the reflection characteristics (Fresnel coefficients) of the target points. The phase term $\exp[-j\omega R_k / c]$ is associated with the path-delay of the echo from the k-th scatterer. The following notation is used:

ω = angular frequency
c = speed of light
R_k = range of the k-th scatterer
N_s = number of scatterers

The assumption of isotropic radiation of each scatterer point means that the k-th phasor contribution $E_{ok}e^{-j\varphi_k}$ in this model is the same in all directions. Only the path delay term given by $2\omega R_k / c$ contributes to the directionality of the scattering pattern. From Eq. (3.8), we note that if the scatterers are fully correlated and have equal amplitudes, the response becomes

$$\langle |E^2| \rangle \approx N_s^2 \langle |E_i|^2 \rangle \tag{3.9}$$

where E_i is the response of the i-th scatterer.

Eq. (3.8) can be written in the form

$$E = \sum_{k=1}^{N_s} E_{0k} e^{-j\theta_k} \tag{3.10}$$

with

$$\theta_k = \varphi_k + \frac{2\omega}{c} R_k \tag{3.11}$$

For equal phase increments $\Delta\theta$ of signals reflected from adjacent scatterers, Eq. (3.10) becomes

$$E = \sum_{k=1}^{N_s} E_{0k} e^{-jk\Delta\theta} \tag{3.12}$$

The scattering pattern is in this case of the type

$$\langle |E|^2 \rangle \approx \left| \frac{\sin N_s \dfrac{\Delta\theta}{2}}{\sin \dfrac{\Delta\theta}{2}} \right|^2 \tag{3.13}$$

On the other hand, if the scatterers are completely uncorrelated, but still have equal amplitudes, we find

$$\langle |E|^2 \rangle \approx N_s \langle |E_i|^2 \rangle \tag{3.14}$$

and the scattering pattern is of the type

$$\langle |E|^2 \rangle \sim \cos\theta \ \cos\theta_2 \tag{3.15}$$

which is the "Lambertian" law for diffuse scattering (Figure 3.1c). Finally, if the scatterers are partially correlated we get

$$\langle |E|^2 \rangle \approx \sum_{k=1}^{N_s} \langle |E_i|^2 + 2 \sum_{i=1}^{N_s - 1} \sum_{k=1}^{N_s - i} \langle E_k E_{k+i}^* \rangle \tag{3.16}$$

The auto and cross correlations of the individual scatterers depend on the fluctuation statistics of the amplitudes and phases of the signals E_i (due to propagation, or relative movement of radar-scatterer). Depending on the values of these correlations, the total average power $\langle |E|^2 \rangle$ [cf. Eqs (3.8), (3.9) and (3.14)] will be in the interval $[N_s, N_s^2]$. The effect of partial correlation is to "modulate" the scattering pattern as shown in Eq. (3.16). This "modulation" can result in broadening of the beam (for example due to the radar-scatter motion), or in filling of the nulls of the pattern (due to path-propagation fluctuations, or multi-path).

b. Facets model

If the radius of curvature of a scatterer is much larger than the wavelength, we may replace it with a finite smooth surface (facet). The incident wave induces an electric field upon the surface of a facet. The pattern of the scattered field will have a form similar to the pattern radiated by an antenna with the same size of the facet. The facet model and the dependence of the scattering process on the size of the facet (compared to wavelength) are illustrated in Figs 3.2 and 3.3:

Fig. 3.2 Representation of a rough surface as a collection of facets

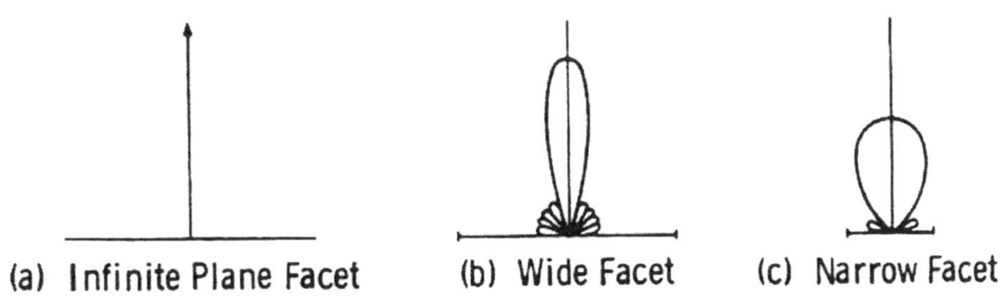

Fig. 3.3 Re-radiation patterns of various sizes of facets at normal incidence. (Reproduced with permission of Artech House, F.T. Ulaby, "Microwave Remote Sensing", Vol. II, Fig. 11.16, Ch. 11, 1982)

It is interesting to note that large facets make larger contributions to the signal if they are "properly" oriented. But their major radiation is in the specular direction away from the radar. For this reason, smaller facets, that are less directional, may give significant contributions to the backscattered signal. To illustrate the dependence of the backscattered signal on these geometrical factors, let us assume a plane wave incident on a facet of size L_f and permittivity ε. The incident field at the surface is given by [*Elachi*, 1987]

$$E_i(x) = E_0 e^{-jkx\sin\vartheta_i} \tag{3.17}$$

Where $k = 2\pi/\lambda$ is the wave-number and ϑ_i is the incident angle with respect to normal. The reflected field has the form

$$E_r(x) = \rho E_0 e^{-jkx\sin\vartheta_i} \tag{3.18}$$

Where ρ is the Fresnel reflection coefficient. The scattered field is proportional to

$$E_s \approx \mathbb{F}\left(\vartheta_i, \vartheta_{s_i}, L_f\right)\rho E_0 \tag{3.19}$$

Where $\mathbb{F}$ is the radiation pattern as a function of the incident angle ϑ_i and the radiation angle ϑ_{s_i} for an antenna of length L_f. The facet model assumes that the total field is the summation of the fields from each facet. The total contribution to the received signal depends on the orientation (slope α_i) of each facet with respect to the radar (incident direction). Only the facets with slope appropriate to beam the signal back to the radar contribute maximally to the total received signal. The total field is proportional to [cf. *Elachi*, 1987]

$$E_s \approx \rho E_0 \cdot \sum_{i=1}^{N} \left\{ \mathbb{F}\left[(\vartheta_i + \alpha_i), (\vartheta_{s_i} + \alpha_i), L_{f_i} \right] \right\} \tag{3.20}$$

Where α_i is the slope, ϑ_i the incident angle, ϑ_s the radiation (scattering) angle and L_{f_i} the size, of the i-th facet. An illustration of re-radiation from different facet-slopes is given in Fig. 3.4.

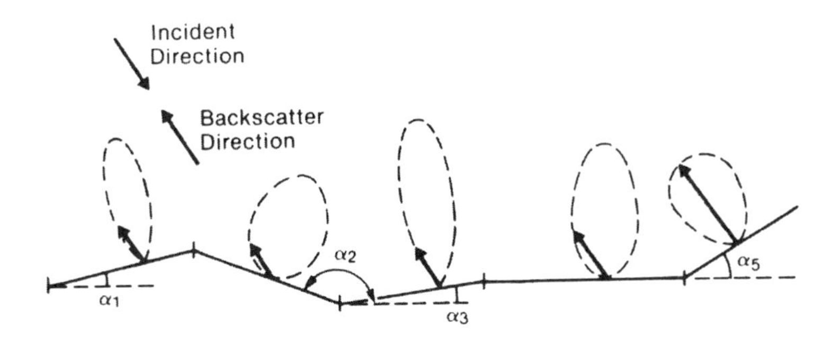

Fig. 3.4 Re-radiation patterns from various facets.
(The waves in the direction of the arrows add together to form the backscattered wave)
©1987 IEEE (Reproduced with permission of IEEE, C. Elachi, "Spaceborne Radar Remote Sensing: Applications and Techniques", Fig. 2.6, Ch. 2, 1987)

c. Bragg resonance model

At large angles of incidence ϑ (small grazing angles $\pi/2-\vartheta$), the facets model may become inadequate. From the point of view of small grazing angles, the facet planes may lie in "shadow zones" and do not contribute to scattering. Only a few points (the highest points of the discontinuities between two facets) would act as effective point scatterers. An example of this geometry is ship-borne radar surveillance. At small grazing angles, we may have as dominant (scattering) mechanism the scattering from tiny capillary waves developing on top of the larger gravitational waves. The capillary waves have sizes in the millimeter to centimeter range in the same order of magnitude as the (microwave) wavelength. This is opposite to the facet model based on facet planes larger than the wavelength. This scattering mechanism can be described with the Bragg model. In this model, the random surface is divided into its Fourier spectral components. The scattering is mainly due to the spectral components of the waves that are in resonance with the wavelength of the incident field (see Fig. 3.5).

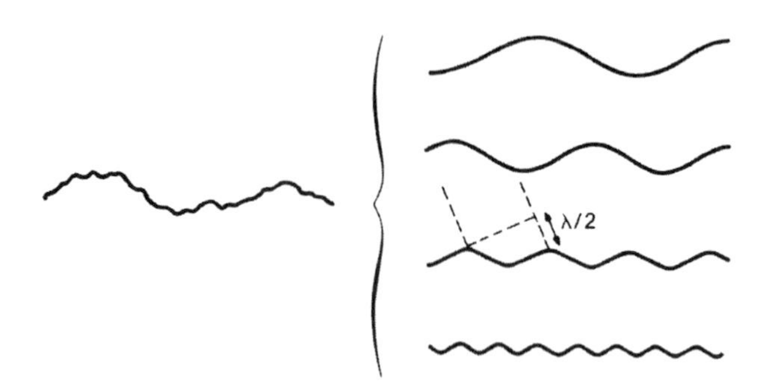

Fig. 3.5 Bragg scattering. (Field resonance with Fourier components of the wave) ©1987 IEEE (Reproduced with permission of IEEE, C. Elachi, "Spaceborne Radar Remote Sensing: Applications and Techniques", Fig. 2.7 Ch. 2, 1987)

To illustrate the Bragg resonance phenomenon, we consider a sinusoidal component of the surface spectrum and an incident plane wave at an angle of incidence ϑ. The wavelength of the surface component is L_0 and the radar wavelength is λ. The received voltage is given by the phase-coherent sum of the voltages from the individual components of the waves [*Ulaby*, 1982]

$$V_r = \sum_{n=1}^{N} V_0 e^{-j2kR_0} \cdot e^{-j2kn\Delta R} \tag{3.21}$$

Here,

V_0 = amplitude of the received signal
k = $2\pi/\lambda$ wavenumber
R_O = distance from the radar to the scatterer
ΔR = path-differential from the source to each successive wavecrest
N = total number of wavelengths of the resonant component within the illuminated area

Following the same procedure used to derive Eq. (3.13), the sum of Eq. (3.21) is found proportional to [*Ulaby*, 1982]:

$$V_r \approx \frac{\sin\left[k(N+1)\Delta R\right]}{\sin\left[k\Delta R\right]} \tag{3.22}$$

The Bragg resonance condition is (see Fig.3.6) given by:

$$k\Delta R = \frac{2\pi}{\lambda}\Delta R = m\pi \quad ; \qquad m=1,2,..... \tag{3.23}$$

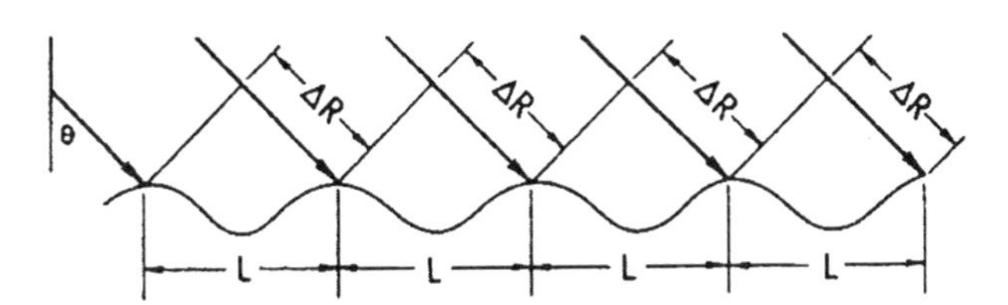

Fig. 3.6 Bragg scattering (in phase addition when $\Delta R = m\lambda/2$)
(Reproduced with permission of Artech House, F.T. Ulaby, "Microwave Remote Sensing", Vol. II, Fig. 11.20, Ch. 11, 1982)

The Eq. (3.23) can be written in terms of the spatial wavelength L_0 and the angle of incidence ϑ:

$$L_0 = m\frac{\lambda}{2}\sin\vartheta \quad ; \qquad m=1,2.... \tag{3.24}$$

The first term (m=1) leads to the strongest scattering.

3.1.2 Wave-scatterer (object) interaction

In the case of an object embedded in a medium, the object can be detected remotely by the radar if its permittivity is different from that of the medium, i.e., if we have "electric contrast." The physical modelling of permittivity is of great importance in order to predict this contrast. An example of an application of this type of modelling is the ground penetrating radar (GPR) used for mine detection, civil engineering applications, oil search, etc. [*Finkelshtein,* 1984, 1994; *Yarovoy*, 1998; *Cerniakov*, 1997]. An embedded object can be modelled, for example, using the model of multiple point scatterers described for the case of surface scattering (see Sec. 3.1.1).

Research activities have been conducted [*Brancaccio*, 1998, 2000; *Pierri*, 1999, 2000] in inverse scattering (non-linear) modelling to reconstruct the shape of unknown scatterers, for GPR applications. Results of this research show the possibility to discriminate small cylindrical shaped objects separated by 20-40 cm in the ground. Research on antennas (bow-tie type) has recently been developed to allow optimal ground penetration with GPR [*Lestari*, 2000].

3.1.3 Wave-medium (volume) scattering

Consider the geometrical configuration shown in Fig. 3.7. If the lower medium is inhomogeneous, or is composed of a mixture of materials with different dielectric properties, then a portion of the waves scattered by the inhomogeneities may cross the boundary surface into the upper medium. Scattering takes place within the volume in the lower medium (volume scattering).

Volume scattering is caused mainly by dielectric discontinuities within the volume. In general, the spatial locations of the discontinuities are random. The scattering depends on the dielectric discontinuities inside the medium, the density of the embedded inhomogeneities (or the variance of the dielectric fluctuations for a random medium) and the geometric size of the inhomogeneities compared to radar wavelength. The latter is a statistical quantity for random media, e.g. the correlation length of the dielectric fluctuations.

The mechanism of volume scattering causes a distribution of the energy of the transmitted wave into directions other than the incident, giving rise to an angular spectrum and resulting in an attenuation of the on-going wave within the medium (scattering loss). The propagating wave inside the medium also experiences a loss due to conduction. The total loss (the sum of the scattering and conduction losses) is referred to as extinction.

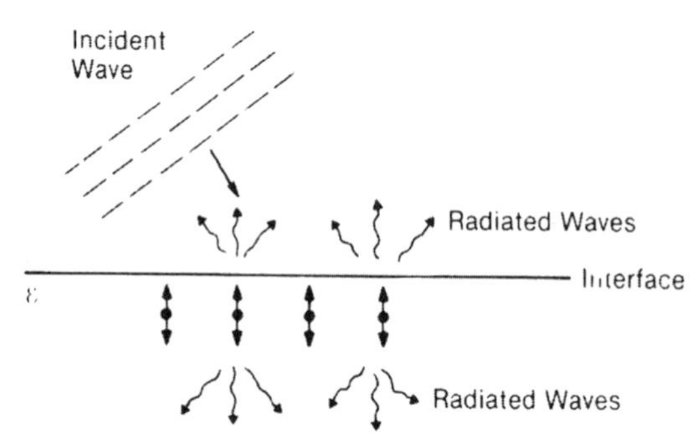

Fig. 3.7 Mechanism of volume scattering

Last decades, several models have been introduced blending surface and volume scattering [*Theocarov*, 1985; *Andreev*, 1985; *Chain*, 1989; *Kuznetsov*, 1992]; e.g., an intermediate layer sandwiched between two homogeneous layers. The interaction of electromagnetic radiation with a medium is described as volume scattering. Various models of vegetation [*Liang*, 1993; *Karam*, 1992] are handled in this manner.

The loss due to conductivity can be estimated with simple considerations on the permittivity of the medium based on physical medium properties described in Sec. 3.1. The imaginary part of the permittivity represents the capability of the medium to absorb electromagnetic energy and transform it to another type of energy (e.g. heat). Consider, for example, a plane wave propagating in a lossy homogeneous medium [*Elachi*, 1987], viz.,

$$E = E_0 e^{j\sqrt{\varepsilon_r} k_0 x} \quad ; \quad \varepsilon_r = \varepsilon' - j\varepsilon'' \quad ; \quad k_0 = \omega / c = 2\pi / \lambda \tag{3.25}$$

If we assume that $\varepsilon'' << \varepsilon'$, then

$$\sqrt{\varepsilon_r} = \sqrt{\varepsilon' - j\varepsilon''} \cong \sqrt{\varepsilon'} - j\frac{\varepsilon''}{2\sqrt{\varepsilon'}} \tag{3.26}$$

and

$$E \cong E_0 e^{-\alpha_a x} e^{j\sqrt{\varepsilon'} k_0 x} \tag{3.27}$$

with the attenuation given by

$$\alpha_a = \frac{\varepsilon'' k_0}{2\sqrt{\varepsilon'}} = \frac{\pi \varepsilon''}{\lambda \sqrt{\varepsilon'}} \tag{3.28}$$

The power of the wave can be written as a function of x as follows:

$$P(x) = P(0) e^{-2\alpha_a x} \tag{3.29}$$

The penetration depth L_p is defined as the distance at which the power reduces to the value $P(0)e^{-1}$, corresponding to a loss of $4.3dB$. We have, then [*Elachi*, 1987],

$$L_p = \frac{1}{2\alpha_a} = \frac{\lambda\sqrt{\varepsilon'}}{2\pi\varepsilon''} \tag{3.30}$$

This can be expressed as a function of the medium loss tangent $(\tan\delta = \varepsilon''/\varepsilon')$, so [*Elachi,* 1987]:

$$L_p = \frac{\lambda}{2\pi\sqrt{\varepsilon'}\tan\delta} \tag{3.31}$$

It can be seen from Eq. (3.31) that penetration depth decreases with the increase of frequency. The loss tangent of the natural surfaces depend on temperature, humidity and salinity and it varies over a wide range of values. For pure ice, dry soil and permafrost it is less than 10^{-2}. For wet soil, sea ice and vegetation, it is around 10^{-1}.

The loss tangent increases and penetration decreases, with the presence of liquid water present in the medium. For example [*Daniels,* 1996], for sandy dry soil ($\varepsilon' \sim 5, \quad \tan\delta \sim 0.01$), the penetration depth [cf. Eq. (3.31)] is $L_p \sim 7\lambda$; for sandy wet soil ($\varepsilon' \sim 20, \quad \tan\delta \sim 0.1$), the penetration depth is reduced to $L_p \sim 0.3\lambda$.

The penetration of the field in the medium is inversely proportional to the effective conductivity, which, in turn, is proportional to the loss tangent. For a medium characterized by a large effective conductivity, the field is almost completely reflected. This is the case for sea water at microwave frequencies.

3.1.4 Effects on the polarization state of an electromagnetic wave

A physical model of a medium is important because its permittivity (electrical properties) and other physical properties (inhomogeneities, discontinuities, etc.) affect the polarization state of an electromagnetic wave. For example, the electric properties of the sea surface cause horizontally polarized waves to be better reflected than vertically polarized ones [*Stratton,* 1941]. For a particular angle of incidence on a smooth surface (i.e., Brewster's angle), the signal level of a vertically polarized wave can vanish completely (the reflection coefficient equals zero). So, we have a "polarizing" effect; only the horizontal component of a field incident at the Brewster's angle is reflected.

Propagation modelling is important in the study of the effects of scattering, or multiple reflections, on the polarization of electromagnetic waves. For example, in the case of bi-static radar a two-ray propagation path model is used to describe the direct and reflected (indirect) paths between the transmitter and the receiver. The propagation over these two paths induces an interference pattern at the receiver. The received field is proportional to

$$R_F \approx 1 + \rho \cdot e^{-j\varphi} \cdot e^{-j\Delta\vartheta} \tag{3.32}$$

where

R_F = resultant field (ratio with the direct field)
ρ = reflection coefficient (amplitude)
φ = phase of the reflection coefficient
$\Delta\vartheta$ = phase difference between direct and reflected paths

In the case of reflection from sea surface (good conductor), the magnitude of the reflected wave is much smaller for vertical polarization than for horizontal polarization. This reduces the amplitude of the vertically polarized sea-reflected wave. The major contribution to the resultant field is given via the direct path. In this case (vertical polarization), the nulls of the interference pattern are reduced compared to those for horizontal polarization. With horizontal polarization, we have significant contribution from both direct and reflected paths, resulting in interference patterns which can have deeper nulls.

Effects on polarization arise also from the degree of roughness of a reflecting surface. In the limiting case of a very rough surface, the direct propagation path is dominant with respect to the reflected path for both polarizations (vertical or horizontal). The interference pattern in this situation has no deep nulls.

In general, the complex reflection coefficient (amplitude and phase) is different for the vertical and horizontal components of the electromagnetic field. This produces the effect of rotation or "depolarizing" of the reflected electric field. Changes in the phase of the reflection coefficient add to or subtract from the phase difference $\Delta\vartheta$ between the direct and reflected paths. This may cause variations in the interference pattern as well.

3.1.5 System design aspects

Physical modelling is important whenever the problem of remote sensing necessitates knowledge of the dependence of the scattered field characteristics on frequency, polarization and propagation conditions. A minimum number of different remote sensing fields should carefully be selected in order to optimize the process of recognition of various types of surfaces (scatterers) with an adequate degree of accuracy. Criteria of optimality may be found by a comparison with classes of concretely solved problems. However, physical modelling of an investigated object should always be at the core of the investigations. This approach, in our point of view, precisely motivates different methods and techniques of remote sensing and answers the question "why remote sensing at wavelength λ_2 has to be done even when remote sensing data at wavelength λ_1 is available." It is because new additional information is obtained and the relationship between data at λ_1 and λ_2 can be crucial in investigating the problem. An important motivation for physical modelling is to get a deeper insight into the core of the pattern recognition problem. Such insight may uncover new features of an investigated object and, thus, is fundamental.

3.2 Mathematical modelling

Once a physical problem is established (for example, scattering of electromagnetic waves from vegetation-covered ground surfaces), a mathematical model is needed to solve it, e.g. to compute the scattering cross section of the surfaces under investigation.

3.2.1 Description of the mathematical model

The mathematical model consists of the following components:

a) Maxwell's equations governing the electromagnetic fields in air, vegetation, soil or sea.
b) The electromagnetic properties of the ground region (complex permittivity, losses, anisotropy and inhomogeneities).
c) Statistical characterization of the vegetation layer.
d) Statistical characterizations of the ground-vegetation and vegetation-air random rough interfaces.

For the computation of the scattering cross section (matrix), we proceed through the following steps:

1) Consider a physical model of scattering (see Sec. 3.1) that describes the physical problem to be solved. For example, the scattering of electromagnetic waves from a surface (surface scattering model) covered by vegetation (scattering layer model).
2) Define the conditions for the electrical properties (permittivity) and the physical characteristics (roughness, etc.) at the boundaries where the scattering (boundary phenomenon) occurs.
3) Postulate the statistics of the fluctuations of the scatterer's properties. For example, the statistics of surface heights variations (surface scattering) or the statistics of the permittivity fluctuations (volume scattering).

4) Compute the scattering cross section σ_s, i.e., the ratio between the scattered power per unit solid angle and the power density (per unit area) incident upon the scatterer, from Maxwell's equations with the boundary conditions as defined in steps 1, 2 and 3. Different methods are used to compute σ_s depending on the type of scattering (single or multiple) and the size L_s of the scatterers compared to the radio wavelength λ. Use is made of the Kirchhoff method if $\lambda << L_s$, or the "small perturbation" method if $\lambda \geq L_s$ [*Ulaby*, 1982]. For the case of multiple scattering, the "radiative transfer theory" is used; the latter is based on the physics of transport of energy through a medium (volume) composed of particles [*Ulaby*, 1990] or on the higher order perturbation theory [*Ogilvy*, 1991].
5) The result in the Kirchhoff "small perturbation" or radiative transfer methods is the parameter σ_s depending on the statistics postulated for the random fluctuations of the scattering medium.
6) The computation of σ_s is done for all polarization combination (V,H) of the transmitting and receiving antennas. The result is completely described by the Stokes matrix [*Fung,* 1994]. The elements (amplitude and phase) of the Stokes matrix are expressed as functions of the elements of the scattering matrix S defined in Eq. (2.10) and Eq. (2.11) of Sec. 2.2.

The validity of this model can be tested through actual experiments. The scattering cross section σ_s computed on the basis of the mathematical model is compared with measured statistics (amplitude and phase) of the scattered signal. If discrepancies exist, the mathematical model must be modified to account for the experimental data. For example, at some frequency regimes, or small grazing angles (in the case of rough surface scattering), multiple scattering effects must be accounted for. The statistics assumed for the fluctuations of the random medium should be modified, for example, with a "wider" variance of the distribution to account for the multiple reflections and with a "narrower" correlation function to simulate highly decorrelated (incoherent) reflected signals. This modelling of the fluctuations in a random medium can be verified experimentally. For example, for surface scattering we can derive from measurements of the "polarization ratio" [ratio between the reflection coefficients (Fresnel) for vertical polarization and horizontal polarization] the value of the complex permittivity of the medium [*Logvin,* 1998] for a given incident angle of the electromagnetic wave. This polarization ratio is equal to the ratio of the voltages of the signals in the orthogonal channels of the receiver. From the measurements of this voltage ratio for small increments of the incident angle, we can obtain a recording in space of the complex permittivity. Repeating the measurements for each point of the surface under investigation, we obtain a statistical distribution of the permittivity. Using the profiles of the permittivity measured in various directions we can compute

the correlation function. This function is used to correct the mathematical model (e.g. Gaussian or exponential).
A different method to determine the permittivity is based on measurement of the delay of the received electromagnetic wave pulse after propagation in the medium (volume scattering). From a recording of the delay, we can compute the propagation velocity v:

$$v = \frac{c}{\sqrt{\varepsilon_r}} \tag{3.33}$$

This expression can be used to derive the value of the relative permittivity ε_r provided that the latter is real. From the measured profile of ε_r we can derive the correlation function.

The first method of "polarization ratio" has been proven useful [*Bogorodsky,* 1985] for some types of rough surfaces. It has also been shown to be accurate as it is based on relative measurements (voltage ratio). The method based on absolute measurements of propagation speed may be less accurate. So far, we have described a mathematical model that leads to the computation of the scattering cross section σ_s based on theoretical modelling of the statistics describing the fluctuations of the random medium. Examples also have been given illustrating the experimental verification of the theoretical statistical model. If we have available experimental statistics of the scattered signal, we can estimate or recognize the type of scatterer under investigation by comparing the experimental distribution of the signal against known statistics of the signal obtained for given configurations (types) of scatterers or scattering surfaces. This method of computing σ_s requires statistical modelling of the scattered signal (amplitude and phase of the elements of the scattering matrix).

3.2.2 Statistical modelling of the scattered signal

The statistical nature of received signals is a basic fact in all scattering problems, e.g. scattering from a random surface. Received signal fluctuations take place due to variations in the phase relationships of signals from separate object elements moving with respect to a radar, or due to the change of scatterers in a resolution cell during the remote sensing process. The scattering matrix is one of the fullest descriptive statistical characteristics of a remote-sensed surface. The elements of this matrix can have different distribution parameters depending on the type of a remote-sensed surface. In spite of the fact that many publications have been devoted to statistical distributions and processing of signals, this problem requires further investigation. The statistical models of the scattered signals need further verification against experimental data. It is especially a vital problem in remote sensing systems for which there are relatively few data available.

In most publications on statistics in remote sensing, normal distributions or their modifications, like Weibull, Rice, Nagakami and a number of other statistics, are assumed [*Kozlov,* 1992]. The application of these distributions is connected to their relative simplicity in performing signal analysis and the possibility of analytical representations of the results. Discrepancy between a model and a real distribution is especially evident when the statistics of objects under consideration are affected by the different statistics of the background. This differentiation in statistics can be used for extraction of the desired surface characteristics from the background clutter or to distinguish between fluctuating and stationary targets. For example, if a radar system observes a vegetation tree canopy as a function of time, the backscattered signal consists of the sum of a constant echo from the trunks and a fluctuating echo from the leaves and branches as they move in the wind. In this case, the Rician probability density function may be a good model of the statistics of the backscatter from heavily wooded terrain.

Intensive experimental investigations in the field of classification of statistical measurements are being carried out over a wide range of wavelengths, with different polarizations, various surfaces and volumes.

Distributions based on experimental parameters can be modelled accurately well. For example, the K-distribution (although used in other fields before) has been applied to remote sensing problems only recently. The K-distribution is nowadays widely used in the theoretical investigations of scattering problems in Russia [e.g. *Kozlov, Logvin and Lutin*, 1992] and in the Western World, e.g., [*Ulaby and Dobson*, 1989]. This distribution gives the best approximation for reflections from (rough) sea surfaces. The generalized K- distribution can be identified by two components of disturbances. The first component arises from scattering by sea waves structures with a long correlation time. The second component is formed by the scattering from a large number of small elements and it exhibits a lower level of amplitude fluctuations.

As a consequence, it has been shown that the K-distribution is a composite of the χ-distribution with a long correlation time and the Rayleigh distribution with a short correlation time [*Kozlov*, 1992]. Experimental studies have indicated the relevance of the K-distribution model in comparison to the normal, Rayleigh, Weibull and other probability distributions. A more accurate description of the statistical distribution is of prime interest when extracting the desired object characteristics from clutter signals, e.g. those due to the ice formations or oil spots over a rough sea surface. In such cases, the difference in statistics of the signals from the desired objects and from clutter is applied. On the basis of signal analysis, the information due to these differences can be calculated and the process of a useful object acquisition is carried out by means of specified algorithms. The application of the K-distribution for extraction of useful signals with Rayleigh distributions has shown a close fit with expected results.

From a statistical point of view, there are several different approaches for modelling radio wave scattering processes. For example, experimental data such as histograms of reflected signals can be taken as a basis. After that, using a corresponding best-fit criterion, these histograms are tested for adequacy by comparing the results with distributions known from literature.

Another approach can be applied if single measurements are available. In this case, various signal statistics are assumed and conditions under which the real experiment is carried out are modelled. The result of the experiment is then compared with the result obtained by the mathematical modelling process by means of known rules.

A third variant is applied when the corresponding experimental data are not available or the conditions under which these experimental data were obtained are not precisely known. In that case, the whole scattering process is modelled, i.e., the statistics of the incident signal on the surface, the statistics of the surface itself and the interaction process (not physically but mathematically). Then, by means of an enumeration of possible incident wave signals and the statistics of the surfaces, the corresponding data bank of the reflected signal statistics is derived. This data bank can be compared with the data bank of statistics obtained from new experiments.

It is obvious that the different statistics of an incident signal are quite limited and the main part in the scattering process is played by the statistics of the surface. Here, various models for roughness and dielectric constants can be assumed. The problem becomes especially sophisticated for surfaces covered with vegetation (forests, agricultural areas, etc.). Nevertheless, various models of such surfaces are widely used in mathematical modelling.

3.2.3 Measured statistics of scattering matrix coefficients

To see the effects on polarization, we show below a few examples of measured statistics of the amplitude and phase (relative phases of the scattering matrix coefficients) of the scattering coefficients of various types of surfaces, at different frequencies and incident angles and for all polarization combinations [*Ulaby*, 1990]. In Fig. 3.8, the histograms of the measured amplitude of the scattering coefficients for a rock surface are shown in all polarization combinations.

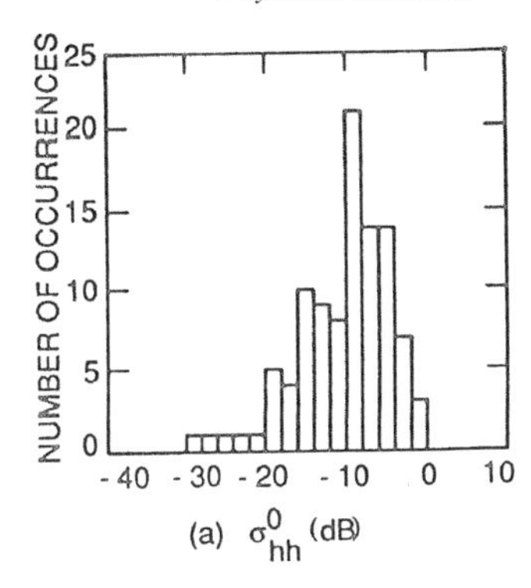

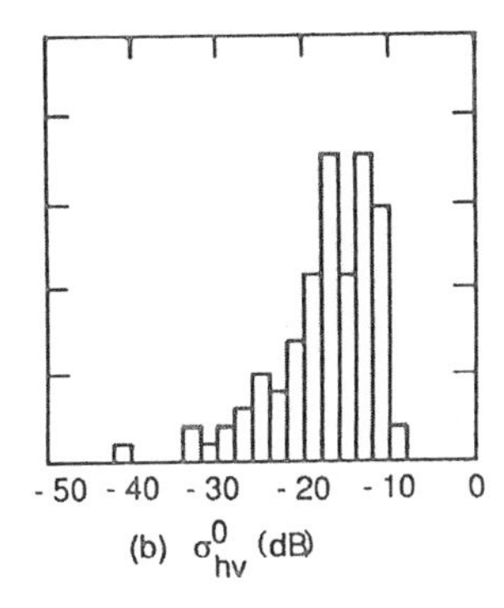

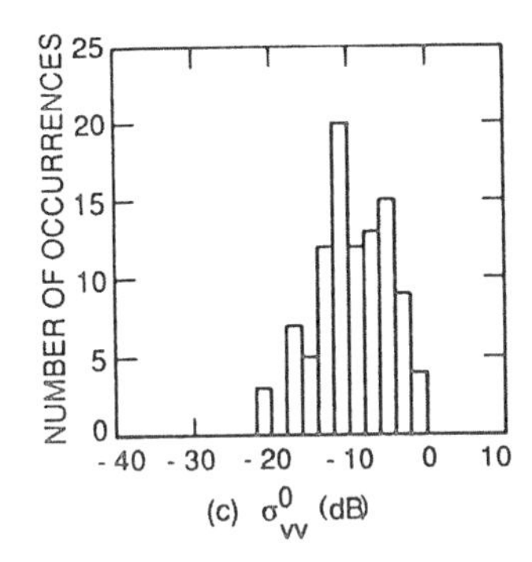

Fig. 3.8 Measured distribution of amplitudes (a) σ^0_{hh}, (b) σ^0_{hv}, (c) σ^0_{vv}, for a rock surface (Reproduced with permission of Artech House, F.T. Ulaby, "Radar Polarimetry for Geoscience Applications", Fig. 5.29, Ch. 5, 1990)

The frequency of the electromagnetic wave is 35 GHz. The incident angle is 60° (relative to normal). We notice that the mean values of the co-polar amplitude components σ^0_{hh} and σ^0_{vv} are larger than those for the cross-polar components σ^0_{hv}. The standard deviation of the co-polar components is significantly smaller than that of the cross-polar. The amplitude distributions σ^0_{hh}, σ^0_{hv} resemble an "inverted" Rayleigh form. The distribution of σ^0_{vv} tends to be more of a Gaussian type.

In Fig. 3.9 histograms are shown of the measured relative phases of the scattering matrix elements for a rock surface. The difference $\phi_{hv} - \phi_{hh}$ is uniform, while the difference $\phi_{vv} - \phi_{hh}$ between the co-polar components is Gaussian.

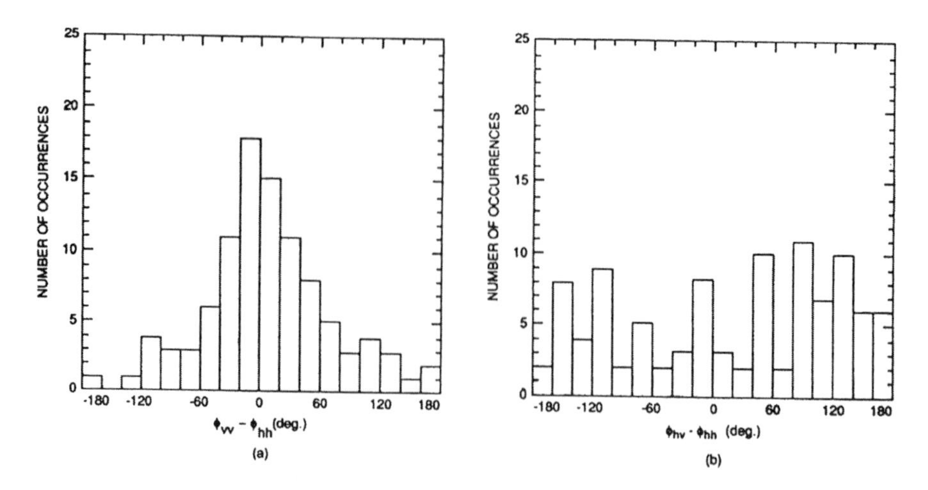

Fig. 3.9 Measured distributions of relative phase for a rock surface: (a) $\phi_{vv} - \phi_{hh}$, (b) $\phi_{hv} - \phi_{hh}$ (Reproduced with permission of Artech House, F.T. Ulaby, "Radar Polarimetry for Geoscience Applications", Fig. 5.30, Ch. 5, 1990)

A histogram of the measured relative phase (co-polar term $\phi_{vv} - \phi_{hh}$) for a rough sand surface is shown in Fig. 3.10. The distribution for this type of surface resembles a Maxwellian one.

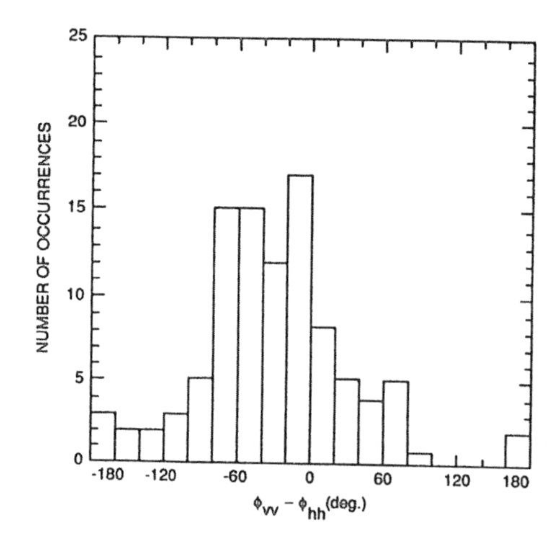

Fig. 3.10 Measured distribution of $\phi_{vv} - \phi_{hh}$ for a rough sand surface. (Reproduced with permission of Artech House, F.T. Ulaby, "Radar Polarimetry for Geoscience Applications", Fig. 5.31, Ch. 5, 1990)

All cases described above refer to measurements done at a frequency of 35 GHz and an angle of incidence of 60°. Changing the frequency from 35 GHz to the L-band and with an angle of incidence of 50°, the measured amplitude statistics for grass surface [*Ulaby*, 1990] are very similar to those obtained at millimeter wavelengths. The comparison of the phases is more interesting. The histograms of the measured relative phases at L-band vary from the case at millimeter waves. The most significant difference is noticed in Fig. 3.11 for the co-polar difference $\phi_{vv} - \phi_{hh}$, which is uniform compared to the Gaussian form of the case at 35 GHz. This indicates that phase statistics from polarimetric measurements of the scattering matrix coefficients can provide additional useful information on the scattering surface.

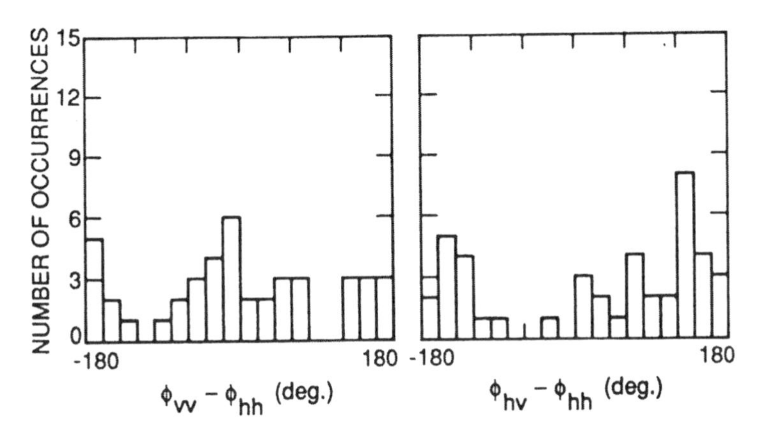

Fig. 3.11 Relative phase distribution for grass surface measured at L-band for incident angle of 50°. (a) $\phi_{vv} - \phi_{hh}$, (b) $\phi_{hv} - \phi_{hh}$ (Reproduced with permission of Artech House, F.T. Ulaby, "Radar Polarimetry for Geoscience Applications", Fig. 5.34, Ch. 5, 1990)

3.2.4 Coherent-incoherent scattering

At the conclusion of this descriptive chapter on modelling of scattering, we would like to make a few comments on the coherence-incoherence characteristics of scattering.

Coherent-incoherent scattering has important physical effects on the scattered signal. In Sec. 3.1.4, on physical scattering modelling, examples were given of scattering from multiple sources producing interference patterns at the receiver. One important consideration was that a very rough surface could cancel the contribution of the reflected (incoherent) component that is diffused by scattering. The major contribution to the received signal would then be the coherent part of the signal due to the direct propagation path from the transmitter to the receiver. For a less rough surface, the incoherent component experiences a lower order of scattering and contributes to the total signal, producing an interference pattern in combination with the coherent part. The most pronounced interference pattern (deep nulls) is formed in the case of a smooth reflecting surface. The horizontal polarization in the example considered on sea water was expected to contribute more than the vertical one to the total (coherent + incoherent) received field.

The coherent-incoherent concept is also important for the mathematical modelling of scattering where polynomial expansions are postulated for the field composed of a coherent component summed to additional (incoherent) higher order "perturbation" terms. It is shown [*Ogilvy*, 1991] that if we use higher order (higher than first order) perturbation terms, the diffused (incoherent) component due to scattering from a rough surface affects also the coherent component of the signal. This is not true if we only use the first order approximation of scattering [*Ogilvy,* 1991]. It would be therefore valuable for the verification of this theory, to compare measured statistics of fields (coherent part) with the statistics of the total field (coherent + incoherent) computed by the theoretical modelling of perturbations.

We consider at this point a very simple case to illustrate the problem. We have a signal composed of a strong specular reflection component (coherent) and a diffused (incoherent) component due to scattering by multi-path as characterized in Fig. 3.12.

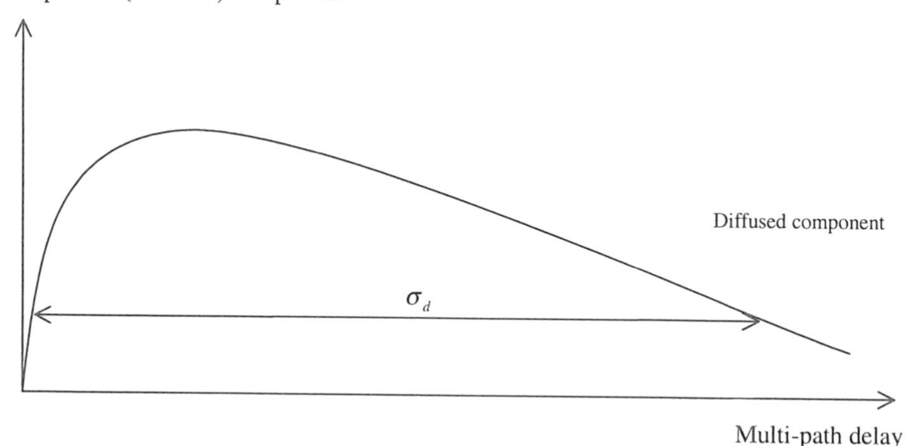

Fig. 3.12 Specular and diffused components (delay power spectrum)

The effects of the rough surface on the specular component are shown in Fig. 3.13:

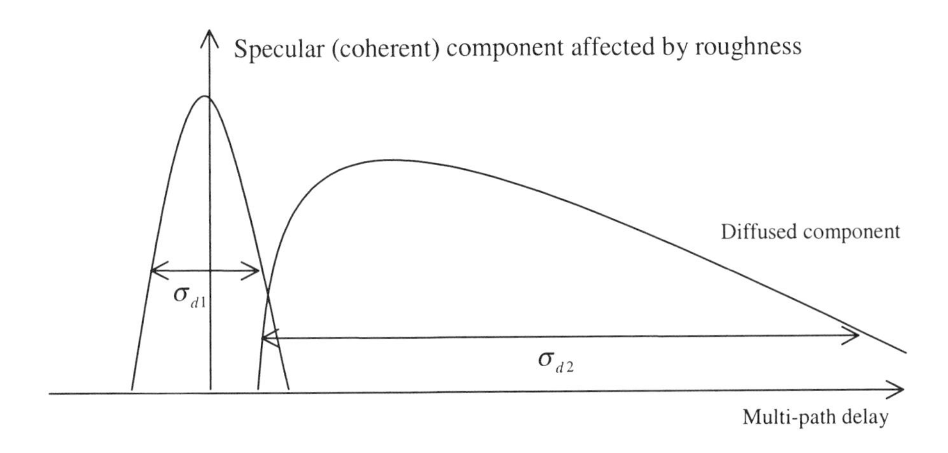

Fig. 3.13 Specular and diffused components (delay power spectrum)

In this case we have an increased total multi-path dispersion (σ_{d1} + σ_{d1}) compared to the case (σ_d) of Fig. 3.12.

CHAPTER 4

Summary of Available Scattering Methods

4.1 Introduction

In Chapter 3, we provided a descriptive introduction to the physics of scattering (physical modelling) and the statistics of scattered signals (mathematical modelling). In the last part of that chapter we also discussed the problem of using different approximations in the scattering model with regard to the accuracy of the prediction of scattering. In this chapter, we present an overview of the available scattering methods used to calculate the scattering cross section in random media. We consider two types of scattering: (a) surface scattering and (b) volume scattering. The major difference between the two is the depth of penetration into the medium. For surfaces of lossy media (such as wet soil), the scattering originates at the surfaces and the volume effects are usually ignored. For volume scattering media, such as vegetation and snow cover, the penetration may be significant and scattering within the volume may become the dominant scatter contribution.

For surface scattering use can be made of the small perturbations method and the Kirchhoff method. For volume scattering, we shall discuss briefly the radiative transfer (or transport) theory.

4.1.1 Perturbation theory of scattering

Lord Rayleigh proposed for the first time (end of 19th century) the method of small perturbations for the description of wave scattering from a surface separating two media. This method has been developed further by a number of other authors. The fullest description of this method can be found in [*Bass and Fuks,* 1972]. We present here an overview of the method of small perturbations. The method of small perturbations is used under the following conditions:

$$P_g \approx \left(2k\sigma_h \cos\theta_1\right)^2 << 1 \tag{4.1}$$

$$\frac{\sigma_h^2}{2L} << 1 \tag{4.2}$$

where

P_g = Rayleigh parameter
k = wavenumber
σ_h^2 = $\langle \xi^2 \rangle$ is the variance of a zero-mean height profile
h = $\xi(x,y)$ of a scattering surface $S(\xi)$
L = the radius of the spatial correlation of the random function $h = \xi(x,y)$
θ_1 = angle of incidence (with respect to the normal to the average surface S_M, assumed to be the $z=0$ plane)

The total wave field $U(\vec{r})$ at position $\vec{r}$ in the presence of a scatterer may be written as the sum of an incident and a scattered field:

$$U(\vec{r}) = U^{inc}(\vec{r}) + U^{sc}(\vec{r}) \tag{4.3}$$

A scattered wave field is represented in the form of a series $U = U_0 + U_1 + U_2 + \ldots$, where U_0 is the reflected field from an undisturbed (averaged) boundary and U_n are disturbances (small corrections to U_0 caused by the roughness) with $U_n \sim \xi^n$. The roughness effect on a scattered field can be taken into account by introducing effective currents on an averaged surface. Such currents depend on the magnitude of surface perturbations.

- **Small perturbations: first-order theory**

If the conditions of small perturbations [cf. Eqs (4.1) and (4.2)] are applicable, the rough scattering surface $S(\xi)$ deviates from the mean surface S_M (assumed to be the plane $z=0$) by a quantity smaller than the wavelength λ.

From the knowledge of the total scattered field on the plane $z=0$, the scattered field at some distance from the surface can be computed using the Helmholtz integral formula [*Ogilvy, 1991*]. If the rough surface has a zero mean value, i.e., $\langle \xi \rangle = 0$, it can be shown that to the first order approximation,

$$\langle U_1(r) \rangle = 0 \tag{4.4}$$

This indicates that first order perturbation predicts no change to the scattered coherent field. That is, the coherent field scattered from a slightly rough surface is the same as that from a smooth surface. However, we do have a contribution to the diffused scattered field. The latter contributes to the averaged scattered intensity via the relationship

$$< I_1 > \; = \; < |U_1|^2 \rangle \tag{4.5}$$

To calculate the diffused scattered intensity, we consider the case of an incident plane wave of the form

$$U^{inc}\left(\vec{r}\right) = e^{j\vec{k}_{inc}\cdot\vec{r}} \tag{4.6}$$

From the geometry shown in Fig. 4.1, the incident and scattered wave vectors are given [*Ogilvy*, 1991] as follows:

$$\vec{k}_{inc} = k\left(\hat{x}\sin\theta_1 - \hat{z}\cos\theta_1\right) \tag{4.7}$$

$$\vec{k}_{sc} = k\left(\hat{x}\sin\theta_2\cos\theta_3 + \hat{y}\sin\theta_2\sin\theta_3 + \hat{z}\cos\theta_2\right) \tag{4.8}$$

where $\hat{x}, \hat{y}, \hat{z}$ are unit vectors along the Cartesian co-ordinate axes x, y, z, respectively and k is the wave number.

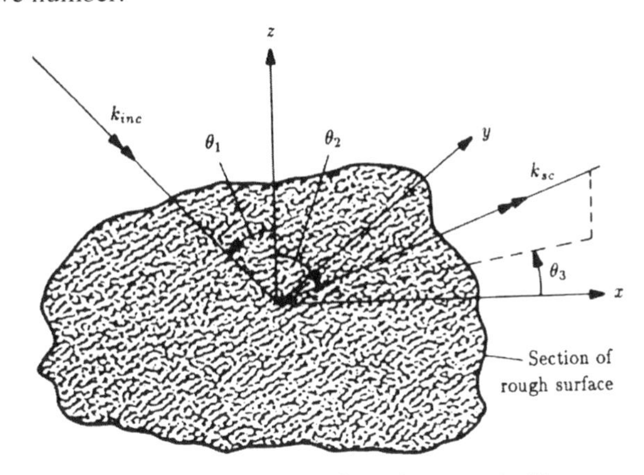

Fig. 4.1 Scattering geometry for a plane wave incidence
(Reproduced with permission of IOP Publishing Ltd, J.A. Ogilvy,
"Theory of Wave Scattering from Random Rough Surfaces, Fig. 3.1, Ch. 3, 1991)

For stationary surfaces, the averaged scattered intensity is given by [*Ogilvy*, 1991]

$$\langle I_{(1)}\left(\vec{r}\right)\rangle = \frac{4k^4\cos^2\theta_1\cos^2\theta_2}{r^2} A_M P\left(v_x, v_y\right) \tag{4.9}$$

where A_M is the area of the mean plane of the scattering surface and $P(v_x, v_y)$ is the surface power spectrum defined by

$$P(v_x, v_y) = \frac{\sigma_h^2}{4\pi^2} \int_{-\infty}^{+\infty} C(\vec{\rho}_d) e^{j(v_x \rho_{d_x} + v_y \rho_{d_y})} d\vec{\rho}_d \tag{4.10}$$

where

$C(\vec{\rho}_d)$ = Correlation function of the surface height fluctuations at different surface points separated by the vector $\vec{\rho}_d$

ρ_{d_x}, ρ_{d_y} = Cartesian components (x, y) of the vector $\vec{\rho}_d$

v_x, v_y = Cartesian components (x, y) of the unit difference vector $\vec{k}_{inc} - \vec{k}_{sc}$, given by $v_x = k(\sin\theta_1 - \sin\theta_2 \cos\theta_3)$, $v_y = -k \sin\theta_2 \sin\theta_3$ (4.11)

It can be seen from Eq. (4.9) that only the components in the x and y directions of the vector difference between incidence and scattering determine the strength of the scattered intensity in all directions. This result is known as "selective scattering" [*Ogilvy*, 1991].

Various autocorrelation functions have been modelled in the literature by various authors [*Ogilvy,* 1991; *Ulaby and Dobson,* 1989]. The most frequently used functions that compare well with experiments are of Gaussian or exponential form. The "width" of the correlation function determines the region where we consider the scattering to be correlated. For the Gaussian correlation function, we have specifically

$$C(\rho_d) = e^{-\left(\frac{\rho_d}{L}\right)^2} \tag{4.12}$$

corresponding to homogeneous, stationary and isotropic fluctuations (dependent only on the magnitude ρ_d of the vector $\vec{\rho}_d$). The correlation distance equals L. At this distance, the correlation is $(1/e)^2$ of its maximum value.

The surface power spectrum computed from the Fourier transform [Eq. (4.10)] of the correlation function Eq. (4.12) is given by

$$P(v_x, v_y) = \frac{\sigma_h^2 L^2}{4\pi} e^{-\frac{L^2}{4}\left[v_x^2 + v_y^2\right]} \tag{4.13}$$

The quantity $v_x^2 + v_y^2$ is determined from Eq. (4.11) as a function of the geometry of scattering (angles $\theta_1, \theta_2, \theta_3$) described in Fig. (4.1). For example, in case of backscattering in the azimuthal plane (x, z) i.e. $\theta_3 = 0$, we compute $v_x^2 + v_y^2 = 4\sin^2\theta_1$ and the surface power spectrum of Eq. (4.13) is given by

$$P\left(v_x, v_y\right) = \frac{\sigma_h^2 L^2}{4\pi} e^{-(kl\sin\theta_1)^2} \tag{4.13a}$$

For the standard deviation and the correlation distance of the surface fluctuations, various models can be adopted, depending on the nature of the surfaces, for example, sea surface in the microwave region. At low grazing angles and moderate wind speed, the small amplitude capillary waves (formed on top of the large gravitational waves) may be the dominant cause of scattering. The standard deviation can be characterized as a function of wind speed statistics.

The correlation distance (length) is a function of the size of the waves characterized as small turbulent "eddies" (scale of turbulence). The scale of turbulence is a statistical quantity that depends on the velocity of the eddies (forced by wind speed) and the superficial tension of the fluid. The tension is a function of pressure and temperature variations at the interface between the sea and the troposphere [*Massey,* 1968, *Landau,* 1979].

By these statistical considerations on scattering, we aim to emphasize that for the prediction-design of a polarimetric radar we would need a method to calculate the surface statistics required for the computation of the diffused scattered intensity [cf. Eq. (4.9)]. No such theoretical method seems to be available. Experimental statistics are available for some cases and were reported in the previous chapter.

- **Depolarization effects of scattering**

Incorporating polarization effects (vector-valued, first-order theory), the averaged diffused intensity [cf. Eq. (4.9)] for any polarization (p, q) of the transmitter and receiver is given by [*Ogilvy*, 1991]:

$$\langle I_{p,q}^{(1)} \rangle = \frac{4k^4 \cos^2\theta_1 \cos^2\theta_2 \cdot [\Phi_{p,q}^P(\theta_1, \theta_2, \theta_3)]^2}{r^2} A_M P\left(v_x, v_y\right); \quad p, q = \mathrm{V,H} \tag{4.14}$$

Here $\Phi^{P}_{p,q}(\theta_1,\theta_2,\theta_3)$ is an angular factor dependent on the polarization (horizontal H or vertical V) of the incident (p) and scattered (q) waves. This polarization factor is a function of the scattering angles $\theta_1,\theta_2,\theta_3$ (see Fig. 4.1) and the permittivity of the scattering surface. Mathematical expressions of the four polarization factors (HH, HV, VH and VV) are given in [*Ogilvy*, 1991].

This first order perturbation theory predicts no depolarization for scattering in the azimuthal plane (x, z) $(\theta_3 = 0)$, i.e., $\left\langle I^{(1)}_{p,q} \right\rangle_d = 0 \quad \text{for } p \neq q$. It is important to notice that the first order perturbation predicts depolarization effects on the field scattered from a (slightly) rough surface for scattering out of the azimuthal plane.

An example of a depolarization effect by a rough surface is illustrated in Fig. 4.2b. A comparison is made to the case of a smooth surface in Fig. 4.2a.

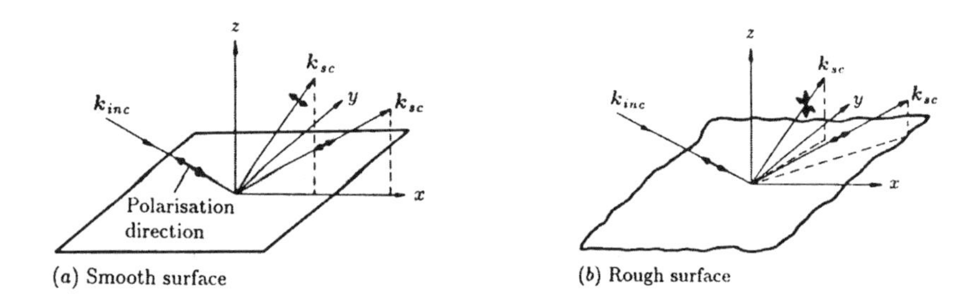

Fig. 4.2 (a) Reflection from a smooth surface; (b) Depolarization of vector waves by a rough surface (Reproduced with permission of IOP Publishing Ltd, J.A. Ogilvy "Theory of Wave Scattering from Random Rough Surfaces", Fig. 5.1, Ch. 5, 1991)

If the incident wave vector is in the (x, z) plane $(\theta_3 = 0)$, the wave reflected from a smooth surface will be in the same plane (x, z) as the incident wave (Fig. 4.2a). If, instead, the surface is rough (Fig. 4.2b), the scattered wave vectors are not necessarily all on the (x, z) plane (wave vectors are scattered by the surface height variations in the y-direction). Depolarization will then occur, depending on the degree of roughness and boundary conditions at the surface. Roughness will affect the relative polarization components (V or H), causing a rotation of the polarization of the scattered field.

- **Higher order perturbation: modified theory**

The second order perturbation theory [retaining terms of order ξ^2 in Eq. (4.3)] predicts a change in the coherent field scattered from a rough surface. The second order perturbation term $\langle U_2 \rangle$ is not zero in this case. It is proportional to the variance and the correlation function of the surface fluctuations [*Bass,* 1979]. The diffused scattered field intensity is given by

$$< I_2 >=< U_2 U_2^* > \tag{4.15}$$

This expression contains the 4th moment of the surface fluctuations and the so-called three-point and four-point surface correlations, which are not known in general, or are very difficult to model. This illustrates the mathematical complexity when higher terms, above the first order, have to be considered.

An alternative perturbation theory that circumvents this complexity is derived by [*Lysanov*, 1970]. In this theory, the total field is written as the sum

$$U(r) =< U(r) > + U_d(r) \tag{4.16}$$

where $\langle U(r) \rangle$ denotes the coherent field component and U_d the diffused field component. The coherent component is assumed of the form

$$< U(r) >= U^{inc} + \rho_r U_0 \tag{4.17}$$

where U^{inc} is a plane incident wave, U_0 is the scattered wave from a smooth surface and ρ_r is the reflection coefficient of the scattered wave. U_0 can be written as

$$U_0 = \rho_0 U^{inc} \tag{4.18}$$

with ρ_0 the Fresnel reflection coefficient for a smooth surface.

The expression for ρ_r in Eq. (4.17) gives the correction to the Fresnel reflection coefficient due to the rough surface. For Dirichlet boundary conditions (expanded to the first order in ξ), isotropic scattering and small perturbations conditions, the coefficient ρ_r is given by [*Ogilvy*, 1991]

$$\rho_r \approx [1 - 2\left(k\sigma_h \cos\theta_1\right)^2] \tag{4.19}$$

where σ_h is the standard deviation of surface height fluctuations and θ_1 denotes the angle of incidence (with respect to normal).

It should be noted from Eqs (4.17) and (4.19) that the modified theory of small perturbations predicts a change in the coherent scattered field, due to surface roughness.

Other corrections due to roughness have been derived and confirmed by experiments [*Ament*, 1953], [*Beard*, 1961]; specifically, we mention the exponential model

$$\rho_r = e^{-2\left(k\sigma_h \cos\theta_1\right)^2} \tag{4.20}$$

or this expression multiplied by the factor $I_0\left[2\left(k\sigma_h \cos\theta_1\right)^2\right]$, where I_0 is the modified Bessel function of zero order [*Miller*, 1984]. For $k\sigma_h \cos\theta_1 << 1$, we obtain the expression for ρ_r given in Eq. (4.19).

It is shown by [*Ogilvy*, 1991] that this modified second-order perturbation theory leads to an expression of the average diffused intensity proportional to the intensity calculated in the first-order approximation:

$$\langle I_{p,q}^{(2)} \rangle = \eta \left\langle I_{p,q}^{(1)} \right\rangle ; \quad p, q = \mathrm{V,H} \tag{4.21}$$

The coefficient of proportionality is given by

$$\eta = \left(1 + \rho_r\right)\left(1 + \rho_r^*\right)/4 \tag{4.22}$$

In general, ρ_r is a complex quantity. In our approximation of small perturbations it is a real quantity, as given in Eq. (4.19), or Eq. (4.20). It should be noted that in our approximation the correction ρ_r due to roughness is not dependent on polarization. For the coherent component of the scattered wave we consider the exponential model for the reflection coefficient [cf. (4.20)]. The coherent scattered intensity is given by

$$I_{coh} = I_{sp} \cdot e^{-P_g} \tag{4.23}$$

where the exponential term in Eq. (4.23) gives a reduction in the amplitude of the specular component due to roughness. The "roughness" parameter P_g is the Rayleigh parameter defined in Eq. (4.1) and I_{sp} is the intensity scattered from a smooth surface (specular component), given by

$$I_{sp} = \left(U^{inc} \cdot \rho_0\right)^2 \tag{4.24}$$

To see the effects of both coherent and diffused components of the scattered wave, we examine the case of back-scattering.

The average intensity of the diffused component per unit solid angle, for any polarization (p,q) of the transmitter and receiver, is derived from Eqs (4.13a), (4.14), (4.21) and (4.22), and is given by

$$< I_{d_{p,q}}^{(2)} >= \left[\frac{\pi}{4}(1+\rho_r)(1+\rho_r^*)\cos^2\theta_1 \cdot \left(\Phi_{p,q}^P\right)^2 \left(\frac{L}{\lambda}\right)^2 e^{-\left(\frac{2\pi L \sin\theta_1}{\lambda}\right)^2} \right] P_g \tag{4.25}$$

as a function of the Rayleigh parameter P_g. For scattering in the azimuthal plane $(\theta_3 = 0)$ the polarization-dependent term $\Phi_{p,q}^P$ is zero for $p \neq q$ (zero cross-polarization components) and for p = q, this term is equal to the Fresnel reflection coefficient for a "p-polarized" or "q-polarized" field (co-polarization components).

Eq. (4.25) shows a linear dependence between the diffused intensity and P_g. The slope $(\tan\beta)$ of the line is given by the quantity in the square parenthesis of Eq. (4.25). The slope (rate of change) of the backscatter diffused intensity is shown in Fig. 4.3 as a function of P_g for three different ratios of L/λ keeping the other parameters constant. It can be seen that the slope decreases for larger values of L/λ.

It can be also noticed from Eq. (4.25) that the backscattered intensity is very small at low angles of incidence with the scattering surface (zero for $\theta_1 = \pi/2$). The intensity becomes significant for $\theta_1 << \pi/2$ (around normal incidence). The backscatter intensity increases with the reflection coefficient ρ_r and with the "roughness" σ_h in the Rayleigh parameter P_g.

With regard to the effects of polarization [term $\Phi^{P}_{p,q}$ in Eq. (4.25)], the following has been established [*Ulaby*, 1982; *Ogilvy*, 1990] for
p = V (vertical polarization)
q = H (horizontal polarization)
surface roughness-wavelength ratio $\sigma_h / \lambda \sim 0.01 - 0.03$:

(1) For $L/\lambda < 1$:
- For angle of incidence in the interval $0 < \theta_1 < \pi/2$, the HH intensity component is smaller than the VV component. The HH component has a faster angular "drop-off" than the VV component.
- At low grazing angles, in the interval $60^\circ < \theta_1 < 90^\circ$, the VV component is much higher than the HH component (~0).

(2) For $L/\lambda \sim 1$:
- The VV and HH components have approximately the same intensity and angular "drop-off" with increase of the angle of incidence.
- At low grazing angles in the interval $40^\circ < \theta_1 < 90^\circ$, both VV and HH components are zero.

The cross-polarized backscatter components VH and HV, predicted by the second order perturbation theory of [*Fung*, 1968] and [*Valenzuela*, 1967], are not zero. The VH or HV components have an angular "drop-off" (as the incidence angle increases) faster than the VV component and slower than the HH component [*Ulaby*, 1982].

The effect of increase in permittivity is an increase of the cross-polar VH components. With regard to the co-polar, the effect of increase permittivity is a slower angular "drop-off" for the VV component and a faster "drop-off" for the HH component [*Ulaby*, 1982].

The total backscattered intensity (see Fig. 4.3) may be written as

$$< I_{sc} > = I_{coh} + < I_d > \tag{4.26}$$

where the coherent component is given by Eq. (4.23) and the diffused component by Eq. (4.25), both as a function of the Rayleigh parameter P_g. By inspection of these two components, we note the following two limiting cases:
(a) For $P_g = 0$ (no roughness), the diffused component is zero and only the coherent (specular) component is present.

(b) For $P_g \to \infty$ (large roughness), the coherent (specular) component vanishes. The contribution of the diffused component increases giving rise to incoherence (multi-path) in the received scattered wave. The depolarization effects due to roughness and phase-delays caused by multi-path are also present. They can be computed from the polarization factor Φ_{pq} in Eq. (4.25) for the diffused term.

As far as dependence on correlation-wavelength ratio, this theory predicts the following:

(c) For $L/\lambda = 0$, no diffused component is present; the coherent component is dominant.

(d) For $L/\lambda \to \infty$, the diffused component decreases and approaches zero because of the exponentially decreasing term in Eq. (4.25).

These considerations are illustrated in Fig. 4.3.

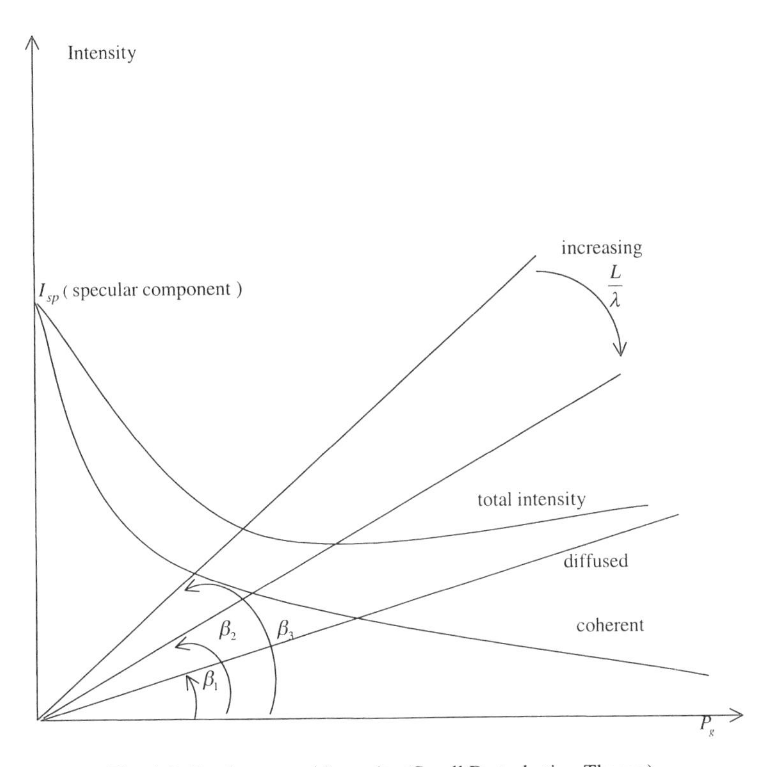

Fig. 4.3 Total scattered intensity (Small Perturbation Theory)

Where

$$\frac{dI}{dP_g} = \tan\beta \approx \frac{\left(\frac{L}{\lambda}\right)^2}{e^{\left(\frac{L}{\lambda}\right)^2}} \text{ ; for three values of } \beta\text{: } \beta_1, \beta_2, \beta_3 \;\; \theta_1 \neq \pi/2 \;\; \text{[cf. Eq. 4.25]}$$

- **Multiple scattering**

The method described sofar ignores the effects of multiple scattering from a rough surface. This is the mechanism by which energy is not transferred directly from the incident wave into the scattered wave. The incident wave may become "trapped" within the undulations of a rough surface by interacting several times with different parts of the surface before it is scattered away from the surface. This process is shown in Fig. 4.4.

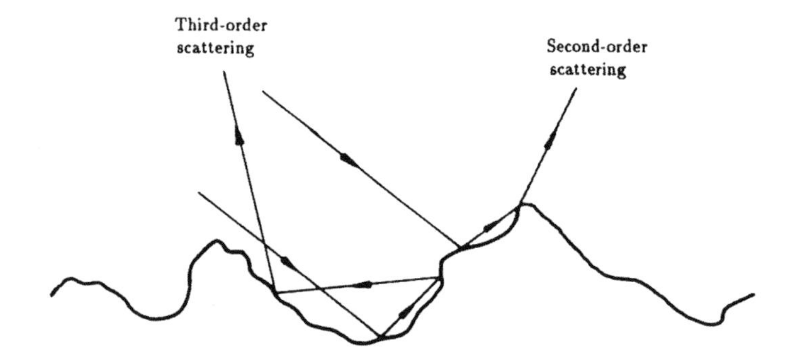

Fig. 4.4 Illustration of multiple scattering from a rough surface
(Reproduced with permission of IOP Publishing Ltd, J.A. Ogilvy,
"Theory of Wave Scattering from Random Rough Surfaces", Fig. 6.1, Ch. 6, 1991)

The "order" of multiple scattering is the number of reflections of the electromagnetic field from the undulations of the rough surface. The multiple scattering effects become more significant as the surface gradient increases, e.g., either by an increase of the surface height fluctuations (σ_h), or as the angles of incidence and scattering increase (with respect to the normal). The gradient of a rough surface is a statistical quantity; specifically,

$$Gradient \approx \gamma \frac{\sigma_h}{L} \tag{4.27}$$

where σ_h is the standard deviation of surface heights, L is the surface correlation length and γ is a coefficient of proportionality.

The latter depends on the model assumed for the random fluctuations of the surface. If the random process is Gaussian, zero-mean, with correlation function of Gaussian form, the coefficient γ is equal to $\sqrt{2}$. If the correlation is exponential, γ is equal to one [*Ulaby,* 1982].

The most significant effects of multiple scattering are the depolarization of the scattered field and the distortion of an electromagnetic pulse shape. The pulse energy distribution is also spread-out. This effect increases with an increase of the number of reflections "n" and the order *"l"* of scattering. The distribution is of Poisson type (see Fig. 4.5) [*Uscinski*, 1977].

Based on the previous considerations made on the physics of multiple scattering, a phenomenological description of the order "l" is as follows:

$$l \approx k \frac{\sigma_h}{L} \theta_i \Delta x \tag{4.28}$$

Here, σ_h / L is proportional to the gradient of the rough surface defined in Eq. (4.27), θ_i is the angle of incidence with respect to the normal and Δx denotes the size of the region of surface where multiple scatter occurs. The latter can be limited by the electromagnetic pulse width or by the antenna beam (foot-print).

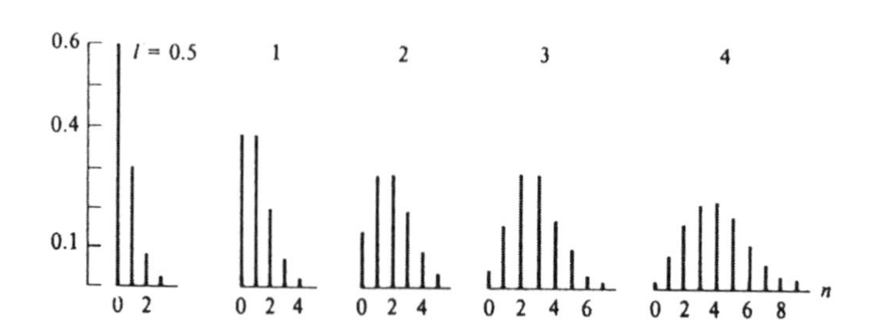

Fig. 4.5 Distribution of energy over different orders of scattering
(for increasing propagation distance through the random medium)
(Reproduced with permission of The Mc Graw-Hill Companies, B.J. Uscinski,
"The Elements of Wave Propagation in Random Media", Fig. 5-4, Ch. 5, ©Mc Graw-Hill Co. 1977)

The effect on pulse shape for increasing order of scattering "l" is shown in Fig. 4.6. It should be noted that the pulse transmitted at $\tau = 0$ is elongated in time-delay due to multiple scattering [*Shishov,* 1968, 1973].

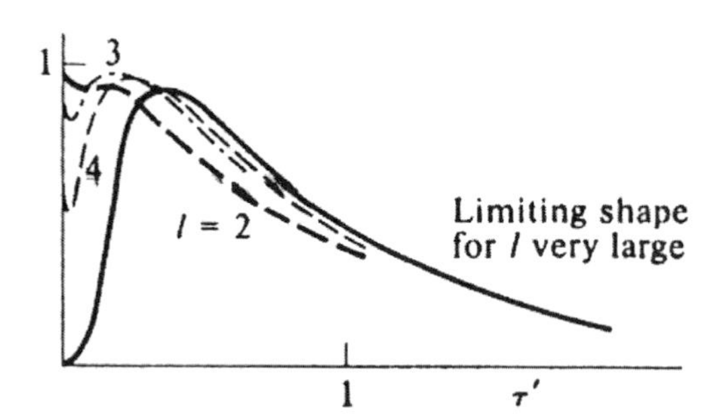

Fig. 4.6 Pulse shape versus delay spread
(Reproduced with permission of The Mc Graw-Hill Companies, B.J. Uscinski,
"The Elements of Wave Propagation in Random Media", Fig. 5-5, Ch. 5, ©Mc Graw-Hill Co. 1977)

The effect of multiple scattering on the average reflected field $\langle U \rangle$ is taken into account by the modified perturbation theory. The total field is written as

$$U = U^{inc} + U_0 + U_P \tag{4.29}$$

Here, U^{inc} denotes the incident (planar) field, U_0 is the field scattered by a (known) smooth plane in the absence of undulations and U_P is the contribution of the undulations; the latter is to be determined.

Modified theories of perturbations take into account the effects of multiple scattering on the magnitude of the average reflected field $\langle U \rangle$. Field reflection from a random surface is the same as from a plane boundary but with an effective surface impedance that depends on the wavelength λ and on the radiation direction [*Bruchovecky,* 1985; *Zhuk*, 1990].

Within the framework of the theory of small perturbations [size of undulations (scatterers) < wavelength)] various models are used to account for the effects of undulations (multiple-scattering medium). One such model considers a rough surface composed of many undulations regarded as a distribution of dipole-type sources of varying strengths and orientations. The term U_P in Eq. (4.29) is calculated under the following conditions [*Ogilvy,* 1991]:

(a) The incident wave is planar
(b) Isotropic scattering from the surface
(c) The average size "*a*" of an undulation is small compared to wavelength
(d) The average separation "*d*" between two adjacent undulations is small compared to wavelength

The results of the model [*Ogilvy*, 1991] show that the multiply scattered field tends to be equal to the field in the absence of multiple scattering when $d >> a$, i.e., when the separation between undulations is much larger than the size of the undulations, or in the (obvious) case when $a = 0$ (no undulations).

Another model [*Ogilvy,* 1991] for "embedded" undulations-scatterers (with separation between scatterers much larger than the size of each scatterer), is based on the assumption that the field reflection from a random surface is the same as from a planar boundary, but with an effective surface impedance $Z(k)$ that depends on wave length, direction of radiation and number of scatterers. The total field is written as the sum:

$$U(r) = U^{inc}(r) + \sum_{n=1}^{N_s} U_n(r - r_n) \tag{4.30}$$

The summation in Eq. (4.30) describes the contribution from all scatterers, with r_n the location of the n-th scatterer and $U_n(r - r_n)$ the field at position r due to a scatterer at the position r_n.

[*Twersky*, 1957] has calculated the total field scattered from an array of N_s identical undulating scatterers, with random positions along the x-direction, under the following assumptions:

(a) the scatterers are "elementary" cylinders of length dx
(b) the probability of a cylinder being within the interval $(x, x+dx)$ is independent from the other scatterers and is given by p(x)dx where $p(x) = \rho_a / N_s$, ρ_a being the average number of scatterers per unit area on the surface
(c) each undulation is excited only by the coherent field scattered from the other undulations (the diffused field is neglected)
(d) scattering between undulations is in the far-field

This model predicts scattered field amplitudes with multiple scattering effects given by the ratio [*Beckmann and Spizzichino*, 1987; *Ogilvy*, 1991]

$$r_{msc} = \frac{1}{1 + \rho_a A_a} \tag{4.31}$$

where A_a is the surface area of each undulating-scatterer. This ratio is a multiplicative factor of the scattered field in the absence of multiple scattering. By increasing ρ_a and/or A_a of the scatterers, the amplitude of the scattered field decreases. This model has been proven useful in predicting scatter for small grazing angles of incidence. In this geometry, the small perturbation theory predicts very small scattered fields.

4.1.2 Kirchhoff theory of scattering (short wavelength limit)

For large surface roughness compared to the radio wavelength $(\sigma_h / \lambda >> 1)$ and for near normal incidence ($\theta_1 = 0$) the Rayleigh condition [cf. Eq. (4.1)], i.e., $\sigma_h \cos\theta_1 / \lambda << 1$, may not be satisfied and the small perturbation theory may not be applicable. With the Kirchhoff method, the scattered field is written in terms of the tangential field on the rough surface and can be calculated even for large $\sigma_h \cos\theta_1 / \lambda$. The surface field is approximated by the field that would be present if the rough surface were replaced by a planar surface tangent to the point of interest (scatter point), i.e., the surface at each point becomes a local tangent plane. This approximation is valid if the radius of curvature R_c at every point on the surface is much larger than the wavelength:

$$R_c >> \lambda \tag{4.32}$$

The radius of curvature is a statistical quantity of the rough surface. For zero-mean Gaussian surface fluctuations, with a correlation function of Gaussian form, the radius of curvature R_c is given by [*Ulaby*, 1982]

$$R_c \approx \frac{L^2}{3\sigma_h} \tag{4.33}$$

where L is the correlation length of surface fluctuations and σ_h is the standard deviation of surface height fluctuations.

From Eq. (4.32), with R_c given in Eq. (4.33), the condition of a large Rayleigh parameter, $P_g = \sigma_h \cos\theta_1 / \lambda >> 1$, yields the following range of validity:

$$\frac{\lambda}{\cos\theta_i} << \sigma_h << \frac{L^2}{3\lambda} \tag{4.34}$$

The latter, in turn, results in the relationship

$$\frac{L^2}{3\lambda} >> \frac{\lambda}{\cos\theta_1} \Rightarrow \frac{L}{\lambda} >> \sqrt{\frac{3}{\cos\theta_i}} \quad \Rightarrow \frac{\mathrm{L}}{\lambda} >> 1 \tag{4.35}$$

We give a numerical example in order to compare the Kirchhoff method with the small perturbation theory. We consider two limiting cases of incidence:

a) For near normal incidence (specifically, $\theta_1 = 1^\circ$), λ =0.03m and correlation length L=1m, the Kirchhoff method is valid in the range [cf. Eq. (4.34)]

$$0.03m << \sigma_h << 10m \tag{4.36}$$

Eq. (4.35) is satisfied because $L/\lambda \simeq 30$. On the other hand, for applying the small perturbation theory we have

$$\frac{\sigma_h}{\lambda}\cos\theta_1 << 1 \quad \Rightarrow \quad \sigma_h << 0.03 \tag{4.37}$$

meaning that for this range of σ_h the condition $\sigma_h / L << 1$ is valid.

b) For incidence far from normal (specifically, $\theta_1 = 89^\circ$) and for the values of λ and L used in case (a), the Kirchhoff method is applicable in the range

$$2m << \sigma_h << 10m \tag{4.38}$$

whereas the small perturbation theory is valid for

$$\sigma_h << 2m \tag{4.39}$$

The ranges of σ_h for the validity of the Kirchhoff method and the small perturbation theory are shown in Fig. 4.7 for cases (a) and (b).
From the example illustrated in Fig. 4.7, it is interesting to note that at an angle of incidence near the normal $(\theta_1 = 1^\circ \text{ deg.})$, the Kirchhoff method can correctly predict scattering for a wide range of "small" and "large" roughness (σ_h). At a low grazing angle $(\theta_1 = 89^\circ \text{ deg.})$, the Kirchhoff method can predict scattering for a limited range of large roughness (σ_h).

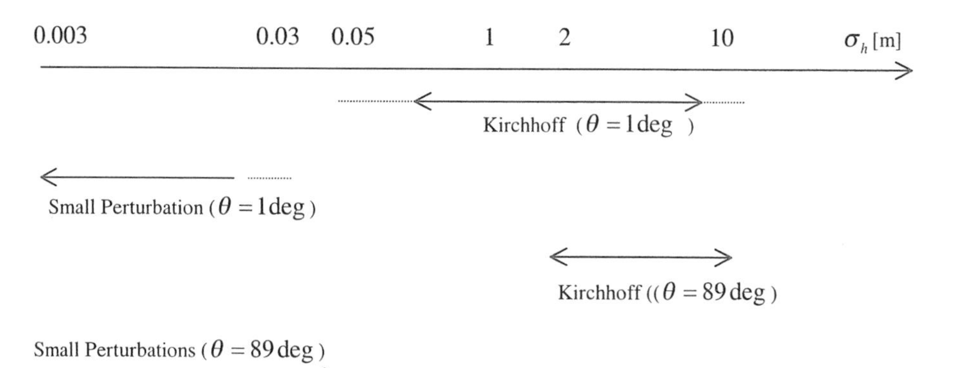

Fig. 4.7 Limits of surface roughness (σ_h) for applying Kirchhoff method and Small Perturbation Theory, $\lambda = 0.03m$, $L = 1m$

The Kirchhoff theory is based on the approximation of a scattering surface with a tangent plane at each (scatterer) point of the surface. The conditions for its validity are $\lambda << L$ and $\lambda << \sigma_h$, where L is the correlation length, σ_h is the standard deviation of surface fluctuations and λ is the wavelength. This tangent plane approximation is used to calculate the field scattered from the surface. The interaction between the radiation and a local surface can be described by the Fresnel formulas for the reflection coefficients [*Stratton*, 1941]. The total field scattered from the rough surface is determined as a result of summing the reflected fields from all elements of the surface. In the tangent plane approximation, the integration is made over projected area elements of the rough surface on the mean plane (see geometry in Fig. 4.8).

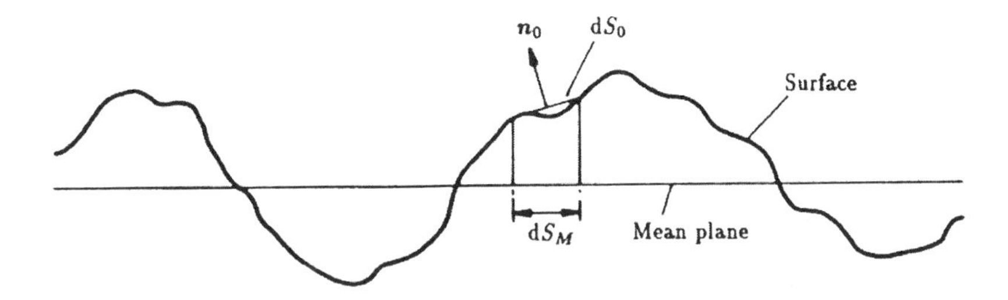

Fig. 4.8 Projection of an area element of the rough surface on the mean plane (Reproduced with permission of IOP Publishing Ltd, J.A. Ogilvy. "Theory of Wave Scattering from Random Rough Surfaces", Fig. 4.2, Ch. 4, 1991)

The computation of the scattered field can be made, for example, with the use of the Helmholtz integral theorem [*Beckmann and Spizzichino,* 1987; *Ogilvy*, 1991]. The scattered field is calculated in the far field of the surface under the assumption that multiple scattering and local diffraction, or shadowing effects, are negligible. The scattering coefficient is computed by [*Beckmann and Spizzichino,* 1987] as the ratio

$$\rho_{sc} = \frac{U_{sc}(\vec{r})}{U_0(\vec{r})} \tag{4.40}$$

where $U_{sc}(\vec{r})$ is the wave scattered field at a distance $\vec{r}$ from the scatterer and $U_0(\vec{r})$ the field reflected in the direction of specular reflection ($\theta_1 = \theta_2$, in Fig. 4.1).

The scattering coefficient is computed in the case of a monochromatic plane wave incident upon a (rough) scattering surface $\xi(x, y)$ [*Beckmann and Spizzichino*, 1987]:

$$\rho_{sc} = \frac{1}{4L_x L_y \cos\theta_1} \int_{-L_x}^{L_x} \int_{-L_y}^{L_y} \left(a \tan\beta_x + c \tan\beta_y - b\right) \cdot e^{j\vec{v}\cdot\vec{r}} dxdy \tag{4.41}$$

Here

L_x, L_y	=	sizes (length) of the scatterer in x and y dimensions, respectively
θ_1	=	angle of incidence with respect to the normal
$\tan\beta_x$, $\tan\beta_y$	=	slopes of tangent to the surface at point of incidence in the planes(x, z) and (x, y), respectively

The vector $\vec{v}$ in Eq. (4.41) is the difference between the incident wave vector and scattered wave vector, $\vec{k}_{inc} - \vec{k}_{sc}$, given by [*Beckmann and Spizzichino*, 1987]

$$\vec{v} = k\left[\left(\sin\theta_1 - \sin\theta_2\cos\theta_3\right)\hat{x} - \left(\sin\theta_2\sin\theta_3\right)\hat{y} - \left(\cos\theta_1 + \cos\theta_2\right)\hat{z}\right] \tag{4.42}$$

where the angles θ_1, θ_2, θ_3 are defined in Fig. (4.1) referred to the "mean" rough surface and $\hat{x}, \hat{y}, \hat{z}$ are the orthogonal unit vectors of the Cartesian reference system in (Fig. 4.1).

The position vector $\vec{r}$ appearing in Eq. (4.41) is given by

$$\vec{r} = x\hat{x} + y\hat{y} + \xi(x, y)\hat{z} \tag{4.43}$$

The parameters a, b and c are functions of the reflection coefficient and the geometry of scattering [see Fig. 4.1], given by:

$$a = (1-\rho)\sin\theta_1 + (1+\rho)\sin\theta_2\cos\theta_3 \tag{4.44}$$

$$b = (1+\rho)\cos\theta_2 - (1-\rho)\cos\theta_1 \tag{4.45}$$

$$c = (1+\rho)\sin\theta_2\sin\theta_3 \tag{4.46}$$

Here, the (Fresnel) reflection coefficient is function of the "local" angle of incidence on the rough surface and it depends, therefore, on the slope of the surface at the point of incidence [*Beckman and Spizzichino*, 1987]. The Fresnel coefficient is computed as a function of the polarization state of the wave [*Stratton*, 1941].

It can be noticed that, if the quantity within the parenthesis in the integral of Eq. (4.41) does not vary in the region of scatter $A = L_x L_y$ and under the assumption that $A >> \lambda^2$ (e.g. for X-band frequencies ($\lambda = 0.033m$) the scattering area A should be much larger than $10^{-3} m^2$), the scattering coefficient can be approximated by

$$\rho_{sc} \sim \frac{F_{p,q}\left(\rho, \beta_x, \beta_y, \theta_1, \theta_2, \theta_3\right)}{L_x L_y} \int_{-L_x}^{L_x} \int_{-L_y}^{L_y} e^{j\vec{v}\cdot\vec{r}} dxdy \qquad (4.47)$$

where $F_{p,q}\left(\rho, \beta_x, \beta_y; \theta_1, \theta_2, \theta_3\right)$ is a polarization-dependent term function of the reflection coefficient ρ, the surface gradients β_x, β_y and the angles $\theta_1, \theta_2, \theta_3$ [see Eqs (4.44), (4.45) and (4.46)]. It is seen [*Beckmann and Spizzichino*, 1987; *Ogilvy*, 1991] that a wave scattered in the plane of incidence (x, z) is not depolarized if the incident wave is polarized either vertically or horizontally. In this case ($\theta_3 = 0$), the parameter c in Eq. (4.46) is zero. The slope of the rough surface in the azimuth plane (x, y) [cf. Eq. (4.41)] is in this case irrelevant (cannot affect the scattered wave). We therefore do not expect depolarization, e.g., transfer of scattered energy from the plane (x, z) to the plane (x, y).

From the integral of Eq. (4.47), it can be noticed that the scattered field can have a "degree" of coherence (given by the pattern sinX/X) as a result of scattering from a two-dimensional "array" of scatterers on the surface area $L_x L_y$. This result shows that this theory can be useful for the computation of scattering from rough surfaces by modelling the surface as an aggregate of facets (see Chapter 3), which radiate electromagnetic energy. Each facet represents an array of scatterers that contributes to the radiation-scattering pattern (of type sinX/X). A mathematical model for sea waves based on an aggregate of facets with pattern given by Eq. (4.41) (in one dimension) is given in [*Pusone*, 2000]. In this model, the (Fresnel) reflection coefficient is computed for finite conductivity of the surface, as a function of the sea-wave slope at the point of incidence for each facet. The slope of each facet is characterized as a function of the speed of the sea particles driven by wind force. The size of each facet is determined by the correlation distance of the speed fluctuations of the sea particles. The model predicts backscatter from rough sea surfaces, at low grazing angles. The total backscattered intensity is computed in the model by the modulus square of the sum of the scattering coefficients of the facets for a sea wave of a given height (sea-state). The results of this model compare well with measurements of backscatter from the sea surface at horizontal and vertical polarizations at grazing angles smaller than 5 degrees [*Levanon*, 1988; *Nathanson*, 1987] and for wind speeds of 5m/s and 10m/s.

The sea waves model of [*Pusone*, 2000] has recently been extended to compute sea backscatter Doppler spectrum [*Pusone*, 2001]. In this model the Doppler spectrum is evaluated as a function of sea state (windspeed) and sea properties (viscosity) at X-band frequencies at low grazing angels. By this model we can retrieve information on sea state and sea properties from the knowledge of the Doppler spectrum of sea backscatter.

In [*Beckmann and Spizzichino*, 1987; *Ogilvy, 1991*], the scattering intensity for coherent and diffused (incoherent) scattering is computed for a known geometry (e.g., periodic) of the surface and for random scattering rough surfaces. We give here a general overview of the results obtained for random surfaces. The average scattered intensity is calculated from the scattering coefficient for any pair of orthogonal polarization states (p, q) of the transmitter and receiver. For coherent scatter we have

$$\langle I_{p,q}\rangle_{coh} = \langle \rho_{sc}\rangle\langle \rho_{sc}\rangle^* = |\langle \rho_{sc}\rangle|^2 \tag{4.48}$$

and for diffused (incoherent) scatter we obtain

$$\langle I_{p,q}\rangle_d = \langle \rho_{sc}\rho_{sc}^*\rangle - \langle \rho_{sc}\rangle\langle \rho_{sc}\rangle^* \tag{4.49}$$

The exact scattering coefficient ρ_{sc} is found from Eq. (4.41) and its approximate value from Eq. (4.47). In the latter case, the average scattering coefficient $\langle \rho_{sc}\rangle$ is given by

$$\langle \rho_{sc}\rangle = \frac{\langle F_{p,q}(\rho,\beta_x,\beta_y,\theta_1,\theta_2,\theta_3)\rangle}{L_x L_y}\int_{-L_x}^{L_x}\int_{-L_y}^{L_y} e^{j(v_x x+v_y y)}\langle e^{jv_z\xi(x,y)}\rangle dxdy \tag{4.50}$$

where v_x, v_y, v_z are the components of the vector defined in Eq. (4.42). The quantity $\langle e^{jv_z\xi(x,y)}\rangle$ is the characteristic function (Fourier transform) of the probability density function of the surface height ξ.

For the case of a normally distributed random surface, with zero mean and standard deviation σ_h, the characteristic function of the distribution is given explicitly by

$$\left\langle e^{jv_z\xi}\right\rangle = e^{-\frac{\sigma_h^2 v_z^2}{2}} \tag{4.51}$$

The average scattered coherent intensity is calculated from Eqs (4.48), (4.50) and (4.51) given by:

$$\left\langle I_{p,q}\right\rangle_{coh} = \left[F_{p,q}\right]^2 \cdot \left[\frac{\sin v_x L_x}{v_x L_x} \cdot \frac{\sin v_y L_y}{v_y L_y}\right]^2 \cdot e^{-\sigma_h^2 v_z^2} \tag{4.52}$$

The coherent intensity is polarization dependent through the factor $\left[F_{p,q}\right]^2$, which is a function of the reflectivity properties (reflection coefficient) and the angles $\theta_1, \theta_2, \theta_3$ (see Fig. 4.1).

For the computation of the diffused component (average complex conjugate cross-product of the scattering coefficient $\left\langle \rho_{sc}\rho_{sc}^*\right\rangle$ in Eq. (4.49)) we must know the joint distribution of two surface height random variables. The average diffused intensity is calculated under the assumption of very rough surfaces ($\sigma_h v_z >> 1$) for the case of a two-dimensional Gaussian distribution of two surface heights random variables ξ_1, ξ_2 correlated by a Gaussian correlation function for isotropic scatter [Eq. (4.12)]. For any polarization pair (p, q), the result of the calculation is given by [*Beckmann and Spizzichino*, 1987; *Ogilvy*, 1991]:

$$\left\langle I_{p,q}\right\rangle_d = \frac{\left[U_{p,q}\right]^2 k^2}{\left|C''(0)\right|}\left(\frac{1}{v_z^2\sigma_h^2}\right) e^{-\frac{[v_x^2+v_y^2]}{2v_z^2\sigma_h^2\left|C''(0)\right|}} \tag{4.53}$$

where [*Ogilvy*, 1991]

$$\left|U_{p,q}\right| = \frac{\left[F_{p,q}\right]^2}{(\cos\theta_1 + \cos\theta_2)^2} \tag{4.54a}$$

$$\left|C''(0)\right| = \frac{2}{L^2} \quad \text{(L: correlation distance of surface height fluctuations)} \tag{4.54b}$$

σ_h^2: Variance of the surface height fluctuations [Eqs (4.1), (4.51)] and

$$v_x = k\left[\sin\theta_1 - \sin\theta_2 \cos\theta_3\right] \tag{4.55}$$

$$v_y = -k\left[\sin\theta_2 \sin\theta_3\right] \tag{4.56}$$

$$v_z = -k\left[\cos\theta_1 + \cos\theta_2\right] \tag{4.57}$$

It is interesting to note that under the assumption of stationary phase of the scattered wave [*Ulaby*, 1982; 1990], the exponential term in Eq. (4.53) can be expressed as a function of the surface gradients. This assumption is correct if the phase in Eq. (4.41), defined by $\vec{k}\cdot\vec{r}$ [cf. Eqs (4.42) and (4.43)], is slowly varying. It means that only the energy scattered around the points of maximum and minimum phase is taken into account in the integral of Eq. (4.41). From the condition of maxima or minima (phase derivative = 0), we obtain from Eqs (4.42) and (4.43) the expressions:

$$\frac{\partial \hat{z}}{\partial x} = -\frac{v_x}{v_z} = grad\ h_x \tag{4.58}$$

$$\frac{\partial \hat{z}}{\partial y} = -\frac{v_y}{v_z} = grad\ h_y \tag{4.59}$$

Through this approximation, the rough surface is replaced locally at each point by a tangent plane with gradients given by Eqs (4.58) and (4.59).

From Eqs (4.53), (4.55), (4.58) and (4.59) we obtain

$$\left\langle I_{p,q}\right\rangle_d = \frac{\left|U_{p,q}\right|^2 k^2}{2v_z^2}\left(\frac{L^2}{\sigma_h^2}\right) e^{-\frac{[grad\ h_x]^2 + [grad\ h_y]^2}{4\sigma_h^2} L^2} \tag{4.60}$$

We note that if we define the "statistical" gradient of the rough surface [see Eq. (4.27)] by the ratio

$$m_s = \sqrt{2}\frac{\sigma_h}{L} \tag{4.61}$$

we obtain from Eqs (4.60) and (4.61) the following expression for diffused intensity

$$\langle I_{p,q}\rangle_d = \frac{\left[U_{p,q}\right]^2 k^2}{4v_z^2 m_s^2} e^{-\frac{[grad\ h_x]^2+[grad\ h_y]^2}{8m_s^2}} \tag{4.62}$$

This quantity increases with an increase of the "statistical" gradient of the rough surface. This means that the diffused component increases with an increase of the "statistical" fluctuations σ_h of surface height, or with a decrease of the correlation distance L of these fluctuations [cf. Eq. (4.61)].

The total scattered intensity may be written as the sum of the coherent component given in Eq. (4.52) and the incoherent component given in Eq. (4.53).

We examine, next, the case of backscatter for the coherent and diffused scattered.

The coherent component of the backscattered intensity (per unit solid angle) for small surface roughness, within the framework of the tangent plane approximation, i.e., under the assumption that $\sigma_h / \lambda > 1$ and $L/\lambda > 1$, is derived from Eq. (4.52) as a function of the Rayleigh parameter P_g, the reflection coefficient and the incident angle. For coherent backscatter in the azimuthal plane (x, z) ($\theta_3 = 0$) we have

$$\langle I_{p,q}\rangle_{coh} = \rho^2 \cdot \left[\frac{\sin(2kL_x \sin\theta_1)}{2kL_x \sin\theta_1}\right]^2 \cdot e^{-P_g} \tag{4.63}$$

where P_g is the Rayleigh parameter defined in Eq. (4.1). It is clear that the coherent component of the backscattered intensity

- decreases with increasing roughness σ_h in the Rayleigh parameter P_g
- takes its maximum value for $\theta_1 = 0$ (normal incidence) and decreases with θ_1 approaching low grazing angles above the scattering surface
- depends on polarization through the Fresnel reflection coefficient ρ

The polarization dependency is determined by the reflection coefficient ρ. At very low grazing angles ($\theta_1 \sim \pi/2$), the reflection coefficient is the same for vertical and horizontal polarizations. For angles between 50 and 85 degrees the reflection coefficient is higher for horizontal polarization [*Stratton*, 1941].

The expression for the diffused (incoherent) backscatter intensity, derived from Eqs (4.53) –(4.57) assumes the form

$$\left\langle I_{p,q}\right\rangle_d = \frac{\left|F_{p,q}\right|^2}{\cos^4\theta_1}\left(\frac{L}{\sigma_h}\right)^2 e^{-\frac{1}{4}\left(\frac{L}{\sigma_h}\right)^2 \tan^2\theta_1} \tag{4.64}$$

The polarization-dependent term $F_{p,q}$ is zero for $p \neq q$ and is equal to the Fresnel reflection coefficient for a "p-polarized" or "q-polarized" field [*Ulaby*, 1982]. Eq. (4.64) can be rewritten as a function of the Rayleigh parameter P_g:

$$\left\langle I_{p,q}\right\rangle_d = \frac{\left|F_{p,q}\right|^2 k^2}{P_g \cos^2\theta_1} e^{-\frac{k^2 L^2 \sin^2\theta_1}{P_g}} \tag{4.65}$$

It is seen from this expression that the diffused intensity equals zero for the two limiting cases $P_g \to 0$ and $P_g \to \infty$. The diffused intensity must therefore have a maximum between the two limiting values of P_g.

The diffused intensity obtained under the assumption that the surface roughness σ_h is much larger than the radio wavelength λ (very rough surface), is given in Eq. (4.64) as a function of polarization, incident angle and the ratio L/σ_h. It is noticed from Eq. (4.64), that the diffused backscatter intensity

- Decreases with an increase in the incidence angle θ_i (away from normal).
- Decreases with an increase of the ratio L/σ_h, meaning that for a large correlation length or small surface height fluctuations, we have reduced the scattering (diffused energy) effects. Since the ratio L/σ_h is proportional to the inverse of the *rms* slope of the rough surface, we can also say that diffused scattering decreases with a decrease of the slope of the surface. In the limit of zero slope, we have no rough surface and therefore no backscattering (diffused energy).
- Depends on the polarization configuration through the term $F_{p,q}$ (p, q orthogonal polarization states) function of the reflectivity properties of the surface.

In Eq. (4.65) the diffused intensity is expressed as a function of the Rayleigh parameter P_g. It is seen that for a given incident angle θ_1 and correlation distance L, the diffused backscattered intensity increases with an increase of P_g.

The diffused scattered intensity $\langle I_d \rangle$ depends on the statistical quantities σ_h (surface roughness) and L (correlation length). In the tangent plane approximation, σ_h and L are both much larger than the radio wavelength λ. For example, scattering from large sea waves (gravitational waves) at microwaves frequencies can satisfy the conditions $\sigma_h / \lambda >> 1$ and $L / \lambda >> 1$.

Various phenomenological models relate the size of gravitational waves to meteo-rological parameters. For example, the standard deviation σ_h of sea height is related to wind speed "*w*" by the semi-empirical expressions [*Richter*, 1990; *Hitney*, 1994]:

$$\sigma_h = 0.0051 \cdot w^2 \qquad \text{(Phillips spectrum)} \tag{4.66}$$

$$\sigma_h = 0.00176 \cdot w^{5/2} \qquad \text{(Neumann-Pierson spectrum)} \tag{4.67}$$

The correlation length of sea fluctuations can be characterized by the mixing-length theory of Obukov [*Massey,* 1968; *Landau*, 1979; *Mosetti*, 1979]. The mixing length is the region where the fluid particles interact by exchange of energy at the interface between sea and troposphere. As in this mixing region we expect high correlation, the mixing length is taken proportional to the correlation length of the sea fluctuations. Values of this mixing length are in the range of the wavelength at microwave frequencies.

The total backscattered intensity may be written as the sum of the coherent component given in Eq. (4.63) and the incoherent component given in Eqs (4.64) and (4.65). By inspection of these components, we note the following two limiting cases:

(a) When σ_h is very small compared to wavelength (small Rayleigh parameter), or L is very large, the diffused component tends to zero and gives no contribution to the total intensity. The total intensity is given by the coherent component only.

(b) When the roughness is very large compared to wavelength (large Rayleigh parameter), both components (coherent and diffused) tend to zero. The total scattered intensity becomes very small. The correlation length L affects the diffused component only. Large values of correlation length reduce the intensity of the diffused scatter.

These considerations are illustrated in Fig. 4.9, where curves of the coherent, diffused and total (coherent + diffused) intensities are shown, as a functions of the Rayleigh roughness parameter P_g.

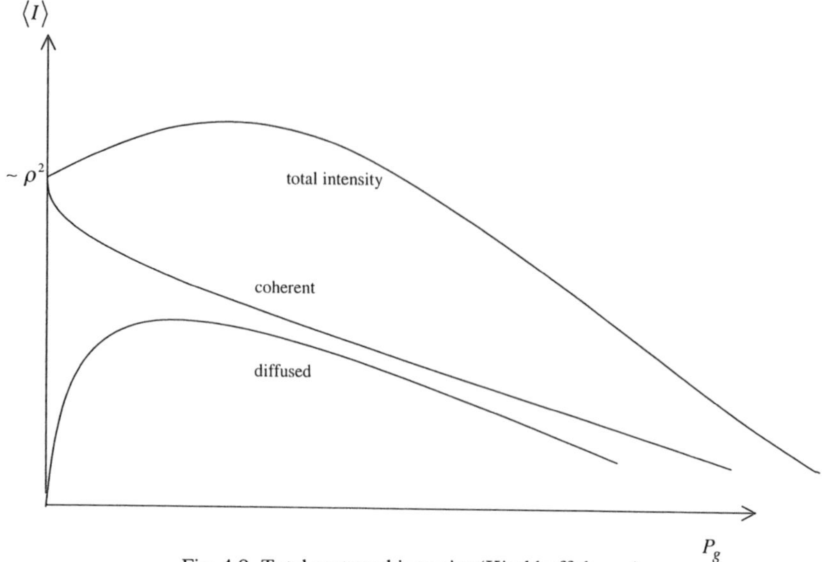

Fig. 4.9 Total scattered intensity (Kirchhoff theory)

Where

$$\lim(diffused): \lim_{\substack{\sigma_h \to 0 (L \to \infty) \\ P_g \to 0}} \frac{\left(\frac{L}{\sigma_h}\right)^2}{e^{\left(\frac{L}{\sigma_h}\right)}} \to 0 \; ; \; \lim(diffused): \lim_{\substack{\sigma_h \to \infty (L \to 0) \\ P_g \to \infty}} \frac{\left(\frac{L}{\sigma_h}\right)^2}{e^{\left(\frac{L}{\sigma_h}\right)}} \to 0$$

[cf. Eqs 4.64 and 4.65]

L: surface height correlation length; σ_h :surface height variance

$$\lim_{P_g \to 0} (coherent) \to \rho^2 \quad (\rho : \text{reflection coefficient}) \qquad \text{[cf. Eq. 4.63]}$$

$$\lim_{P_g \to \infty} (coherent) \to 0 \qquad \text{[cf. Eq. 4.63]}$$

From the intensity of the scattered field we can derive the scattering cross section of a radar object and the scattering matrix elements.

From the (polarization) changes of the scattering elements we can retrieve information on the objects and on the propagation media. The variation of the scattering elements is, in general, random and time dependent. This requires a stochastic analysis of the scattering matrix [*Fung*, 1994].

We can compute the expected values of the scattering cross section (modulus square of the scattering matrix elements) and the correlation products of the scattering matrix elements using the equation [*Ulaby*, 1982; 1990]:

$$\left\langle S_{p,q} S_{m,n}^{*} \right\rangle = \frac{k^2}{16\pi^2} \left\langle I_{p,q} I_{m,n}^{*} \right\rangle \tag{4.68}$$

where k is the wavenumber and

$\left\langle I_{p,q} I_{m,n}^{*} \right\rangle$: Correlation product of the scattered intensity calculated for the four (orthogonal) polarization combinations e.g., vv, vh, hv and hh. The scattered intensity I is given, for example, by the Eq. (4.52) for coherent scatter and by Eq. (4.53) for the diffused scattered component.

$\left\langle S_{p,q} S_{m,n}^{*} \right\rangle$: Correlation products of the scattering matrix elements for four orthogonal polarization combinations per scattering matrix. The scattering cross sections are computed for the polarization configuration p=m; q=n.

- **Multiple scattering: depolarization**

A measure of the effects on polarization is given by the polarization matrix $F_{p,q}$ defined in Eq. (4.64). We have seen in the backscatter case that the co-polar term of the scattered intensity $\left\langle I_{p,p} \right\rangle_d$ is proportional to the Fresnel reflection coefficient for the p-polarization, and the cross-polar term $\left\langle I_{p,q} \right\rangle_d$ for $p \neq q$ is equal to zero. Absence of depolarization can be explained by the fact that multiple scattering has been neglected.

The Kirchhoff theory has been used in a multiple scattering formalism (higher order Kirchhoff theory) [*Fung*, 1981] to predict depolarization from surfaces of finite conductivity. Multiple scattering is considered to occur only between specular points (scatterers). The multiple-scattered field is calculated under the assumption that the surface acts as a distributed source of rays with amplitudes given by the Kirchhoff theory. It is assumed that all scattered components interact with the surface again, but with a shadowing correction applied to the back-scattered wave and to the forward-scattered wave. An iterative series results for each Fourier component of the scattered field, with successive iterations corresponding to increasing orders of multiple scattering [*Ogilvy*, 1991]. In Fig. 4.10, the predicted scattered field intensity is shown for incident and scattered waves of both polarizations. In all cases, multiple scattering enhances the intensity for all scattered directions, the effect becoming more noticeable as the angle of scattering increases away from the normal for a fixed angle of incidence.

The depolarization term, which is zero in the single scattering Kirchhoff theory, becomes nonzero within the setting of the multiple scattering Kirchhoff theory. This theory is valid in the limit of very high frequencies for which the wavelength is much smaller than any change of surface properties.

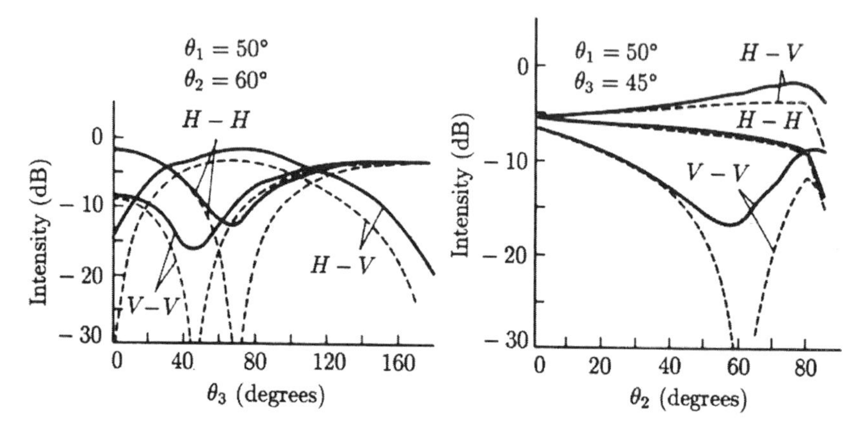

Fig. 4.10 Predicted scattered intensities based on second-order scattering theory (full curves) compared to single scattering theory (broken-lines curves)
(Reproduced with permission of IOP Publishing Ltd, J.A. Ogilvy, "Theory of Wave Scattering from Random Rough Surfaces", Fig. 5.9, Ch. 5, 1991)

Consideration of partial surface shadowing and multiple scattering are serious problems in the Kirchhoff method, especially at grazing incidence. The shadowing corrections used in multiple scattering account for regions of the surface that may be screened by other parts (higher undulations) of the surface. Not all "potential scatterers" are illuminated by the incident wave. The shadowing correction is in general a function of the incident angle θ_1, the scattering angle θ_2 and the *rms* slope gradients (m) of the rough surface. If the shadowing function is indicated by $S(\theta_1, \theta_2, m)$, it satisfies the condition $0 < S < 1$ and it multiplies the scattered field intensity computed without shadowing. Various functions $S(\theta_1, \theta_2, m)$ are used [*Beckmann*, 1965]. They decrease with an increase of the *rms* slope m_s and the angle of incidence (and scattering) away from normal. Examples of shadowing functions are given in Fig. 4.11, where the gradient m_s is indicated with the ratio σ / λ_0

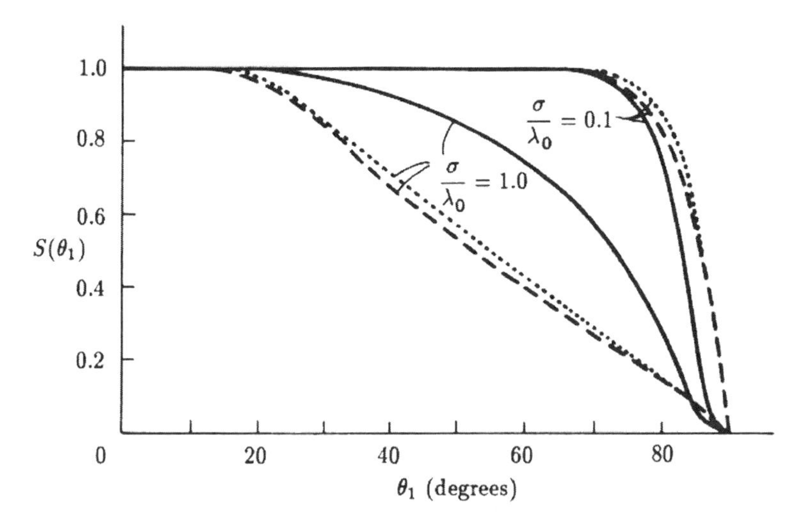

Fig. 4.11 Comparison of three shadowing functions: Beckmann (full curve); Wagner (broken-lines curve); Smith (dotted curve)
(Reproduced with permission of IOP Publishing Ltd, J.A. Ogilvy, "Theory of Wave scattering from Random Rough Surfaces", Fig. 7.6, Ch. 7, 1991)

Results of a numerical simulation [*Bass and Fuks*, 1979; *Ogilvy*, 1991] indicate that shadowing becomes "effective" at large angles (incident and scattering) away from the normal. The effects of shadowing on the number of scatterers (specular points) on a random surface has been also studied by [*Mikhaylovskiy and Fuks*, 1993]. The result of the theoretical studies indicate that the number of "effective" or "bright" scatterers depends on the surface roughness. The number of "bright" scatterers for a given geometry (antenna height) and propagation path decreases for larger standard deviation σ_h of the surface height fluctuations.

The number of "unshadowed" scatterers (specular points) is equal to the number of "correlation distances" enclosed in the surface area illuminated by the antenna beam. By increasing the antenna height, the number of "unshadowed" points increases, and the distribution of these points approaches the Poisson law at long ranges.

We give a numerical example. If the number of potential "unshadowed" scatterers is N_s and S is the shadowing function, the number of effective "bright" scatterers per square meter can be estimated by the product $N_B = \{[N_s]S\}^2$. If we take for example N_s equal to the number of wavelengths in one meter (30 at X-band frequencies), and a shadowing value of S = 0.2 (corresponding in Fig. 4.11 to an incident angle of 85° for an *rms* gradient equal to 1) we obtain $N_B = 36$ effective scatterers per square meter (the unshadowed points are $N_s = 900$ per square meter).

Depolarization in the backscattered direction has also been modelled with a composite roughness model in order to explain observations of scattering from the sea surface [*Fuks*, 1966; 1969]. The surface is modelled in terms of two independent roughness components: a small scale, high-frequency roughness superimposed on a large scale, low-frequency roughness. The effect of the large amplitude roughness is modelled as 'tilt' to the small amplitude scale of roughness. This results into a tilt of the scattering angle that changes the polarization of the scattered wave with respect to the incident wave. A mathematical model of this two-scale roughness has been provided by [*Brown*, 1980]. He considers the surface height function as the sum of two random variables, viz. [*Ogilvy*, 1991]:

$$h(x,y) = h_s(x,y) + h_l(x,y) \tag{4.69}$$

where h_s is the random variable associated with the small-scale component, and h_l is the random variable corresponding to the large-scale component. Assuming these two random components to be Gaussian and independent, he uses a model which combines the Kirchhoff and the small perturbations theories. The Kirchhoff method applies to the large-scale roughness, and the (first-order) perturbation theory accounts for the small-scale roughness. This model can be applied, for example, to scattering from a rough sea surface composed of large-scale roughness (gravitational waves) and small-scale roughness (capillary waves).
In Fig. 4.12, predictions are shown of scattering from the two-scale rough surface, with use of the 'combined' (Kirchhoff + small perturbations) model. The results of the model indicate that scattering from small scale fluctuations is predicted by small perturbations (first-order) at large angles (away from normal), and scattering from large scale fluctuations by the Kirchhoff at small angles (near to normal).

The effect of backscatter from steep sea-waves on polarization are predicted by a model based on two-scale (large-small) roughness developed by [*Voronovich and Zavorotny*, 2000]. Results of predictions indicate large differences between backscatter levels for VV (transmit and receive vertical polarization) and for HH (transmit and receive horizontal polarization), resulting in VV levels higher than HH.

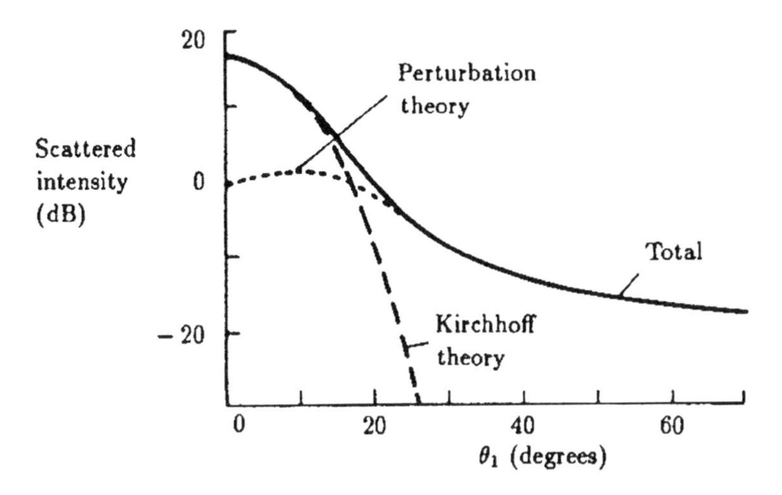

Fig. 4.12 Scattered intensity for a composite rough surface
(Reproduced with permission of IEEE © 1978 (Brown) and IOP Publishing Ltd, J.A. Ogilvy, "Theory of Wave Scattering from Random Rough Surfaces", Fig. 7.3, Ch. 7, 1991)

4.1.3 Other types of scattering modelling

Another interesting method for the calculation of scattering from composite rough surfaces has been developed by [*Voronovich*, 1983; 1985]. By this method, the intensity of the scattered wave is calculated based on an expansion of the logarithm of the scattering matrix as a function of a power-series of the surface heights. For small heights, this method gives results equivalent to those obtained with the approximation by the first terms of the series in the perturbation theory. In the cases of large-scale and low-slope irregularities, the method reduces to the Kirchhoff approximation. This method is applicable for various types of surface irregularities. The scattering coefficients computed by this method for the case of sinusoidal surfaces are shown to compare well with experimental data [*Voronovich*, 1985]. The scattering coefficient of statistically irregular surfaces is also calculated by this method. The scattering coefficient is given by a formula which includes both cases of small and large scale of roughness, with no need to have a parameter separating these scales.

A mathematical model of scattering from multi-path propagation over (composite) rough surfaces has been developed by [*Pusone*, 1999]. The model is based on a phenomenological characterization of the surface roughness as composed of several delayed paths. The model computes the coefficients of the scattering matrix for the case of rough sea surfaces, various sea states (wind velocity) conditions, at low grazing angles. The multi-path delay spread is computed from the scattering coefficient computed for the two cases of large scale roughness (gravitational waves for large wind velocities) and small scale roughness (capillary waves for small wind velocities), at low grazing angles (less than 5 degrees) for two polarizations, vertical and horizontal.

For the computation of the scattering coefficient, the Kirchhoff method for the case of gravitational waves is used. The small perturbation theory is used for capillary waves. Results of the model indicate that the multi-path propagation causes distortion in the radar pulse shape. The distortion is found more pronounced for wide-band radar signals, in the case of scattering from gravitational waves at horizontal polarization.

The multi-path model of [*Pusone*, 1999] has recently been extended to evalute the effects of multi-path by scattering in the polarization of the scattered wave. The "extended" model [*Pusone*, 2001] evaluates the changes of polarization characteristics (e.g. ellipticity) of the scattered wave as a function of sea roughness (sea state). The theoretical results of the model indicate that by increase of wind speed the ellipticity of the polarized wave is decreased.

Considerations of surface partial shadowing and multiple scattering constitute serious problems in scattering modelling, especially at low grazing angles [*Voronovich*, 1996]. An attempt to solve the problem of scattering in a general form was undertaken [*Walsh*, 1987]. The merit of the approach consists in its relative simplicity, the natural derivation of proper boundary conditions and the absence of restrictions on surface roughness characteristics, such as their heights and tilt angle. The problem is reduced to a joint solution of a system of integral equations for a field located directly above and below the surface.

The solutions enable the calculation of a field reflected by the boundary and a field passing into the other medium. The ensuing integral equations have the same nature (but a different form) as the equations derived from the application of Green's theorem. The mathematical complication of these equations is connected with the problem of the calculation of the field actually incident on the surface. Each surface element is illuminated not only by the incident wave but also by fields reflected by adjacent elements (i.e. repeatedly scattered waves). Therefore, the problem is identical to the multiple scattering in a volume. If a real field is found (a field affecting each separate scatterer), then the calculation of the total field becomes trivial.

[*Walsh*, 1987] has proposed simplifications assuming that the spatial Fourier transform of the field and the surface profile is restricted to spatial frequencies for which $k << 2\pi / \Lambda$, where k is the wavenumber and Λ is the (mean) spatial frequency of the surface fluctuations (perturbations). Actually, this approximation means that only smooth disturbances are analyzed and that the incident angle is close to zero. This approximation is close to the Rayleigh hypothesis according to which each surface element only interacts with the incident wave.

In our opinion, the correct description of surface scattering with repeated interaction between the waves and a medium can be carried out by considering this scattering as volume scattering in an electrodynamic dense medium.

4.2 Transport theory: radiative transfer equation

Radiative transfer theory deals with the transport of energy through a medium containing particles. The equation of transfer governs the variation of intensities in a medium that absorbs, emits and scatters radiation [*Barabanenkov*, 1975; *Chandrasekhar*, 1960].

The vector-valued radiative transfer equation is given by [*Ulaby*, 1990; *Tsang*, 2000]

$$\frac{d\vec{I}(\vec{r},\hat{s})}{d\hat{s}} = -\overline{\overline{k}}_e \vec{I}(\vec{r},\hat{s}) - k_{ag}\vec{I}(\vec{r},\hat{s}) + \vec{J}_e + \int_{4\pi} P(\hat{s},\hat{s}')\vec{I}(\vec{r},\hat{s}')d\Omega' \tag{4.70}$$

using the Stokes vector representation, in which $\vec{I}$ is a (4×1) matrix representing the vector specific intensity, $\overline{\overline{k}}_e$ is (4×4) extinction matrix, k_{ag} is a power absorption coefficient, $\vec{J}_e$ is the (thermal) emission vector and $P(s,s')$ is the (4×4) phase function matrix accounting for scattering. The latter gives the ratio between the intensity scattered at direction $\hat{s}$ and incident at direction $\hat{s}'$. Equation (4.70) governs the change in the specific intensity $\vec{I}$ through an elementary volume of unit cross section along a differential distance $(d\hat{s})$, accounting for absorption loss and scattering loss.

To solve the radiative transfer equation, we need to know the phase and extinction matrices. The phase matrix is proportional to the scattering matrix. It is defined by

$$P = N_v < L_m > \tag{4.71}$$

Where N_v is the number of particles per unit volume and $\langle L_m \rangle$ is the average of the modified Mueller matrix. The latter is given by [*Ulaby*, 1990]

$$L_m = \begin{pmatrix} |S_{VV}|^2 & |S_{VH}|^2 & \mathrm{Re}\left(S_{VV}S^*_{VH}\right) & -\mathrm{Im}\left(S_{VV}S^*_{VH}\right) \\ |S_{HV}|^2 & |S_{HH}|^2 & \mathrm{Re}\left(S_{HV}S^*_{HH}\right) & -\mathrm{Im}\left(S_{HV}S^*_{HH}\right) \\ 2\,\mathrm{Re}\left(S_{VV}S^*_{HV}\right) & 2\,\mathrm{Re}\left(S_{VH}S^*_{HH}\right) & \mathrm{Re}\left(S_{VV}S^*_{HH} + S_{VH}S^*_{HV}\right) & -\mathrm{Im}\left(S_{VV}S^*_{HH} - S_{VH}S^*_{HV}\right) \\ 2\,\mathrm{Im}\left(S_{VV}S^*_{HV}\right) & 2\,\mathrm{Im}\left(S_{VH}S^*_{HH}\right) & \mathrm{Im}\left(S_{VV}S^*_{HH} + S_{VH}S^*_{HV}\right) & \mathrm{Re}\left(S_{VV}S^*_{HH} - S_{VH}S^*_{HV}\right) \end{pmatrix} \tag{4.72}$$

where $S_{vv}, S_{vh}, S_{hv}, S_{hh}$ are the scattering elements defined in the Chapter 2.

The total extinction matrix $\overline{\overline{k}}_e$ becomes [*Ulaby,* 1990]

$$\overline{\overline{k}}_e = \begin{pmatrix} -2\,\mathrm{Re}(M_{VV}) & 0 & -\mathrm{Re}(M_{VH}) & -\mathrm{Im}(M_{VH}) \\ 0 & -2\,\mathrm{Re}(M_{HH}) & -\mathrm{Re}(M_{HV}) & \mathrm{Im}(M_{HV}) \\ -2\,\mathrm{Re}(M_{HV}) & -2\,\mathrm{Re}(M_{VH}) & -\mathrm{Re}(M_{VV}+M_{HH}) & \mathrm{Im}(M_{VV}-M_{HH}) \\ 2\,\mathrm{Im}(M_{HV}) & -2\,\mathrm{Im}(M_{VH}) & -\mathrm{Im}(M_{VV}-M_{HH}) & -\mathrm{Re}(M_{VV}+M_{HH}) \end{pmatrix} \tag{4.73}$$

where

$$M_{pq} = \frac{i2\pi N}{k_0} \left\langle S_{mn}\left(\theta_i\,,\,\phi_i\,;\,\theta_i\,,\,\phi_i\,;\,\theta_j\,,\,\phi_j\right)\right\rangle \;\; ; \;\; \text{p,q=V,H} \tag{4.74}$$

With the definition of the phase and the extinction matrices we can solve the radiative transfer equation for a specific particle distribution and given boundary conditions.

The modified scattering (Mueller) matrix L_m [see Eq. (4.72)] is computed using the elements S_{pq} of the scattering matrix obtained for the case of scattering from given particle shapes (e.g., small spheres, cylinders, etc.), depending on the types of remotely sensed objects. The elements of L_m must be averaged in accordance with the specified particle sizes, shapes, or space distributions in order to obtain the phase matrix given in Eq. (4.71). With the matrices P and k_e' known, the radiative transfer Eq. (4.70) may be solved, in general, using numerical methods.

- **A specific example**

To solve the radiative transfer equation (4.70) we need to specify appropriate boundary conditions. From the definitions of power reflectivity and transmissivity, we can obtain the relations between incident-reflected and incident-transmitted Stokes vectors. Using the definitions of the Fresnel reflection and transmission coefficients, Snell's law of refraction, and the conservation of electromagnetic energy (incident energy = reflected + transmitted energy), we obtain the relation between the incident I^i upon a planar boundary from medium 1 to medium 2 and the reflected I^r Stokes vector [*Ulaby*, 1990]:

$$I^r = R_{12}(\theta_1) I^i \tag{4.75}$$

Here $R_{12}(\theta_1)$ is the reflectivity matrix defined as a function of the Fresnel reflection coefficient for vertical polarization and for horizontal polarization [*Ulaby*, 1990].

The relation between the incident I^i and the transmitted I^t Stokes vector is given by

$$I^t = T_{12}(\theta_1) I^i \tag{4.76}$$

Where $T_{12}(\theta_1)$ is the transmissivity matrix defined as a function of the Fresnel transmission coefficient for vertical polarization and for horizontal polarization [*Ulaby*, 1990].

The boundary conditions associated with a rough surface are determined in [*Fung and Eom*, 1981] for scattering from a layer of Rayleigh particles ($a << \lambda$, with a is the size of particles and λ the wavelength) over a rough surface using the Kirchhoff approximation.

The equation of radiative transfer [cf. Eq. (4.70)], with the definition of the phase matrix [cf. Eq. (4.71)] and the extinction matrix given in Eq. (4.73), together with the boundary conditions [cf. Eqs. (4.75), (4.76)], constitutes a complete mathematical formulation of the problem of multiple scattering in a random medium.

Various methods are used to solve the radiative transfer equation. The most frequently used are an iterative technique when the multiple scattering is small, or a discrete eigenanalysis technique when the multiple scattering is more significant. The iterative procedure starts by computing the zero-order solution which ignores scattering. The zero-order solution is used as a source function for computing the first-order solution. This procedure has been applied to solve the problem for the 3-layer model [*Ulaby*, 1990]. This model considers a specific intensity incident onto the upper layer (1) of scatterers. It is assumed that the background permittivity of the mid-layer (2) is the same of upper layer. The bottom boundary of layer (3) is flat and its permittivity is different from that of the mid-layer. No reflection occurs at the top interface. The incident intensity will be reflected at the flat interface between the mid and bottom layers.

Applying the iterative procedure to the transfer equation with the described boundary conditions, the solutions are expanded into series of perturbation orders as follows [*Ulaby*, 1990]:

$$I(\cos\theta_s,\phi_s,z) = I^{(0)}(\cos\theta_s,\phi_s,z) + I^{(1)}(\cos\theta_s,\phi_s,z) + I^{(2)}(\cos\theta_s,\phi_s,z) + \ldots\ldots \quad (4.77)$$

Here θ_s,ϕ_s are the scattering angles and z the depth. The zero-order solution $I^{(0)}$ represents the reduced intensity which attenuates exponentially in the medium and is proportional to [*Ulaby*, 1990]:

$$I_s^{(0)} \approx R_{23}(\cos\theta_s)e^{-\frac{k_e(z+2d)}{\cos\theta_s}} I_0 \quad (4.78)$$

Where $I_s^{(0)}$ is the (zero-order solution) scattered (upward) intensity, I_0 is the incident intensity, R_{23} is the reflectivity at mid-bottom boundary layer, θ_s is the scattering angle, z is the depth variable, d is the thickness of the interface layer and k_e is the total extinction coefficient of a particle (scatterer).

The first-order solution $I^{(1)}$ represents the single scattering solution. It is composed of a term proportional to the reduced intensity (zero-order solution) plus a second term that is a function of the phase scattering matrix P [*Ulaby,* 1990]. It should be noted that the first-order solution does not give depolarization in the back-scattering direction for spherical scatterers.

4.2.1 Polarization synthesis

The solution of the radiative equation can be used to calculate the scattering coefficient of the medium, defined by the ratio:

$$\sigma_{pq} = 4\pi \frac{\left|E_p^s\right|^2}{\left|E_q^i\right|^2} \tag{4.79}$$

Where E_p^s is the scattered field (p-polarized) and E_q^i the incident field (q-polarized). E_p^s is related to the scattered intensity I_p^s by

$$\left|E_p^s\right|^2 = I_p^s \cdot \cos\theta_s \quad , \qquad \text{with } \theta_s \text{ scattering angle} \tag{4.80}$$

And E_q^i is related to the incident intensity I_q^i by

$$\left|E_q^i\right|^2 = I_q^i \tag{4.81}$$

The scattering coefficient is given [from Eq. (4.79), (4.80), (4.81)], by:

$$\sigma_{pq} = 4\pi \cos\theta_s \frac{I_p^s}{I_q^i} \tag{4.82}$$

where I_p^s is the solution of the radiative equation.

PART III

DIAGNOSTICS OF THE EARTH'S ENVIRONMENT USING POLARIMETRIC RADAR MONITORING: FORMULATION AND POTENTIAL APPLICATIONS

CHAPTER 5

Basic Mathematical Modelling for Random Environments

5.1 Introduction

Basic mathematical modelling for random media is provided in this chapter. This background information is needed for the calculation of the fields scattered by vegetation-covered ground surfaces, with specific applications to radar remote sensing.

In remote sensing, the incident field is transformed when it interacts with the earth's surface covered with vegetation. Amplitude and phase relationships of the echo signal carry all the information needed for the solution of inverse problems in pattern recognition of forest, bush, crops, desert, etc., as well as for the evaluation of the vegetation parameters (biomass density, humidity, vegetation cover thickness). However, the extraction of information by solving inverse problems is problematic due to the following facts:

1. In most cases, there is no clear understanding of the "coding" of information during the echo signal formation in the scattering volume. The complexity is caused by the presence of geophysical objects and by processes of interaction between radiation and such objects.

2. A significant loss of information occurs in antenna systems during reception and antenna processing. This loss can be reduced by combining frequency and polarization measurements and/or using a coherent adaptive phased-array radar or a synthetic aperture radar.

Theoretical research [*Rino,* 1988; *Klyatskin,* 1975; *Novikov,* 1964; *Furutsu,* 1963; *Rytov,* 1978; *Landau,* 1982; *Li,* 1992] and experimental research [*Schiffer,* 1979; *Karam,* 1988] is devoted to the investigation of the reflected fields as a basis for the solution of the inverse problem. The experimental research [*Karam*, 1988] deals in particular with propagation (scattering) through (by) vegetation.

The main stages in solving the inverse problem are formulated below:

- Physical processes have to be specified on the basis of model analysis. Distinct features of the echo signal and unambiguous parameters of vegetation models play an important role.

- Adequate mathematical methods for the description of the interaction between radiation and vegetation must be chosen from the available methods, or have to be modified or newly developed. Then, the relationships between the observed parameters of echo signals and the corresponding parameters of vegetation can be calculated.

- Solution schemes for the inverse problem have to be validated by testing and verifying the physical models.

It is pointed out that our approach based on physical models should not be confused with the work in [*Schiffer*, 1979] and [*Karam*, 1988], which is classified as phenomenological.

The difficulty of formulating strict algorithms is the main disadvantage of the phenomenological approach, which is grounded on intuition and ease-of-success, instead of on a strict analysis of the physical situation. However, a theoretical analysis of the interaction of radiation with a complex vegetation model is so laborious that in the general case far-reaching approximations need to be made. The oversimplification introduced by these approximations in forming the final result may become so great that discrimination of a specific vegetation cover by the echo signal may prove infeasible or erroneous.

Here, we attempt a different approach; specifically, grass covers are modelled as a series of cylinders. This means that we do not solve the problem of pattern recognition but the more straightforward problem of evaluation of parameters.

The first step in this approach uses the knowledge that cylinders modelling grass covers have prevailing directions in space. This implies that their ensemble, in accordance with their geometrical anisotropy, must be characterized by an anisotropic effective dielectric constant. The model of such vegetation may be a (horizontally) parallel layered structure of an anisotropic medium situated above the soil surface. However, in this case we encounter several problems characterized by two fundamental questions:

a) Which physical radiation process involved in media interactions should be chosen as dominant when forming distinctions in the echo signal?

b) How can we quantify the tensor components of the effective dielectric constant of the modelled medium connected with the parameters of vegetation (geometry of cylinders and dielectric constant of biomass)?

We discuss first question (a). The interaction effects can be classified into coherent and incoherent. With an ensemble of N randomly positioned scatterers, the intensity of coherently scattered fields in the Born approximation is proportional to N^2, whereas the intensity of incoherent scattering is proportional to N. Taking into account that N is usually large, we notice that the difference in intensities of coherent and incoherent processes is significant. It is, therefore, natural to consider in the first place the effects connected with coherent scattering. In the ensemble of non-specially prepared scatterers, the coherent processes are associated with forward scattering only. The effect of these processes is manifested in the first place in the phase velocity of the waves in the medium. If the medium is anisotropic (as in our case), then the phase velocity becomes dependent on the direction of propagation (in our case on the angle of viewing) and the polarization of the field. For waves with different polarizations, the optical thickness of a layer will be different and an additional phase shift will appear between waves. (This phase shift is determined by the extent of anisotropy of the medium). If this phase shift is determined experimentally and theoretically when calculating the interaction between medium and radiation, then the first part of the solution of the inverse problem concerning the evaluation of the parameters may be obtained.

Theoretical and experimental determination of the phase shift of fields with different polarizations (these fields compose the echo signal) determine the subject of our analysis and are the core of the second and third steps in our approach.

After exposition of the main physical idea (as the basic method for remote determination of grass cover parameters), we discuss problems connected to the second stage: the choice of an adequate mathematical method for the description of the radiation-vegetation interaction. For the description of coherent scattering effects, the approach based on the radiative transfer equations unfortunately cannot be applied. It is necessary to find equations describing the behavior of the coherent averaged field in a randomly inhomogeneous medium. It is well known that in general the description becomes an application of approximate solutions of the Dyson equation [*Karam*, 1988; *Kuznetsov*, 1988]. Here, no comparative analysis of these approximations is undertaken. We shall describe our own approach, which possesses adequate physical transparency and can easily be interpreted in terms of Feynman diagrams [*Tatarski*, 1971].

In the next step, a concrete solution scheme is proposed, allowing an experimental determination of the decisive phase shift of reflected waves with orthogonal polarizations. Waves with circular polarization are considered to be most suitable. This is because the phase velocities of waves with linear polarization (waves with 90 degrees phase difference and equal amplitudes form circularly polarized waves) will result in the appearance of waves with circular polarization, but with two directions of revolution of the $\vec{E}$ vector (right-hand circular and left-hand circular polarizations) in the echo signal. Knowing the relationship between the two components in the echo signal for two orthogonal circular polarizations, it is possible (and relatively simple) to calculate the value of the phase shift of the waves with horizontal and vertical polarizations.

5.2 Space spectrum method

5.2.1 General concepts and relationships

When electromagnetic wave propagation in randomly inhomogeneous media takes place, the main problem is the effect of multiple scattering due to fluctuations in the permittivity.

An iterative method is conceptually simple but not always an optimal description of a real geometry from a computational point of view. Our method allows writing the problem solution in the form of a Born series, i.e. a scattered field can be represented in the form of an infinite sum of fields with single scattering, double scattering, etc. Using this approach, single and double scattering processes are taken into account and the processes of higher order scattering are neglected. Such a truncation of a series is possible when the parameter $v = L/l_f$ is small, where l_f is the photon free path length in the medium and L is the characteristic linear dimension of the scattering area.

When $v \ll 1$, the Born approximation reduces the calculations to a first iteration procedure. Such an approximation is called Born scattering. Our approach takes the effects of multiple scattering into account by partitioning the whole interaction area into small zones. Within these zones, the condition $v_i = L_i / l_f \ll 1$ is met and, therefore, the Born approximation is applicable. The main problem of such an approach is combining the solutions at the boundary of such zones. Some aspects of this problem will be discussed in the sequel.

In a number of problems of practical interest, the region with a randomly inhomogeneous medium can be modelled by a plane layer. In this case, it is advisable to consider infinitesimal thin layers with boundaries parallel to the boundaries of the macrolayer. Each thin layer has an elementary volume and meets the criterion for applying the Born scattering approximation. Within the framework of such a geometry, it is convenient to represent the field in the medium as a superposition of plane waves. Such a field representation becomes the basis of the spectral domain method of multiple scattering [*Rino*, 1988]. Consider, specifically, the Fourier synthesis

$$\tilde{\vec{E}}^{\pm}(\vec{q},z)=\frac{1}{(2\pi)^2}\int_{\Delta A}\vec{E}^{\pm}(\vec{\rho},z)e^{-i\vec{q}\cdot\vec{\rho}}d\vec{\rho} \tag{5.1}$$

Here, ΔA is the area of an elementary layer on the x-y plane, z is the coordinate of an observation point along the axis directed perpendicular to the boundary of the planar layer and $\vec{q}$ is the projection of the wave vector $\vec{k}$ on this boundary plane. The total field $\vec{E}$ is represented in the form of a sum: $\vec{E}=\vec{E}^{+}+\vec{E}^{-}$. The sign " + " corresponds to waves propagating in the direction of the + z axis and the sign " $-$ " corresponds to waves propagating in the opposite direction.

Within an elementary layer i, the electric field is the solution of the inhomogeneous vector-valued Helmholtz equation

$$\vec{\nabla}\times\vec{\nabla}\times\vec{E}-k_0^2\varepsilon_r\vec{E}_i=k_0^2\delta\varepsilon_r\vec{E}_e \tag{5.2}$$

Here, ε_r and $\delta\varepsilon_r$ denote the coherent and incoherent parts of the relative permittivity of the layer, respectively, and $\vec{E}_e$ is the field within the elementary layer and is determined by the waves entering this layer. For the three layers modelling air-vegetation-ground (see Fig. 5.1) the Helmholtz equations are

$$\vec{\nabla}\times\vec{\nabla}\times\vec{E}_1-k_0^2\varepsilon_{r_1}\vec{E}_1=0 \tag{5.2a}$$

$$\vec{\nabla}\times\vec{\nabla}\times\vec{E}_2-k_0^2\varepsilon_{r_2}(\vec{\rho},z;d)\vec{E}_2=0 \tag{5.2b}$$

$$\vec{\nabla}\times\vec{\nabla}\times\vec{E}_3-k_0^2\varepsilon_{r_3}\vec{E}_3=0 \tag{5.2c}$$

where $k_0=\dfrac{\omega}{c}$ is the wavenumber, $\vec{E}_1,\vec{E}_2,\vec{E}_3$ are the electric fields and $\varepsilon_{r_1},\varepsilon_{r_2},\varepsilon_{r_3}$ are the relative (to vacuum) permittivities in the three layers (air, vegetation and ground), respectively. To solve Eqs (5.2a-c) we need to choose appropriate 'realistic' boundary conditions [*Lang*, 1983].

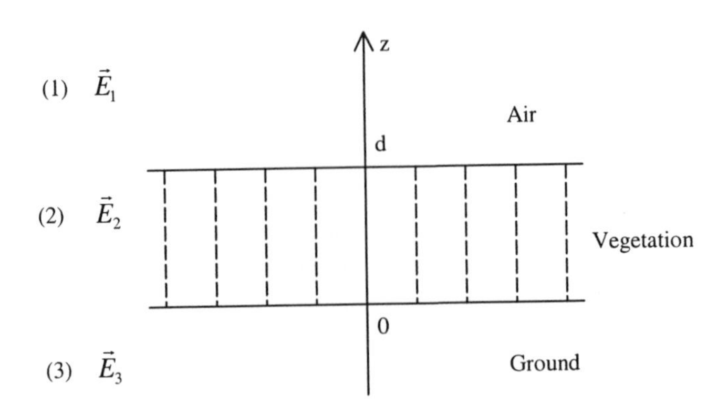

Fig. 5.1 Illustration of a three-layer model: air-vegetation-ground

The field leaving each layer can be represented as

$$\vec{E} = \vec{E}_0 + \vec{E}_s \tag{5.3}$$

where $\vec{E}_0$ is the solution of Eqs (5.2a-c) in the absence of scatterers and $\vec{E}_s$ is the scattered field given by

$$\vec{E}_s = \int \overline{\overline{\Gamma}}(\vec{r},\vec{r}') \cdot \vec{S}(\vec{r}')d\vec{r}' \tag{5.4}$$

$\vec{S}(\vec{r}')$ being the 'source' defined on the right-hand side of equation (5.2), viz.

$$\vec{S}(\vec{r}') = k_0^2 \delta\varepsilon_r \vec{E}_e \tag{5.4a}$$

The dyadic Green's function $\overline{\overline{\Gamma}}(\vec{r},\vec{r}')$ is given by

$$\overline{\overline{\Gamma}}(\vec{r},\vec{r}') = \left[\overline{\overline{1}} + \frac{1}{k_0^2\varepsilon_r}\nabla\nabla\right] G(\vec{r},\vec{r}') \tag{5.5}$$

where $G(\vec{r},\vec{r}')$ is the ordinary Green's function for the scalar-valued Helmholtz equation, given by

$$G(\vec{r},\vec{r}') = \exp\left\{ik_0\sqrt{\varepsilon_r}\,|\vec{r}-\vec{r}'|\right\}/\left(4\pi|\vec{r}-\vec{r}'|\right) \tag{5.6}$$

It should be pointed out that the Green's function is chosen in relation to a specific boundary problem. The choice of (5.3) allows us to meet the boundary conditions for an elementary layer (i.e., the conditions for inhomogeneous media). By doing so, the dyadic Green's function (5.5), can be applied in the calculation of $\vec{E}_s$ with the scalar Green's function given in Eq. (5.6) meeting the Sommerfeld radiation condition. Assuming that the fields entering an elementary layer are known, it is easy to derive finite difference equations for the fields leaving the layer:

$$\tilde{\vec{E}}^{\pm}(\vec{q},z+\Delta z) = \tilde{\vec{E}}^{\pm}(\vec{q},z)e^{ik_z\Delta z} + \tilde{\vec{E}}_s^{\pm} \tag{5.7}$$

Since Δz (the thickness of an elementary layer) is small and the fields at point z can be expressed as series in Δz, we derive the following system of integro-differential Eqs:

$$\pm\frac{d\tilde{E}_\alpha^{\pm}(\vec{q},z)}{dz} = ik_z(\vec{q})\tilde{E}_\alpha^{\pm}(\vec{q},z) + \Gamma_{\alpha\beta}(\vec{q},0)\int\delta\tilde{\varepsilon}_r(\vec{q}-\vec{\mu},z)\left[\tilde{E}_\beta^{+}(\vec{\mu}) + \tilde{E}_\beta^{-}(\vec{\mu})\right]d\vec{\mu} \tag{5.8}$$

Here, $\tilde{E}_{\alpha,\beta}^{\pm}(\vec{q},z)$ is the transverse (with respect to x and y) Fourier transform of the forward and backward electric field corresponding to polarizations α and β, respectively, $\Gamma_{\alpha\beta}(\vec{q},0)$ is a polarization-dependent (coupling) coefficient, $k_z(\vec{q}) = \sqrt{k_0^2\varepsilon_r - k_x^2 - k_y^2}$ and $\delta\tilde{\varepsilon}_r(\vec{q},z)$ denotes the transverse Fourier transform of the fluctuating part of the relative permittivity. The relationships in (5.8) constitute a set of coupled stochastic integro-differential equations expressed in terms of the fluctuations of the medium and the field. The transition from stochastic equations to averaged equations will be discussed in the next subsection.

5.2.2 Stochastic or ensemble averaging

The equations derived in the last section require further processing in order to be used for the description of the behavior of the average fields. For the calculation of the first moment, a problem arises in calculating the value $\left\langle\delta\tilde{\varepsilon}_r\tilde{E}\right\rangle$, where the random field $\tilde{E} = \tilde{E}[\delta\tilde{\varepsilon}_r]$ is a functional of the dielectric constant fluctuations $\delta\tilde{\varepsilon}_r = \delta\tilde{\varepsilon}_r(\vec{q},z)$. To derive a useful formula for $\left\langle\delta\tilde{\varepsilon}_r\tilde{E}\right\rangle$, we consider the following stochastic (or ensemble) averaging [*Klyatskin,* 1975]:

$$f(\vec{r},\eta)=\left\langle \xi(\vec{r})\tilde{E}[\xi+\eta]\right\rangle \tag{5.9}$$

Here, $\eta=\eta(\vec{r})$ is a certain deterministic function. The function $\tilde{E}[\xi+\eta]$ can be expanded in a Taylor series around η as follows:

$$\tilde{E}[\xi+\eta]=\exp\left\{\int d\vec{r}'\xi(\vec{r}')\frac{\delta}{\delta\eta(\vec{r}')}\right\}\tilde{E}[\eta] \tag{5.10}$$

Inserting (5.10) into (5.9) we obtain

$$f(\vec{r},\eta)=\left\langle \xi(\vec{r})\exp\left\{\int d\vec{r}'\xi(\vec{r}')\frac{\delta}{\delta\eta(\vec{r}')}\right\}\right\rangle \tilde{E}(\eta) \tag{5.11}$$

and derive

$$f[\vec{r},\eta]=\frac{\left\langle \exp\left\{\int d\vec{r}'\xi(\vec{r}')(\delta/\delta\eta(\vec{r}'))\right\}\xi(\vec{r})\right\rangle}{\left\langle \exp\left\{\int d\vec{r}'\xi(\vec{r}')(\delta/\delta\eta(\vec{r}'))\right\}\right\rangle}\left\langle \exp\left\{d\vec{r}'\xi(\vec{r}')(\delta/\delta\eta(\vec{r}'))\right\}\right\rangle E(\eta) \tag{5.12}$$

Introducing

$$\Omega[\vec{r},U]=\frac{\left\langle \xi(\vec{r})\exp\left\{i\int d\vec{r}'\xi(\vec{r}')U(\vec{r}')\right\}\right\rangle}{\left\langle \exp\left\{i\int dr'\xi(\vec{r}')U(\vec{r}')\right\}\right\rangle} \tag{5.13}$$

the equation for $f[\vec{r},\eta]$ can be written in the following form:

$$f[\vec{r},\eta]=\Omega\left[\vec{r},\frac{\delta}{i\delta\eta}\right]\left\langle \exp\left\{\int d\vec{r}'\xi(\vec{r}')\frac{\delta}{\delta\eta(\vec{r}')}\right\}\tilde{E}[\eta]\right\rangle \tag{5.13a}$$

Consider next, the expression

$$\Omega\left[\vec{r},\frac{\delta}{i\delta\eta}\right]\tilde{E}[\eta+\xi] \tag{5.13b}$$

Since it is non-random, we obtain formally

$$f[\vec{r},\eta]=\left\langle \Omega\left[\vec{r},\frac{\delta}{i\delta\eta}\right]\tilde{E}[\eta+\xi]\right\rangle \tag{5.14}$$

$\tilde{E}$ depends on η and ξ only in the combination $(\eta+\xi)$, meaning that in the operator Ω the differentiation with respect to η can be changed to a differentiation with respect to ξ. We find [*Klyatskin*, 1975], then, by setting $\eta=0$,

$$\langle \xi(\vec{r})E(\xi)\rangle=\left\langle \Omega\left[\vec{r},\frac{\delta}{i\delta\xi}\right]\tilde{E}(\xi)\right\rangle \tag{5.15}$$

Note that $\Omega[\vec{r},U]$ can be expressed as

$$\Omega[\vec{r},U]=\frac{1}{\Phi[U]}\frac{1}{i}\frac{\delta\Phi}{\delta U(\vec{r})}=\frac{1}{i}\frac{\delta\ln\Phi[U]}{\delta U(\vec{r})}=\frac{1}{i}\frac{\delta\Theta[U]}{\delta U(\vec{r})} \tag{5.16}$$

where

$$\Phi[U]=\left\langle \exp\left\{i\int d\vec{r}'\xi(\vec{r}')U(\vec{r}')\right\}\right\rangle \tag{5.16a}$$

$$\Theta[U]=\ln\Phi[U] \tag{5.16b}$$

$\Theta[U]$ is developed in a series expansion as follows:

$$\Theta[U]=\sum_{k=1}^{\infty}\frac{i^k}{k!}\int\Psi_k(\vec{r},\vec{r}_1,\ldots\vec{r}_k)U(\vec{r}_1)\ldots U(\vec{r}_k)d\vec{r}_1\ldots d\vec{r}_k \tag{5.16c}$$

Differentiating $\Theta[U]$ results into

$$\Omega[\vec{r},U]=\frac{1}{i}\frac{\delta\Theta[U]}{\delta U(\vec{r})}=\sum_{k=1}^{\infty}\frac{i^{k-1}}{(k-1)!}\int\Psi_k(\vec{r},\vec{r}_1,\ldots\vec{r}_{k-1})\prod_{j=1}^{k-1}U(\vec{r}_j)d\vec{r}_j \tag{5.17}$$

Substituting $U\left(\vec{r}_j\right) = \frac{1}{i}\frac{\delta}{\delta\xi\left(\vec{r}_j\right)}$ in Eq. (5.15) yields

$$\left\langle \xi(\vec{r})\tilde{E}(\xi)\right\rangle = \sum_{k=1}^{\infty}\frac{1}{(k-1)!}\int \Psi_k\left(\vec{r},\vec{r}_1,...\vec{r}_{k-1}\right)\left\langle \prod_{j=1}^{k-1}\frac{\delta\tilde{E}[\xi]}{\delta\xi\left(\vec{r}_j\right)}\right\rangle d\vec{r}_j \tag{5.18}$$

In many experimental situations it suffices to assume a Gaussian distributed $\xi(\vec{r})$ with $\Psi_3 = \Psi_4 = ... = 0$, so that only the terms containing $\Psi_1 = <\xi(\vec{r})>$ and $\Psi_2(\vec{r},\vec{r}') = \left\langle \xi(\vec{r})\ \xi(\vec{r}')\right\rangle$ remain. In that case we find

$$\left\langle \xi(\vec{r})\tilde{E}[\xi]\right\rangle = \left\langle \xi(\vec{r})\right\rangle\left\langle \tilde{E}[\xi]\right\rangle + \int\left\langle \xi(\vec{r})\xi(\vec{r}')\right\rangle\left\langle \frac{\delta\tilde{E}[\xi]}{\delta\xi(\vec{r}')}\right\rangle d\vec{r}' \tag{5.19}$$

This relationship was derived for the first time by [*Novikov,* 1964] and [*Furutsu,* 1963]. It follows from this Novikov-Furutsu formula that the average value of the product of two random functions (which have a certain functional interdependence) can be represented in the form of the sum of two components. The first term is equal to the product of their average values and the second term depends on an average value of the variational derivative and the correlation function of the random function $\xi(\vec{r})$.

We make the following remarks concerning the Novikov-Furutsu formula applied to wave scattering in randomly inhomogeneous media:

- If $\xi(\vec{r})$ describes fluctuations in the dielectric constant, the first term of the Novikov-Furutsu formula means that the field is scattered from each inhomogeneity only once. Repeated field scattering from inhomogeneities is described by the second term. This can easily be understood if we take into account the fact that after the first scattering the disturbed field is a carrier of scattering information. The field and the medium become, therefore, statistically dependent.

- The application of the Novikov-Furutsu formula is advantageous as long as an explicit form of the function $\tilde{E}[\xi]$ is unknown. The variational derivative should then be considered as a new functional relationship.

- The variational derivative of the functional $\tilde{E}\left[U\left(\vec{r}\right)\right]$ at the point $\vec{r}=\vec{r}_0$ is defined as

$$\lim_{\Delta\to 0,\max|\delta U|\to 0}\frac{E\left[U+\delta U\right]-E\left[U\right]}{\delta U\left(\vec{r}\right)} \tag{5.20}$$

in which δU is localized (different from zero) in the vicinity Δ around the point $r=r_0$.

5.3 Solutions

5.3.1 Cylinders as vegetation model

Proceeding to the description of the coherent interaction between radiation and vegetation, we introduce a model of the grass cover as shown in Fig. 5.2.

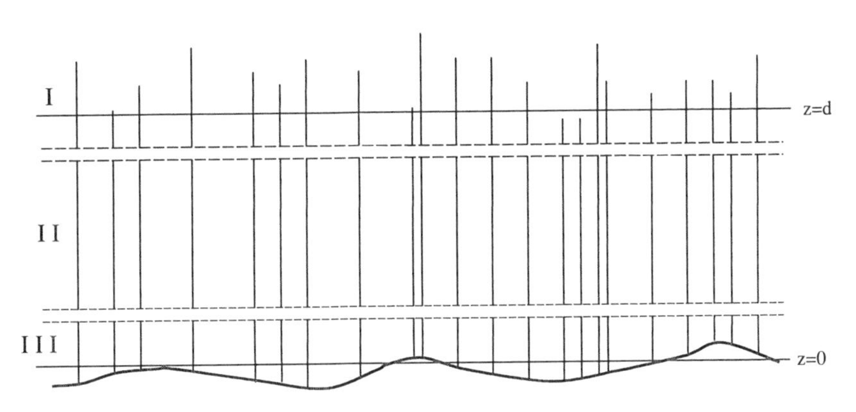

Fig. 5.2 Model of a grass cover

The actual vegetation layer on the earth's surface is divided into three areas:

- The first area is the diffused scattering boundary of the grass cover. Its effect on the coherent reflected signal is insignificant and it will be neglected in the following discussion.

- The second area is the homogeneous layer of the grass cover. Here, the coherent effects of forward scattering mainly prevail. The backward radiation reflected by this layer is incoherent and its effect is negligible in comparison to forward scattering.

- The third area is an earth boundary layer "smoothed" by vegetation. At the present stage of the investigation we assume that the reflection matrix in this region is symmetric.

An introductive description of a polarimetric radar signal propagating through a vegetation layer was given in section 2.2.1. The influence of unwanted factors, such as soil conditions, must be recognized as noise. Separation of information in such situations is a complicated problem. In fact, the scattered field $\tilde{\vec{E}}_s(\vec{q},z)$ has to be represented in the following form:

$$\tilde{\vec{E}}_s(\vec{q},z)=\left\{\hat{\vec{R}}_C^-(\vec{q})+\sum_{n=0}^{\infty}\left[\hat{\vec{T}}_C^-(\vec{q})\left[\hat{\vec{T}}_C^+(\vec{q})\hat{\vec{R}}_C^+(\vec{q})\hat{\vec{T}}_C^-(\vec{q})\hat{\vec{R}}_B(\vec{q})\right]^n\hat{\vec{R}}_B(\vec{q})\hat{\vec{T}}_C^+(\vec{q})\right]\right\}\tilde{\vec{E}}(\vec{q},z) \tag{5.21}$$

Here, the indices B and C refer to the earth and vegetation cover, respectively, and $\tilde{\vec{E}}(\vec{q},z)$ is the incident wave. $\hat{\vec{R}}(\vec{q})$ and $\hat{\vec{T}}(\vec{q})$ are matrix integral operators of the form

$$\begin{cases} \hat{\vec{R}}(\vec{q})=\int d\vec{q}'\cdot\hat{r}(\vec{q},\vec{q}'), \\ \hat{\vec{T}}(\vec{q})=\int d\vec{q}'\cdot\hat{t}'(\vec{q},\vec{q}'); \\ \hat{r}(\vec{q},\vec{q}')=\begin{Vmatrix} r_{11}(\vec{q},\vec{q}') & r_{12}(\vec{q},\vec{q}') \\ r_{21}(\vec{q},\vec{q}') & r_{22}(\vec{q},\vec{q}') \end{Vmatrix};\ \hat{t}'(\vec{q},\vec{q}')=\begin{Vmatrix} t'_{11}(\vec{q},\vec{q}') & t'_{12}(\vec{q},\vec{q}') \\ t'_{21}(\vec{q},\vec{q}') & t'_{22}(\vec{q},\vec{q}') \end{Vmatrix} \end{cases} \tag{5.22}$$

The operator $\hat{\vec{R}}_C^-$ describes the reflection of an electromagnetic wave from the vegetation layer when a field is incident from above and $\hat{\vec{R}}_C^+$ gives the reflection of a wave from the vegetation layer that has been reflected from the soil layer. The operators $\hat{\vec{T}}_C^{\pm}$ describe the "transparency" of the vegetation layer. Integration over $\vec{q}'$ in operators $\hat{\vec{R}}$ and $\hat{\vec{T}}$ describes the field angular spectrum transformation under the condition that there is a single interaction of the wave with the vegetation or soil cover.

It must be emphasized that this single interaction should not be confused with the first Born approximation.
The non-diagonal elements of the matrix operators describe the process of wave depolarization upon reflecting from or passing through the vegetation layer.

Relationship (5.21) shows that the disturbance is both of a multiplicative and additive nature. Signal separation into two parts, each of which is the response to interaction of one system component only (either vegetation or soil surface), is not possible. It should be noted, however, that this difficulty of signal separation does not mean that information extraction (in order to determine the object parameters of the sub-systems) is impossible.

We model the vegetation in the form of cylinders oriented normally to the plane of the averaged earth surface. In other words, we shall not consider the effects of grass which is not perpendicular to the earth plane. For actual experimental situations, this constraint can be quite serious, but then the analysis needs to be extended, with modifications so that the theory can be applied to more complicated spatially oriented cylinders, as well.

5.3.2 Stochastic field equations
In this section we derive a finite set of differential equations for the field in a randomly inhomogeneous medium (formed by an ensemble of spatially oriented scatterers) taking into account the effects of multiple scattering.

We start with the two curl Maxwell`s equations:

$$\left.\begin{aligned} \vec{\nabla}\times\vec{E} &= -\frac{\partial\vec{B}}{\partial t} \\ \vec{\nabla}\times\vec{B} &= \mu_0\vec{j}+\varepsilon_0\mu_0\frac{\partial\vec{E}}{\partial t} \end{aligned}\right\} \tag{5.23}$$

The current density in the medium becomes

$$\vec{j} = \sigma\vec{E}+\frac{\partial\vec{P}}{\partial t} \tag{5.24}$$

Here, it is assumed that the medium conductivity is independent of frequency. The medium polarizability $\vec{P}$ and conductivity σ are local characteristics determined by the local instantaneous internal field. In other words, in describing the random fields, spatial and temporal dispersion is neglected.

Let us assume that on a microscopic level the biomass substance is isotropic, i.e., $\vec{P} = \varepsilon_0 \chi \vec{E}$ and χ and σ are scalars. Assuming a monochromatic field, Maxwell's curl equations (5.23) can be transformed into the vector-valued Helmholtz equation

$$\vec{\nabla}\times \vec{\nabla}\times \vec{E}(\vec{r},\omega) - k_0^{\,2}\,\varepsilon(\vec{r})\vec{E}(\vec{r},\omega) = 0 \qquad (5.25)$$

Here,

$$\varepsilon(\vec{r}) = \begin{cases} 1 & outside \quad the \quad cylinders \\ 1 + i\chi \dfrac{\sigma}{\varepsilon_0 \omega} & inside \quad the \quad cylinders \end{cases} \qquad (5.26)$$

The elementary layer has a small thickness Δz inside the vegetation layer parallel above the average plane-earth interface. The area of this layer (in the plane $z = const$) is filled with an ensemble of randomly located disks which are the intersections of the cylinders modelling the grass cover.

We now consider the interaction between the radiation and the disks. We assume that inside the elementary layer we can neglect the effects of internal radiation. In fact, when modelling each disk by an extremely oblate spheroid, we find that its polarization factor α is a scalar ($\vec{P} = \alpha \vec{E}$) and is very small (α is proportional to Δz). In other words, each disk is a "soft" scatterer in spite of the fact that the average distance between disks is not large. This assumption is justified because for $\Delta z \to 0$ the condition of electrodynamic medium sparseness can still be met [*Rino,* 1988]; that is,

$$\alpha \bar{n} << 1 \qquad (5.27)$$

where $\bar{n}$ is the average concentration of scatterers.

Describing the interaction between the field and each separate scatterer (disk), we use the generalized Rayleigh-Gans approximation [*Klyatskin,* 1975] which is valid if a disk scatterer is much smaller than the wavelength. The modelling of a disk by an extremely oblate spheroid allows us to simplify the problem. The field inside each spheroid may be taken to be uniform if the external field is also uniform.

Following [*Novikov*, 1964] and under the condition $k_0 D\left[\sqrt{\varepsilon}-1\right] << 1$ (in our case $D = \Delta z$), we write for the relationship between the internal and external fields

$$\vec{E}_{\text{int}} = \overline{\overline{d}}\, \vec{E}_{ext} \tag{5.28}$$

$$\text{where} \begin{cases} \overline{\overline{d}} = d_T \overline{\overline{I}} + \left(d_N - d_T\right)\hat{z}\hat{z} \\ d_T = \dfrac{1}{\left(\varepsilon - 1\right) g_T + 1} \\ d_N = \dfrac{1}{\left(\varepsilon - 1\right) g_N + 1} \end{cases} \tag{5.28a}$$

For disks with thickness $2h = \Delta z$ and radius a we know [*Schiffer,* 1979],

$$\left.\begin{aligned} g_T &= \frac{1}{2\left(m^2-1\right)}\left[\frac{m^2}{\sqrt{m^2-1}}\sin^{-1}\left(\frac{\sqrt{m^2-1}}{m}\right)-1\right] \\ g_N &= \frac{m^2}{m^2-1}\left[1-\frac{1}{\sqrt{m^2-1}}\sin^{-1}\left(\frac{\sqrt{m^2-1}}{m}\right)\right] \\ m &= \frac{a}{h} \end{aligned}\right\} \tag{5.29}$$

In our case $\Delta z \rightarrow 0$. All relationships are then substantially simplified. We find $m \rightarrow \infty$, $g_T \rightarrow 0$, $g_N \rightarrow 1$, $d_T \rightarrow 1$, $d_N \rightarrow 1/\varepsilon$ and

$$\overline{\overline{d}} = \begin{pmatrix} 1 & 0 & 0 \\ 0 & 1 & 0 \\ 0 & 0 & 1/\varepsilon \end{pmatrix} \tag{5.30}$$

A similar result can be derived using the Rayleigh approximation for thin cylinders [*Furutsu,* 1963].

It is noted that in earlier research [*Rytov,* 1978] an attempt was made to take into account the interactions within the elementary layer. Scalar waves in a continuous turbulent medium were assumed and a solution by means of iteration was proposed.

The investigations, however, are not fully correct and the conclusion concerning the insufficiency of the Born approximation is questionable.
Following the generalized approach of the spatial spectrum method [*Rino,* 1988], we divide the field in the plane of intersection z = constant into two parts: $\tilde{E}_\alpha^+$ and $\tilde{E}_\alpha^-$. The quantities $\tilde{E}_\alpha^{+(-)}(\vec{q},z)$ describe the angular spectra of the forward and backward wave fields. When constructing the equation for the field in the medium, we assume that the fields entering the elementary layer are known, i.e., the values of $\tilde{E}_\alpha^+(\vec{q},z)$ and $\tilde{E}_\alpha^-(\vec{q},z+\Delta z)$ are specified. We have to calculate the values of the fields leaving the elementary layer, i.e., $\tilde{E}_\alpha^+(\vec{\mu},z+\Delta z)$ and $\tilde{E}_\alpha^-(\vec{\mu},z)$.

The equation describing the interaction between the radiation and the layer is rewritten as follows:

$$\nabla^2\vec{E}+k_0^2\vec{E}=-k_0^2(\varepsilon_r-1)\vec{E}+\nabla(\nabla\cdot\vec{E}) \tag{5.31}$$

We now transform the second term on the right-hand in Eq. (5.31). Such a transformation is needed because it mixes the second derivatives of the field components and thus introduces depolarization. Convolution with the Green's function of the left-hand term in (5.31) will only remove one differentiation. As a result, when applying the first iteration procedure as a solution technique, we have to possess knowledge not only about the field itself but about its spatial derivatives also. This is why some transformations are carried out first. We derive from Maxwell's divergence equation

$$\vec{\nabla}\cdot\vec{D}=\vec{\nabla}\cdot(\varepsilon\vec{E})=0 \tag{5.32}$$

$$\nabla\cdot\vec{E}=-\nabla\cdot\left[(\varepsilon-1)\vec{E}\right]=-\left(\nabla(\varepsilon-1)\cdot\vec{E}\right)-(\varepsilon-1)\nabla\cdot\vec{E} \tag{5.33}$$

which leads to

$$\nabla\cdot\vec{E}=-\frac{1}{\varepsilon}\left(\nabla[\varepsilon-1]\cdot\vec{E}\right) \tag{5.34}$$

This equation demonstrates that $\nabla\cdot\vec{E}$ manifests itself at the boundaries of the areas occupied by the dielectric cylinders.

Next, we introduce the function

$$\vec{F}(\vec{r}) = -k_0^2\left(\varepsilon - 1\right)\vec{E} - \nabla\left(\frac{1}{\varepsilon}\nabla\left[\varepsilon - 1\right]\cdot\vec{E}\right) \tag{5.35}$$

and use it in conjunction with Eq. (5.31). Upon Fourier transformation with respect to the transverse variables x and y, we obtain the inhomogeneous one-dimensional Helmholtz equation

$$\frac{d^2}{dz^2}\tilde{E}_\alpha^\pm(\vec{q},z) + \left(k_0^2 - q^2\right)\tilde{E}_\alpha^\pm(\vec{q},z) = \tilde{F}_\alpha(\vec{q},z) \tag{5.36}$$

where $\tilde{E}_\alpha^\pm(\vec{q},z)$ denotes the α- (rectangular) component of the forward (+) and backward (-) Fourier-transformed electric field component. The (infinite domain) Green's function associated with the ordinary differential equation (5.36) is given by

$$G^\pm(\vec{q},z) = -\frac{i}{2\sqrt{k_0^2 - q^2}}e^{\pm i\sqrt{k_o^2 - q^2}\,z} \tag{5.37}$$

When describing the field scattering within the elementary layer ($\Delta z \to 0$), the outward-radiating fields $\tilde{E}_\alpha^+(\vec{q},z+\Delta z)$, $\tilde{E}_\alpha^-(\vec{q},z)$ can be represented as

$$\tilde{E}_\alpha^+(\vec{q},z+\Delta z) = \tilde{E}_\alpha^+(\vec{q},z)e^{i\sqrt{k_0^2-q^2}\Delta z} + \int_z^{z+\Delta z} dz' G^+(\vec{q},z+\Delta z - z')\tilde{F}_\alpha(\vec{q},z') \tag{5.38}$$

for the electric field leaving the elementary layer at the top side (+z direction) and as

$$\tilde{E}_\alpha^-(\vec{q},z) = \tilde{E}_\alpha^-(\vec{q},z+\Delta z)e^{i\sqrt{k_0^2-q^2}\Delta z} + \int_z^{z+\Delta z} dz' G^-(\vec{q},z - z')\tilde{F}_\alpha(\vec{q},z') \tag{5.39}$$

for the electric field leaving at the bottom side (in the opposite direction).
The relationships (5.38) and (5.39) are similar in structure. Each includes the sum of two components: The first one describes the part of the field passing through an elementary layer without interaction with vegetation. The second component describes the distortion of the incident field which is due either to forward or backward scattering. The thickness of the elementary layer (Δz) is small and on the right-hand sides of (5.38) and (5.39) we can limit our consideration to those components that are proportional to the first power of Δz only.

Recalling the definition of the function $F(\vec{r})$ in Eq. (5.35), we have

$$\tilde{F}(\vec{q},z)=\int d\rho e^{-i\vec{q}\cdot\vec{\rho}}\left[-k_0^2\left(\varepsilon-1\right)\widetilde{E}-\nabla\left(\frac{1}{\varepsilon}\nabla\left[\varepsilon-1\right]\cdot\widetilde{E}\right)\right] \tag{5.40}$$

As a consequence, the integral in Eq. (5.38) can be written explicitly as

$$\begin{aligned}&-k_0^2\int_z^{z+\Delta z}(\varepsilon-1)G^+(\vec{q},z+\Delta z-z')\tilde{\vec{E}}dz'+\\&-\int_z^{z+\Delta z}\int d\rho' e^{-i\vec{q}\cdot\vec{\rho}'}\nabla\left[\frac{1}{\varepsilon}\nabla(\varepsilon-1)\cdot\vec{E}\right]G^+(\vec{q},z+\Delta z-z')dz'\end{aligned} \tag{5.41}$$

At first, we shall carry out the calculation of the second integral. By introducing the auxiliary vector $\vec{L}$ as

$$\vec{L}=-\int_z^{z+\Delta z}dz'G^+(\vec{q},z+\Delta z-z')\left\{\int d\rho' e^{-i\vec{q}\cdot\vec{\rho}'}\nabla\left[\frac{1}{\varepsilon}\nabla(\varepsilon-1)\cdot\overline{\overline{d}}\cdot\vec{E}_{ext}(\vec{\rho}',z')\right]\right\} \tag{5.42}$$

where $\overline{\overline{d}}$ is the matrix operator defined in Eq. (5.30). After integration by parts in Eq. (5.42), we obtain

$$\vec{L}=\vec{L}_1+\vec{L}_2 \tag{5.43}$$

where

$$\vec{L}_1=\int d\rho'\left[-\frac{1}{\varepsilon}\nabla(\varepsilon-1)\cdot\overline{\overline{d}}\cdot\vec{E}_{ext}(\vec{\rho}',z')e^{-i\vec{q}\cdot\vec{\rho}'}G^+(\vec{q},z+\Delta z-z')\Big|_z^{z+\Delta z}\right] \tag{5.44}$$

$$\vec{L}_2=\int_{\Delta V}\frac{1}{\varepsilon}\nabla(\varepsilon-1)\cdot\overline{\overline{d}}\cdot\vec{E}_{ext}(\vec{\rho}',z')\cdot\nabla\left[e^{-i\vec{q}\cdot\vec{\rho}'}G^+(\vec{q},z+\Delta z-z')\right]dV' \tag{5.45}$$

In the limit $\Delta z\to 0$,

$$\vec{L}_1=0 \tag{5.46}$$

Next, we use the property

$$\nabla\left[e^{-i\vec{q}\cdot\vec{\rho}'}G^{+}\left(\vec{q},z-z'\right)\right]=-i\vec{k}e^{-i\vec{q}\cdot\vec{\rho}'}G^{+}\left(\vec{q},z-z'\right) \tag{5.47}$$

where

$$\vec{k}=\left\{\vec{q},k_z\left(\vec{q}\right)\right\} \tag{5.48}$$

$$\vec{q}=\left(k_x,k_y\right) \tag{5.48a}$$

$$k_z\left(\vec{q}\right)=\sqrt{k_0^2-q^2} \tag{5.49}$$

$$q^2=k_x^2+k_y^2 \tag{5.49a}$$

We also note that

$$\frac{1}{\varepsilon}\nabla\left(\varepsilon-1\right)=-\frac{\varepsilon-1}{\varepsilon}\delta_S\vec{n}\left(\vec{V}'\right) \tag{5.50}$$

where δ_S is the Dirac (delta) function specified at the surface S as a discontinuity in the dielectric constant. S characterizes the layer structure [*Karam*, 1988] and $\vec{n}$ is the unit vector normal to the surface S.

Substituting Eqs (5.47) and (5.50) into Eq. (5.45) yields

$$\vec{L}_2=i\vec{k}\int_{\Delta V}dV\frac{\varepsilon-1}{\varepsilon}e^{-i\vec{q}\cdot\vec{\rho}'}G^{+}\left(\vec{q},z-z'\right)\delta_S E_n(\rho',z') \tag{5.51}$$

where E_n is the normal component of the internal field on the boundary S. This expression can be rewritten as

$$\vec{L}_2=i\frac{\varepsilon-1}{\varepsilon}\frac{e^{i\sqrt{k_0^2-q^2}\,z}}{2\sqrt{k_0^2-q^2}}\vec{k}\sum_j\oint_{S_j}d\vec{S}e^{-i\vec{k}\cdot\vec{r}'}\vec{n}\cdot\vec{E}_{\text{int}}\left(\vec{r}'\right) \tag{5.52}$$

where, $\vec{E}_{\text{int}}=\overline{\overline{d}}\cdot\vec{E}_{ext}$. To transform the surface integral to a volumetric integral, it is necessary to take into account that

$$\nabla' \cdot \left(e^{-i\vec{k}\cdot\vec{r}'}\vec{E}(r')\right) = -i\vec{k}\cdot e^{-i\vec{k}\cdot\vec{r}'}\vec{E}(\vec{r}') + e^{-i\vec{k}\cdot\vec{r}'}\nabla'\cdot\vec{E}(\vec{r}') \tag{5.53}$$

where $\nabla' \cdot \vec{E}(\vec{r}')$ is determined from Eq. (5.34). As a result, we obtain

$$\vec{L} = \frac{\varepsilon - 1}{\varepsilon}\left[-i\frac{e^{i\sqrt{k_0^2 - q^2}z}}{\sqrt{k_0^2 - q^2}}\vec{k}\sum_j \int dV' e^{-i\vec{k}\cdot\vec{r}'}\left(-i\vec{k}\right)\cdot\vec{E}_{\text{int}}(\vec{r}') + \vec{L}_1\right] \tag{5.54}$$

The volume integral can be transformed using the relationship

$$\begin{aligned} A &\equiv \frac{-i\vec{k}}{(2\pi)^2}\cdot\sum_j \int dV' e^{-i\vec{k}\cdot\vec{r}'}\int d\vec{\mu}\tilde{\vec{E}}_{\text{int}}(\vec{\mu},z')e^{i\vec{\mu}\cdot\vec{\rho}'} = \\ &= \frac{-i\vec{k}}{(2\pi)^2}\cdot\sum_j \int d\vec{\mu}\int d\vec{\rho}'\int dz' e^{-i(\vec{q}-\vec{\mu})\cdot\vec{\rho}'}\tilde{\vec{E}}_{\text{int}}(\vec{\mu},z')e^{-ik_z(\vec{q})z'} \end{aligned} \tag{5.55}$$

Let $\vec{\rho} = \vec{\rho}' - \vec{\rho}_j$. Then,

$$\begin{aligned} A &\equiv \frac{-i\vec{k}}{(2\pi)^2}\cdot\sum_j \int d\vec{\mu}\int d\vec{\rho}\int dz' e^{-i(\vec{q}-\vec{\mu})\cdot\vec{\rho}}\tilde{\vec{E}}_{\text{int}}(\vec{\mu},z')e^{-ik_z(\vec{q})z'}e^{-i(\vec{q}-\vec{\mu})\cdot\vec{\rho}_j} \\ &= -\frac{i\vec{k}}{2\pi}\cdot\sum_j \int dz' \exp\{-ik_z(\vec{q})z'\}\int d\vec{\mu}\tilde{\vec{E}}_{\text{int}}(\vec{\mu},z')e^{-i(\vec{q}-\vec{\mu})\cdot\vec{\rho}_j}\int_0^R d\rho J_0\left(|\vec{q}-\vec{\mu}|\rho\right)\rho \end{aligned} \tag{5.55b}$$

where $\vec{\rho}_j$ is the transverse coordinate of each disk center and R is the disk radius. Finally, Eq. (5.42) is approximately recasted into the form

$$\vec{L} = \frac{1}{2\pi}\frac{(\varepsilon - 1)}{\varepsilon}\Delta z\, G^+(\vec{q}, z - z')e^{-ik_z(\vec{q})z}\int d\vec{\mu}\,\vec{k}\cdot\tilde{\vec{E}}_{\text{int}}(\vec{\mu},z)\sum_j e^{-i(\vec{q}-\vec{\mu})\cdot\vec{\rho}_j}\int_0^R d\rho\rho J_0\left(|\vec{q}-\vec{\mu}|\rho\right) \tag{5.56}$$

Returning to Eq. (5.37), the approximation over the interval

$$G^+(\vec{q}, z - z') \approx G^-(\vec{q}, z + \Delta z - z') \tag{5.57}$$

is justified because the integration over z' is carried out over an infinitesimally small interval. Using the mean-value theorem for the integration over z', we replace the integration by a multiplicative factor Δz and obtain

$$\vec{L} = \frac{ik_0^2}{k_z(\vec{q})}\frac{\varepsilon - 1}{\varepsilon}\int d\vec{\mu}\Delta\varepsilon(\vec{q}-\vec{\mu})\, d_{\alpha\beta}\left\{\tilde{E}_\beta^+(\vec{\mu},z) + \tilde{E}_\beta^-(\vec{\mu},z)\right\}\Delta z \tag{5.58}$$

where we have used the relationship $\left(\bar{\bar{d}}\cdot\tilde{\bar{E}}_{ext}\right)_\alpha = d_{\alpha\beta}\left(\tilde{E}_\beta^+ + \tilde{E}_\beta^-\right)$. The term $\Delta\varepsilon(\vec{q}-\vec{\mu})$ is due to the random characteristics of the location and the shape of the scatterers. We know that

$$\Delta\varepsilon(\vec{\rho}) = \sum_{j=1}^{N}\Delta\varepsilon_j(\vec{\rho}) \tag{5.59}$$

Carrying out the two-dimensional transformation over $\vec{\rho}$ in (5.59) yields

$$\Delta\varepsilon(\vec{q}-\vec{\mu}) = \sum_{j=1}^{N}\frac{1}{(2\pi)^2}\int e^{-i(\vec{q}-\vec{\mu})\cdot\vec{\rho}}\Delta\varepsilon_j(\vec{\rho})\,d\vec{\rho} \tag{5.60}$$

Assuming that all cylinders have equal radius R, we write

$$\Delta\varepsilon_j(\vec{\rho}) = \Delta\varepsilon_0(\vec{\rho}-\vec{\rho}_j) \tag{5.61}$$

where

$$\Delta\varepsilon_0(\vec{\rho}) = \begin{cases}(\varepsilon-1), & when\ |\vec{\rho}|\le R\\ 0, & when\ |\vec{\rho}| > R\end{cases} \tag{5.62}$$

so that

$$\Delta\varepsilon(\vec{q}-\vec{\mu}) = \Delta\varepsilon_0(\vec{q}-\vec{\mu})\sum_{j=1}^{N}e^{-i(\vec{q}-\vec{\mu})\cdot\vec{\rho}_j} = \frac{1}{2\pi}\int_0^R d\rho\rho J_0(|\vec{q}-\vec{\mu}|\rho)\sum_{j=1}^{N}e^{-i(\vec{q}-\vec{\mu})\cdot\vec{\rho}_j} \tag{5.63}$$

Carrying out similar calculations for the other integral expressions in (5.38) and (5.39), we finally arrive at the following finite-difference system of stochastic equations:

$$\tilde{E}_{\alpha}^{+}(\vec{q}, z+\Delta z) = \tilde{E}_{\alpha}^{+}(\vec{q}, z) e^{ik_Z(\vec{q})\Delta Z} + i\frac{k_0^2\delta_{\alpha\beta} - k_\alpha k_\beta}{2k_Z(\vec{q})}$$
$$\times \sum_j \int d\vec{\mu} e^{-i(\vec{q}-\vec{\mu})\cdot\vec{\rho}_j} \Delta\varepsilon_0(\vec{q}-\vec{\mu}) d_{\beta\gamma} \left\{\tilde{E}_{\gamma}^{+}(\vec{\mu}, z) + \tilde{E}_{\gamma}^{-}(\vec{\mu}, z)\right\} \Delta z \tag{5.64}$$

and

$$\tilde{E}_{\alpha}^{-}(\vec{q}, z) = \tilde{E}_{\alpha}^{-}(\vec{q}, z+\Delta z) e^{ik_Z(\vec{q})\Delta Z} + i\frac{k_0^2\delta_{\alpha\beta} - k_\alpha k_\beta}{2k_Z(\vec{q})}$$
$$\times \sum_j \int d\vec{\mu} e^{-i(\vec{q}-\vec{\mu})\cdot\vec{\rho}_j} \Delta\varepsilon_0(\vec{q}-\vec{\mu}) d_{\beta\gamma} \left\{\tilde{E}_{\gamma}^{+}(\vec{\mu}, z) + \tilde{E}_{\gamma}^{-}(\vec{\mu}, z)\right\} \Delta z \tag{5.65}$$

The contribution of the total layer is formed by the summation. The summation sign is omitted in the following. When $\Delta z \to 0$, these two equations can be transformed into a system of integro-differential equations:

$$\pm\frac{d}{dz}\tilde{E}_{\alpha}^{\pm}(\vec{q}, z) = ik_Z(\vec{q})\tilde{E}_{\alpha}^{\pm}(\vec{q}, z) + i\frac{k_0^2\delta_{\alpha\beta} - k_\alpha k_\beta}{2k_Z(\vec{q})}$$
$$\times \sum_j \int d\vec{\mu} e^{-i(\vec{q}-\vec{\mu})\cdot\vec{\rho}_j} \Delta\varepsilon_0(\vec{q}-\vec{\mu}) d_{\beta\gamma} \left\{\tilde{E}_{\gamma}^{+}(\vec{\mu}, z) + \tilde{E}_{\gamma}^{-}(\vec{\mu}, z)\right\} \tag{5.66}$$

In Eqs (5.64) – (5.66), the Einstein summation over β and γ is implied.

The integral in Eq. (5.56) describes the interactions of the angular spectrum components in the vegetation layer and the re-partitioning of the field energy in an angular spectrum.

Equation (5.66) resembles the radiative-transfer equation [cf. *Ulaby*, 1990]. However, there are some distinctive aspects which should be pointed out. In the first place, the relationships (5.66) are derived from wave equations, while the radiative-transfer equations are phenomenological (Transport theory can also be rigorously derived from basic principles via wave-kinetic techniques (theory of gases) of Boltzmann [*Lifshitz*, 1979] and neutron diffusion [*Sommerfeld,* 1956]). Secondly, Eqs (5.66) describe the fields themselves and not in quadratic or "power-like" quantities. That is why the principle of field superposition can be applied in (5.66), while the corresponding integral in the radiative-transfer equations requires incoherence of the scattering processes. Equation (5.66) is stochastic; in the next section, we shall undertake an ensemble averaging in order to find the equations for the averaged fields.

5.3.3 Averaged stochastic equations describing scattering from extended scatterers: first-order approximation

The fields in Eq. (5.66) depend on the spatial positions and orientations of the cylinders:

$$\tilde{E}_{\alpha}^{\pm}(\vec{q},z) = \tilde{E}_{\alpha}^{\pm}\left(\vec{q},z\middle|\{\vec{\rho}_i\}\right) \tag{5.67}$$

The averaged fields are found by averaging over the ensemble of cylinder realizations; specifically,

$$\left\langle \tilde{E}_{\alpha}^{\pm}\left(\vec{q},z\middle|\{\vec{\rho}_i\}\right)\right\rangle \equiv \left\langle \tilde{E}_{\alpha}^{\pm}(\vec{q},z)\right\rangle = \int \prod_{i=1}^{N} d\vec{\rho}\, W(\vec{\rho}_1,\vec{\rho}_2,...\vec{\rho}_N)\cdot \tilde{E}_{\alpha}^{\pm}\left(\vec{q},z\middle|\{\vec{\rho}_i\}\right) \tag{5.68}$$

where $W(\vec{\rho}_1,\vec{\rho}_2,...\vec{\rho}_N)$ denotes the joint probability density function.

Averaging of the second term on the right-hand side of equation (5.66) yields

$$i\frac{k_0^2\delta_{\alpha\beta} - k_\alpha k_\beta}{2k_z(\vec{q})}\sum_{j=1}^{N}\int d\vec{\mu}\Delta\varepsilon_0(\vec{q}-\vec{\mu})\langle Q_j\rangle \tag{5.69}$$

where

$$\langle Q_j\rangle = \left\langle e^{-i(\vec{q}-\vec{\mu})\cdot\vec{\rho}_j} d_{\beta\gamma}\left[\tilde{E}_{\gamma}^{+}\left(\vec{\mu},z\middle|\{\vec{\rho}_j\}\right) + \tilde{E}_{\gamma}^{-}\left(\vec{\mu},z\middle|\{\vec{\rho}_j\}\right)\right]\right\rangle \tag{5.70}$$

Hypothesizing statistical independence of scatterers and invoking the property of permutation symmetry for the field, we obtain

$$\tilde{E}_{\gamma}^{\pm}\left(\vec{\mu},z\middle|\vec{\rho}_1...\vec{\rho}_S...\vec{\rho}_K...\vec{\rho}_N\right) = \tilde{E}_{\gamma}^{\pm}\left(\vec{\mu},z\middle|\vec{\rho}_N...\vec{\rho}_K...\vec{\rho}_S...\vec{\rho}_1\right) \tag{5.71}$$

Eq. (5.70) can be further simplified. The averaging in (5.70) is executed over all $\vec{\rho}_S$. When averaging is restricted to the position of cylinder j and not over the realizations, we deal with a spatial Fourier transformation. For a given S_0 (which is the area covered with cylinders), the distribution density $W(\vec{\rho}_j) = 1/S_0$ holds. With this, Eq. (5.70) can then be written as

$$\langle Q_j \rangle = \frac{d_{\beta\gamma}}{S_0} \left\{ \tilde{E}_\gamma^+ (\vec{\mu}, z | \vec{q} - \vec{\mu}) + \tilde{E}_\gamma^- (\vec{\mu}, z | \vec{q} - \vec{\mu}) \right\} \tag{5.72}$$

Substitution of (5.72) in (5.69) and under the assumption that $\langle Q_j \rangle$ is independent of *j*, Eq. (5.70) becomes

$$i\Pi_{\alpha\beta} \frac{N}{S_0} \int d\vec{\mu} \, \Delta\varepsilon_0 (\vec{q} - \vec{\mu}) d_{\beta\gamma} \left\{ \tilde{E}_\gamma^+ (\vec{\mu}, z | \vec{q} - \vec{\mu}) + \tilde{E}_\gamma^- (\vec{\mu}, z | \vec{q} - \vec{\mu}) \right\} \tag{5.72a}$$

with

$$\Pi_{\alpha\beta} = \frac{k_0^2 \delta_{\alpha\beta} - k_\alpha k_\beta}{2\sqrt{k_0^2 - q^2}} \tag{5.72b}$$

In (5.72a), we proceed with the limit $N \to \infty, S_0 \to \infty, N / S_0 = n$ (i.e., the position density of the cylinders has a finite value). In this case, the averaged equation corresponding to (5.66) is given by

$$\begin{aligned} &\pm \frac{d}{dz} \left\langle \tilde{E}_\alpha^\pm (\vec{q}, z) \right\rangle - ik_Z \left\langle \tilde{E}_\alpha^\pm (\vec{q}, z) \right\rangle - i\Pi_{\alpha\beta} n d_{\beta\gamma} \int d\vec{\mu} \, \Delta\varepsilon_0 (\vec{q} - \vec{\mu}) \\ &\times \left\{ \tilde{E}_\gamma^+ (\vec{\mu}, z | \vec{q} - \vec{\mu}) + \tilde{E}_\gamma^- (\vec{\mu}, z | \vec{q} - \vec{\mu}) \right\} = 0 \end{aligned} \tag{5.73}$$

It should be pointed out that equation (5.73) is similar to that for the averaged field in a cloud of discrete scatterers [*Landau*, 1982]. Complete correspondence holds when $d_{\beta\gamma} \Delta\varepsilon_0 (\vec{q} - \vec{\mu}) \to \chi_{\beta\gamma}$, where $\chi_{\beta\gamma}$ is the polarization tensor of a separate discrete scatterer.

The system of Eq. (5.73) contains not only the desired average fields $\left\langle \tilde{E}_\alpha^\mp (\vec{q}, z) \right\rangle$ but also the fields $\tilde{E}_\alpha^\pm (\vec{\mu}, z | \vec{q} - \vec{\mu})$, of which the physical meaning has not yet been explained. $\tilde{E}_{\alpha}^{\pm}(\vec{\mu}, z|\vec{q}-\vec{\mu})$ is determined by the dependence of the random field on the coordinates of the scatterers $E_{\alpha}^{\pm}(\vec{r}, |\{\vec{\rho}_i\})$.

Note, again, that $E_{\alpha}^{\pm}(\vec{r})$ is the field entering the elementary layer and that the dependence on $\{\vec{\rho}_i\}$ is thus a "memory effect" connected to earlier scatterings. In the absence of such a "memory," we obtain

$$\left\langle e^{-i(\vec{q}-\vec{\mu})\cdot\vec{\rho}_j}\left[\tilde{E}_{\gamma}^{+}+\tilde{E}_{\gamma}^{-}\right]\right\rangle \approx (2\pi)^2\delta(\vec{q}-\vec{\mu})\left[\left\langle \tilde{E}_{\gamma}^{+}(\vec{\mu},z)\right\rangle+\left\langle \tilde{E}_{\gamma}^{-}(\vec{\mu},z)\right\rangle\right] \tag{5.74}$$

Using this expression, Eq. (5.73) becomes

$$\pm\frac{d}{dz}\left\langle \tilde{E}_{\alpha}^{\pm}(\vec{q},z)\right\rangle = ik_z\left\langle \tilde{E}_{\alpha}^{\pm}(\vec{q},z)\right\rangle+4i\pi^2 n\Delta\varepsilon_0(0)\Pi_{\alpha\beta}d_{\beta\gamma}\left[\left\langle \tilde{E}_{\gamma}^{+}(\vec{q},z)\right\rangle+\left\langle \tilde{E}_{\gamma}^{-}(\vec{q},z)\right\rangle\right] \tag{5.75}$$

System (5.75) describes the propagation of an average field in a continuous anisotropic medium with effective dielectric constant characterized by the tensor $\varepsilon_{\alpha\gamma}^{eff}$. We now want to calculate this tensor. By subtracting and summing the forward and backward averaged fields $\left\langle \tilde{E}_{\alpha}^{\pm}(\vec{q},z)\right\rangle$, we find

$$\left.\begin{aligned}\frac{d}{dz}\left(\left\langle \tilde{E}_{\alpha}^{+}(\vec{q},z)\right\rangle-\left\langle \tilde{E}_{\alpha}^{-}(\vec{q},z)\right\rangle\right) &= ik_z(\vec{q})\left\langle \tilde{E}_{\alpha}^{\Sigma}(\vec{q},z)\right\rangle+4i\pi^2\Pi_{\alpha\beta}d_{\beta\gamma}n\Delta\varepsilon_0(0)\left\langle \tilde{E}_{\gamma}^{\Sigma}(\vec{q},z)\right\rangle\\ \frac{d}{dz}\left\langle \tilde{E}_{\alpha}^{\Sigma}(\vec{q},z)\right\rangle &= ik_z(\vec{q})\left(\left\langle \tilde{E}_{\alpha}^{+}(\vec{q},z)\right\rangle-\left\langle \tilde{E}_{\alpha}^{-}(\vec{q},z)\right\rangle\right)\end{aligned}\right\} \tag{5.76}$$

where $\left\langle \tilde{E}_{\gamma}^{\Sigma}\right\rangle = \left\langle \tilde{E}_{\gamma}^{+}\right\rangle+\left\langle \tilde{E}_{\gamma}^{-}\right\rangle$. By differentiation of the second equation of (5.76) with respect to z and substituting the result into the first equation, we subsequently find

$$\frac{d^2}{dz^2}\left\langle \tilde{E}_{\alpha}^{\Sigma}\right\rangle+\left\{k_z^2(\vec{q})\delta_{\alpha\beta}+8\pi^2 k_z(\vec{q})\Pi_{\alpha\beta}d_{\beta\gamma}n\Delta\varepsilon_0(0)\right\}\left\langle \tilde{E}_{\gamma}^{\Sigma}(\vec{q},z)\right\rangle = 0 \tag{5.77}$$

Comparison of Eq. (5.77) with the standard one-dimensional Helmholtz equation

$$\frac{d^2}{dz^2}E_\alpha + \left(k_0^2\varepsilon_{\alpha\beta} - q^2\delta_{\alpha\beta}\right)E_\beta = 0 \tag{5.78}$$

leads to the following effective permittivity for the grass layer:

$$\varepsilon_{\alpha\gamma}^{eff} = \delta_{\alpha\gamma} + 8\pi^2 k_0^2 n\Delta\varepsilon_0(0)k_z(\vec{q})\Pi_{\alpha\beta}d_{\beta\gamma} \tag{5.79}$$

With the choice of reference coordinate system such that the wavevector $\vec{k}$ is positioned on the (y,z) plane, the permittivity tensor is proportional to the following matrix:

$$\hat{A} = \begin{pmatrix} k_0^2 & 0 & 0 \\ 0 & k_0^2 - k_Y^2 & -k_Y k_z / \varepsilon \\ 0 & -k_Y k_z & k_0^2 - k_z^2 / \varepsilon \end{pmatrix} \tag{5.80}$$

This describes an anisotropic medium or an "artificial dielectric" with permittivity ε.

We now analyze the propagation of vertically and horizontally polarized waves through such a medium. In the first case, the electrical field is specified by

$$\vec{E}_h = \begin{pmatrix} 1 \\ 0 \\ 0 \end{pmatrix} \tag{5.81}$$

Using Eqs (5.79) and (5.80), the effective permittivity of the grass layer for horizontally polarized waves equals

$$\varepsilon_h^{eff} = 1 + 2\pi^2 n\Delta\varepsilon_0(0) \tag{5.82}$$

For vertically polarized waves, the field vector is specified by

$$\vec{E}_v = \begin{pmatrix} 0 \\ \cos\theta \\ -\sin\theta \end{pmatrix} \tag{5.83}$$

and k_Y and k_Z are given by $(-k_0 \sin\theta)$ and $(k_0 \cos\theta)$, respectively. It is easily shown that

$$\hat{A}\vec{E}_v = k_0^2\left(\cos^2\theta + \sin^2\theta/\varepsilon\right) \tag{5.84}$$

With (5.84), the effective permittivity of the grass layer for vertically polarized waves equals

$$\varepsilon_v^{eff} = 1 + 2\pi^2 n\Delta\varepsilon_0(0)\left(\cos^2\theta + \sin^2\theta/\varepsilon\right) \tag{5.85}$$

Comparing formulas (5.82) and (5.85), it is clear that the medium is anisotropic. The anisotropy manifests itself when changing the incidence angle θ. A plot of the relationship

$$\Delta\varepsilon_{h,v}^{eff} = \varepsilon_h^{eff} - \varepsilon_v^{eff} = 2\pi^2 n\Delta\varepsilon_0(0)\frac{\varepsilon-1}{\varepsilon}\sin^2\theta \tag{5.86}$$

is presented in Fig. 5.3. It is evident that at normal incidence($\theta = 0$), there is no anisotropy. Anisotropy increases with an increase in the concentration of cylinders (stems of grass) and the value of dielectric constant of biomass ε and is proportional to the cross-sectional area of the stems ($\Delta\varepsilon_0(0) \sim \pi R^2$).

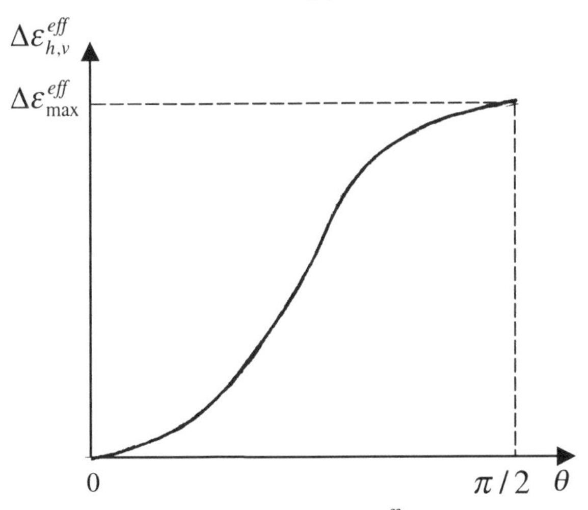

Fig. 5.3 Dependence of $\Delta\varepsilon_{h,v}^{eff}$ and θ

Using (5.86), the additional phase shift between the waves with horizontal and vertical polarization is readily obtained when these waves pass through a layer of grass with thickness H:

$$\Delta\varphi = \Delta\varepsilon_{h,v}^{eff} H \tag{5.87}$$

Next, we will discuss the physical meaning of the approximation leading to equation (5.75) and the validity of the ensuing results. As pointed out earlier, the "coding" of information by a scattering object takes place when a field interacts with it. After such an interaction, the field carries this information in amplitude-phase relationships of its angular spectrum. This means that within the layer of vegetation the field depends in particular on the geometry of the vegetation layer, i.e., it depends on the spatial orientation of the stems of grass. Eq. (5.75) implies that there is no correlation between the field and the position of scatterer j and that only a part of the field has been considered, which assumes no interaction between the stems. In summary, of all possible models of wave scattering only those were applied that did not involve repeated scattering.

The processes of scattering not described by (5.75) are represented in Fig. 5.4. The approximation considered in this section is therefore called the first approximation.

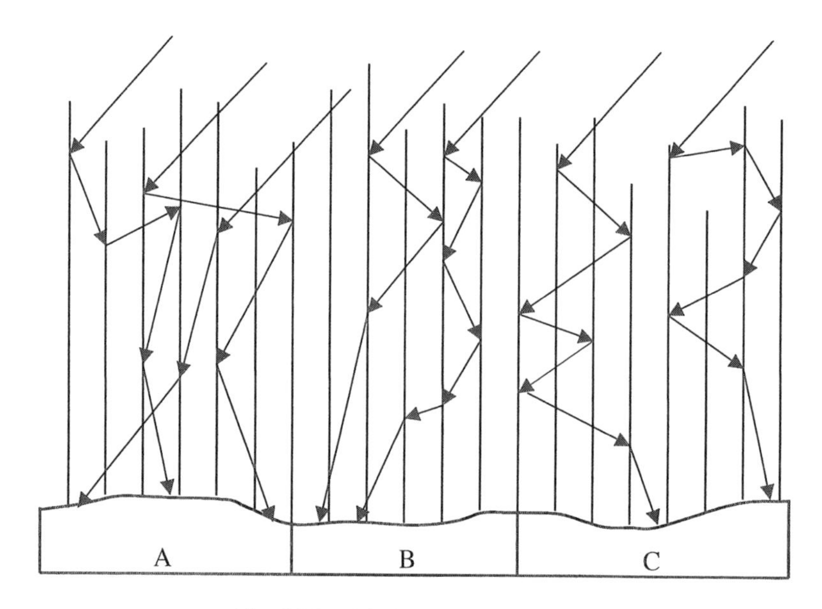

Fig. 5.4 Model of scattering processes

5.3.4 Use of field equations derived from the first approximation
Eq. (5.73) illustrate the well-known problem of field averaging in connection with a random medium. This problem is further discussed when applied to vegetation models, but now for the quantity $\tilde{E}_\alpha^\pm(\vec{\mu},z|\vec{q}-\vec{\mu})$. We multiply the system of stochastic Eq. (5.66) with $\exp\{-i\vec{\chi}\cdot\vec{\rho}_k\}$ and then conduct the averaging by a procedure analogous to that used in the previous section. This results in the expression

$$\begin{aligned}&\pm\frac{d}{dz}\tilde{E}_\alpha^\pm(\vec{q},z|\vec{\chi})-ik_Z(\vec{q})\tilde{E}_\alpha^\pm(\vec{q},z|\vec{\chi})-\\&-in\Pi_{\alpha\beta}d_{\beta\gamma}\int d\vec{\mu}\cdot\Delta\varepsilon_0(\vec{q}-\vec{\mu})\left[\tilde{E}_\gamma^+(\vec{\mu},z|\vec{q}-\vec{\mu};\vec{\chi})+\tilde{E}_\gamma^-(\vec{\mu},z|\vec{q}-\vec{\mu};\vec{\chi})\right]=\\&=in\Pi_{\alpha\beta}d_{\beta\gamma}\int d\vec{\mu}\cdot\Delta\varepsilon_0(\vec{q}-\vec{\mu})\left[\tilde{E}_\gamma^+(\vec{\mu},z|\vec{q}-\vec{\mu}+\vec{\chi})+\tilde{E}_\gamma^-(\vec{\mu},z|\vec{q}-\vec{\mu}+\vec{\chi})\right]\end{aligned} \tag{5.88}$$

This system of equations as in the case of Eq. (5.73) is not-closed. The question then arises whether it can be closed and, if so, which approximations should be applied.

When considering Eq. (5.88), the left-hand side is similar to the left-hand part of Eq. (5.73). It is, therefore, important to focus our attention on the differences of these systems, especially in the right-hand parts. To do so, we execute in (5.88) an inverse Fourier transform over the vector $\vec{\chi}$; this operation yields:

$$\begin{aligned}&in\Pi_{\alpha\beta}d_{\beta\gamma}\int d\vec{\mu}\Delta\varepsilon_0(\vec{q}-\vec{\mu})\sum_{+,-}e^{i(\vec{q}-\vec{\mu})\cdot\vec{\rho}_j}\int d\vec{\chi}e^{i\vec{\chi}\cdot\vec{\rho}_j}\tilde{E}_\gamma^\pm(\vec{\mu},z|\vec{q}-\vec{\mu}+\vec{\chi})=\\&=in\Pi_{\alpha\beta}d_{\beta\gamma}\int d\vec{\mu}\Delta\varepsilon_0(\vec{q}-\vec{\mu})\sum_{+,-}e^{i(\vec{q}-\vec{\mu})\cdot\vec{\rho}_j}\tilde{E}_\gamma^\pm(\vec{\mu},z|\vec{\rho}_j)=\\&=in\Pi_{\alpha\beta}d_{\beta\gamma}\int d\vec{\mu}\Delta\varepsilon_j(\vec{q}-\vec{\mu})\left(\tilde{E}_\gamma^+(\vec{\mu},z|\vec{\rho}_j)+\tilde{E}_\gamma^-(\vec{\mu},z|\vec{\rho}_j)\right)\end{aligned} \tag{5.89}$$

A comparison of Eqs (5.89) and (5.64) leads to the physical interpretation that this field component describes a distortion to the average field, which appears when an additional cylindrical scatterer is placed at point $\vec{\rho}_j$. The distortion itself is propagating in the medium and undergoes multiple scattering from other cylinders.

Using this physical explanation, the approximate relationship (5.75) becomes a more exact one by introducing the additional field disturbance caused by the source (5.89). It leads to

$$\delta E_\alpha^\pm(\vec{\mu},z|\vec{q}-\vec{\mu})=E_\alpha^\pm(\vec{\mu},z|\vec{q}-\vec{\mu})-(2\pi)^2\delta(\vec{q}-\vec{\mu})E_\alpha^\pm(\vec{\mu},z) \tag{5.90}$$

This relationship needs to be explained in more detail. Starting with a vegetation model consisting of N randomly positioned cylinders, we carry out the procedure of field averaging over (N - 1) cylinders. In order to carry out such a partial averaging of the field, we write the obvious relationship

$$\left\langle E_\alpha\left(\vec{r}\middle|\left\{\vec{\rho}_j\right\}\right)\right\rangle_{j\neq N}=E_\alpha\left(\vec{r}\middle|\vec{\rho}_N\right)=\left\langle E_\alpha(\vec{r})\right\rangle+\delta E_\alpha\left(\vec{r},\vec{\rho}_N\right) \tag{5.91}$$

Now we carry out a Fourier transform over the coordinates of cylinder N, with (5.90) as a result.

The field distortions can be derived by multiplying Eq. (5.73) with $(2\pi)^2\delta(\vec{\chi})$ and by subtracting it from (5.88). The resulting expression can be written in the following form:

$$\begin{aligned}&\pm\frac{d}{dz}\left[\delta\tilde{E}_\alpha^{\pm}\left(\vec{q},z\middle|\vec{\chi}\right)\right]-ik_Z\left(\vec{q}\right)\left[\delta\tilde{E}_\alpha^{\pm}\left(\vec{q},z\middle|\vec{\chi}\right)\right]-\\&-in\Pi_{\alpha\beta}d_{\beta\gamma}\int d\vec{\mu}\cdot\Delta\varepsilon_0\left(\vec{q}-\vec{\mu}\right)\left[\delta\tilde{E}_\gamma^{+}\left(\vec{\mu},z\middle|\vec{q}-\vec{\mu};\vec{\chi}\right)+\delta\tilde{E}_\gamma^{-}\left(\vec{\mu},z\middle|\vec{q}-\vec{\mu};\vec{\chi}\right)\right]=\\&=in\Pi_{\alpha\beta}d_{\beta\gamma}\int d\vec{\mu}\cdot\Delta\varepsilon_0\left(\vec{q}-\vec{\mu}\right)\left[\tilde{E}_\gamma^{+}\left(\vec{\mu},z\middle|\vec{q}-\vec{\mu}+\vec{\chi}\right)+\tilde{E}_\gamma^{-}\left(\vec{\mu},z\middle|\vec{q}-\vec{\mu}+\vec{\chi}\right)\right]\end{aligned} \tag{5.92}$$

The boundary conditions for this set of equations are

$$\left.\begin{aligned}\delta\tilde{E}_\alpha^{+}\left(\vec{q},H\middle|\vec{\chi}\right)=0\\\delta E_\alpha^{-}\left(\vec{q},0\middle|\vec{\chi}\right)=\int d\vec{\mu}R_{\alpha\beta}\left(\vec{q},0\middle|\vec{\mu}\right)\delta\tilde{E}_\beta^{+}\left(\vec{\mu},0\middle|\vec{\chi}\right)\end{aligned}\right\} \tag{5.93}$$

Here, $R_{\alpha\beta}\left(\vec{q},H\middle|\vec{\mu}\right)$ is the complex matrix coefficient of the earth's surface reflection. This coefficient describes the transformation of the wave from the state $(\beta,+,\vec{\mu})$ to the state $(\alpha,-,\vec{q})$.

Equation (5.92) characterizes field distortions, but similarly to Eqs (5.73) and (5.88) we are dealing with a non-closed system. It is possible to associate (5.92) to (5.73) and (5.75) by using the approximation

$$\delta E_\alpha^{\pm}\left(\vec{\mu},z\middle|\vec{q}-\vec{\mu};\vec{\chi}\right)\approx(2\pi)^2\delta E_\alpha^{\pm}\left(\vec{\mu},z\middle|\vec{\chi}\right) \tag{5.94}$$

This approximation physically means that the field is not doubly scattered (see Fig. 5.4).

5.3.5 Spatial dispersion effects: the second approximation

In order to investigate the effect of multiple scattering, we apply approximation (5.94) to (5.92). The two equations can be united in one using the new unknown function

$$\delta E_{\alpha}^{\Sigma} = \delta \tilde{E}_{\alpha}^{+}\left(\vec{q}, z|\vec{\chi}\right) + \delta \tilde{E}_{\alpha}^{-}\left(\vec{q}, z|\vec{\chi}\right) \tag{5.95}$$

In fact, by summing and subtracting Eqs (5.92) and using the approximation (5.94) we obtain

$$\frac{d}{dz}\left\{\delta \tilde{E}_{\alpha}^{+}\left(\vec{q}, z|\vec{\chi}\right) - ik_{z}(\vec{q})\delta \tilde{E}_{\alpha}^{-}\left(\vec{q}, z|\vec{\chi}\right)\right\} - 2ik_{z}\left(\vec{q}\right)\delta \tilde{E}_{\alpha}^{\Sigma} - 2in\Pi_{\alpha\beta}d_{\beta\gamma}(2\pi)^{2}\delta \tilde{E}_{\gamma}^{\Sigma}\Delta\varepsilon_{0}(0) =$$
$$= 2in\Pi_{\alpha\beta}d_{\beta\gamma}\int d\vec{\mu}\Delta\varepsilon_{0}\left(\vec{q} - \vec{\mu}\right)\delta \tilde{E}_{\gamma}^{\Sigma}(\vec{\mu}, z \,|\, \vec{q} - \vec{\mu} + \vec{\chi}) \tag{5.96}$$

and

$$\frac{d}{dz}\delta \tilde{E}_{\alpha}^{\Sigma} - ik_{Z}\left(\vec{q}\right)\left\{\delta \tilde{E}_{\alpha}^{+}\left(\vec{q}, z|\vec{\chi}\right) - \delta \tilde{E}_{\alpha}^{-}\left(\vec{q}, z|\vec{\chi}\right)\right\} = 0 \tag{5.97}$$

Carrying out the differentiation over z in (5.97) and substituting the result into Eq. (5.96) gives

$$\frac{d^{2}}{dz^{2}}\delta \tilde{E}_{\alpha}^{\Sigma} + 2\left\{k_{Z}^{2}\left(\vec{q}\right)\delta_{\alpha\gamma} + (2\pi)^{2}k_{Z}\left(\vec{q}\right)\Pi_{\alpha\beta}d_{\beta\gamma}n\Delta\varepsilon_{0}(0)\right\}'\delta \widetilde{E}_{\gamma}^{\Sigma} =$$
$$= -2k_{Z}(\vec{q})n\Pi_{\alpha\beta}d_{\beta\gamma}\int d\vec{\mu}\Delta\varepsilon_{0}\left(\vec{q} - \vec{\mu}\right)\delta \tilde{E}_{\gamma}^{\Sigma}(\vec{\mu}, z \,|\, \vec{q} - \vec{\mu} + \vec{\chi}) \tag{5.98}$$

Eqs (5.97) and (5.98) have to be true for any cross-section of the vegetation layer and therefore also at its boundaries. These equations have to fulfill the boundary conditions in a similar way as (5.93) for (5.92).

Starting with the upper boundary of the layer (z = H), it follows from (5.93) and (5.97) that

$$\frac{d}{dz}\left\{\delta \tilde{E}_{\alpha}^{\Sigma}\right\} + ik_{Z}\left(\vec{q}\right)\left\{\delta \tilde{E}_{\alpha}^{\Sigma}\right\} = 0 \tag{5.99}$$

The lowest boundary of the layer is the earth's surface. In general, the relationship between incident and reflected fields at this boundary is substantially complicated. However, here we can apply simplifications in $\langle R_{\alpha\beta}(\vec{q},\vec{\mu})\rangle$, the reflection coefficient of the earth's random surface for the field strength and not for the intensity. The absence of factors quadratic in the field allows us to simplify the integral relation (5.93). If the earth's surface can be described by the method of small perturbations (the conditions of this method are well known [*Ulaby*, 1982, *Beckmann and Spizzichino*, 1987]), then the expression for the reflection coefficient can be written as

$$\langle R_{\alpha\beta}(\vec{q},\vec{\mu})\rangle = \delta(\vec{q}-\vec{\mu})\begin{Vmatrix} R_1(\vec{q}) & 0 \\ 0 & R_2(\vec{q}) \end{Vmatrix} \tag{5.100}$$

i.e., only the coherent reflection component for the field in the vegetation layer remains and the diffused scattering can be neglected because of its incoherence.

If the surface perturbations are substantial, then the coherent component is absent, meaning that in Eq. (5.100) $R_1(\vec{q}) = R_2(\vec{q}) = 0$. The small-roughness approximation holds surprisingly well for more general surfaces and turns out to be less restrictive.

From (5.93) and by using (5.100), we find

$$\delta\tilde{E}_{\alpha}^{-}(\vec{q},z|\vec{\chi}) = R(\vec{q})\delta\tilde{E}_{\alpha}^{+}(\vec{q},z|\vec{\chi}) \tag{5.101}$$

Equations (5.95) and (5.101) give

$$\delta\tilde{E}_{\alpha}^{+}(\vec{q},z|\vec{\chi}) - \delta\tilde{E}_{\alpha}^{-}(\vec{q},z|\vec{\chi}) = \frac{1-R(\vec{q})}{1+R(\vec{q})}\{\delta\tilde{E}_{\alpha}^{\Sigma}\} \tag{5.102}$$

With this relationship, the final expression of the boundary condition for field distortions at the lower boundary of the layer becomes

$$\frac{d}{dz}\{\delta\tilde{E}_{\alpha}^{\Sigma}\} - ik_Z(\vec{q})\frac{1-R(\vec{q})}{1+R(\vec{q})}\{\delta\tilde{E}_{\alpha}^{\Sigma}\} = 0 \tag{5.103}$$

We now consider the solution of the stated boundary-value problem. Denoting the right-hand side of (5.98) as $M_{\alpha}(z)$, the general solution of this equation can be written as

$$\delta \tilde{\vec{E}}^{\Sigma} = e^{i\overline{\overline{K}}(\vec{q})z} \cdot \vec{l}_1 + e^{i\overline{\overline{K}}(\vec{q})z} \cdot \vec{l}_2 + \overline{\overline{K}}(\vec{q}) \cdot \int_0^H \sin\left[\overline{\overline{K}}(\vec{q})(z-z')\right] \cdot \vec{M}(z')dz' \tag{5.104}$$

Here, we use the notation

$$\overline{\overline{K}}(\vec{q}) = K_{\alpha\gamma} = \left\{k_Z^2(\vec{q})\delta_{\alpha\gamma} + k_Z(\vec{q})\Pi_{\alpha\beta} d_{\beta\gamma} n\Delta\varepsilon_0(0)\right\} \tag{5.105}$$

The exponential with the matrix $\overline{\overline{K}}(\vec{q})$ is approximated by

$$e^{i\overline{\overline{K}}(\vec{q})z} = \overline{\overline{I}} + i\overline{\overline{K}}(\vec{q})z - \frac{1}{2!}\overline{\overline{K}}(\vec{q}) \cdot \overline{\overline{K}}(\vec{q})z^2 - \frac{i}{3!}\overline{\overline{K}}(\vec{q}) \cdot \overline{\overline{K}}(\vec{q}) \cdot \overline{\overline{K}}(\vec{q})z^3 + \ldots, \tag{5.106}$$

where

$$\overline{\overline{K}}(\vec{q}) \cdot \overline{\overline{K}}(\vec{q}) = \sum K_{\alpha\beta}(\vec{q}) K_{\beta\gamma}(\vec{q}) \ldots, \quad etc.$$

and $\overline{\overline{I}}$ is unit matrix operator. Introducing coefficients $\vec{C}_1$ and $\vec{C}_2$ such that

$$\left(\overline{\overline{K}}(\vec{q}) + k_Z(\vec{q})\overline{\overline{I}}\right)\vec{C}_1 = \left(\overline{\overline{K}}(\vec{q}) - k_Z(\vec{q})\overline{\overline{I}}\right)\vec{C}_2 \tag{5.107}$$

we find from Eqs (5.99) and (5.103) that

$$\left\{\overline{\overline{K}}(\vec{q}) - k_Z(\vec{q})\frac{1-R}{1+R}\overline{\overline{I}}\right\} \cdot e^{i\overline{\overline{K}}(\vec{q})H} \cdot \vec{C}_1 - \left\{\overline{\overline{K}}(\vec{q}) + k_Z(\vec{q})\frac{1-R}{1+R}\overline{\overline{I}}\right\} \cdot e^{-i\overline{\overline{K}}(\vec{q})H} \cdot \vec{C}_2 = $$
$$= \int_0^H dz' \overline{\overline{\Phi}}(z') \cdot \vec{M}(z') \tag{5.108}$$

Here,

$$\overline{\overline{\Phi}}(z') = k_Z(\vec{q})\frac{1-R}{1+R}\sin\left[\overline{\overline{K}}(\vec{q})(H-z')\right] + i\overline{\overline{K}} \cdot \cos\left[\overline{\overline{K}}(\vec{q})(H-z')\right] \tag{5.109}$$

Solving the system of linear Eqs (5.107) and (5.108) with respect to the vectors $\vec{C}_1$ and $\vec{C}_2$ and substituting the derived expressions into (5.104) results in

$$\delta\tilde{\vec{E}}^{\Sigma}\left(\vec{q},z\middle|\vec{\chi}\right)=\int_{0}^{H}\overline{\overline{\Psi}}\left(z,z',\vec{q}\right)\cdot\vec{M}\left(z',\vec{q},\vec{\chi}\right)dz' \tag{5.110}$$

The dependence of $\vec{M}$ on $\vec{q}$ and $\vec{\chi}$ follows from (5.98). This expression can be simplified by approximating (5.75). In this case, the integral over $\vec{\mu}$ is trivially evaluated, due to the presence of a Dirac delta function:

$$\vec{M}\left(z',\vec{q},\vec{\chi}\right)\approx-(2\pi)^{2}k_{Z}\left(\vec{q}\right)\Delta\varepsilon_{0}\left(\vec{\chi}\right)\overline{\overline{\Pi}}\cdot\overline{\overline{d}}.\tilde{\vec{E}}\left(\vec{q}+\vec{\chi},z'\right) \tag{5.111}$$

where

$$\overline{\overline{\Pi}}\overline{\overline{d}}=\Pi_{\alpha\beta}d_{\beta\gamma} \tag{5.111a}$$

We subtracted $\delta\vec{E}^{\Sigma}\left(\vec{q},z\middle|\vec{\chi}\right)$ for evaluation of the part we removed in (5.75) when proceeding from (5.73) to (5.77). Carrying out the calculations for (5.73) (the calculations are similar to those yielding (5.98)), we obtain

$$\begin{aligned}&\frac{d^{2}}{dz^{2}}\left\{\tilde{\vec{E}}^{\Sigma}\right\}+\left\{k_{Z}^{2}\left(\vec{q}\right)\overline{\overline{I}}+k_{Z}\left(\vec{q}\right)\overline{\overline{\Pi}}.\overline{\overline{d}}n\Delta\varepsilon_{0}\left(0\right)\right\}\cdot\tilde{\vec{E}}^{\Sigma}=\\&=-(2\pi)^{2}k_{Z}(\vec{q})\overline{\overline{\Pi}}\cdot\overline{\overline{d}}\int d\vec{\mu}\Delta\varepsilon_{0}\left(\vec{q}-\vec{\mu}\right)\delta\tilde{\vec{E}}^{\Sigma}\left(\vec{\mu},z\middle|\vec{q}-\vec{\mu}\right)\end{aligned} \tag{5.112}$$

We make some substitutions in Eq. (5.110) before calculating the right part of (5.112). The vector $\vec{\chi}$ is arbitrary and consequently in (5.110) we carry out the substitutions

$$\left.\begin{aligned}\vec{q}&\rightarrow\vec{\mu}\\\vec{\chi}&\rightarrow\vec{q}-\vec{\mu}\end{aligned}\right\} \tag{5.113}$$

We, then, multiply the expression by $-(2\pi)^{2}k_{Z}\left(\vec{q}\right)\overline{\overline{\Pi}}.\overline{\overline{d}}\Delta\varepsilon_{0}\left(\vec{q}-\vec{\mu}\right)$ and perform an integration over $\vec{\mu}$. As a result, we obtain

$$\left(4\pi^{2}\right)^{2}k_{Z}\left(\vec{q}\right)\overline{\overline{\Pi}}\left(\vec{q}\right)\cdot\overline{\overline{d}}\cdot\int_{0}^{H}dz'\left\{d\vec{\mu}\left(\Delta\varepsilon_{0}\left(\vec{q}-\vec{\mu}\right)\right)^{2}k_{Z}\left(\vec{\mu}\right)\overline{\overline{\Pi}}\left(\vec{\mu}\right)\cdot\overline{\overline{d}}\cdot\overline{\overline{\Psi}}\left(z,z',\vec{\mu}\right)\right\}\cdot\tilde{\vec{E}}^{\Sigma}\left(\vec{q},z'\right) \tag{5.114}$$

Placing (5.114) in the right part of (5.112) we get an equation close to the previous one for $\bar{\vec{E}}^{\Sigma}(\vec{q}, z)$ in which the effects of multiple scattering are included up to second-order (which should not be confused with double scattering!).

It is clear that the higher-order returns result in the appearance of spatial dispersion effects. The resulting equation for $\vec{E}^{\Sigma}(\vec{q}, z)$ is integro-differential.

Equations analogous to (5.112) and (5.114) can be derived from Maxwell`s equations in the case of an isotropic medium that has a spatially non-local nature. The equation for the electric field is given in this case by

$$\nabla^2 \vec{E} + k^2 \vec{E} + 4\pi k^2 \int \chi(\vec{r}, \vec{r}') \vec{E}(\vec{r}') d\vec{r} = \vec{\nabla}(\vec{\nabla} \cdot \vec{E}) \tag{5.115}$$

Equation (5.114) and the integral part of (5.115) constitute an addition to the medium polarizability caused by spatial dispersion. The underlying physics can be easily understood: an average field at a specified point $\vec{r}$ depends on the inhomogeneities surrounding this point.

We have derived an effect due to spatial dispersion when solving the boundary value problem. The method of the dispersion equation is most suitable when analyzing such problems. This will be considered in the next section.

5.3.6 Spatial dispersion in a grass layer

The permittivity matrix of a medium determines the relationship between the electric displacement density $\vec{D}(\omega, \vec{k})$ and the electric field intensity $\vec{E}(\omega, \vec{k})$:

$$D_i(\omega, \vec{k}) = \varepsilon_0 \varepsilon_{ij}(\omega, \vec{k}) E_j(\omega, \vec{k}) \tag{5.116}$$

As mentioned earlier, the summation has to be carried out over repeated indices. The sign of summation is omitted for simplification of the notation. The dependence of the relative permittivity matrix $\varepsilon_{ij}(\omega, \vec{k})$ on ω determines the frequency (time) dispersion and its dependence on the wave vector $\vec{k}$ gives the spatial dispersion of the electromagnetic field in the medium. The tensor $\varepsilon_{ij}(\omega, \vec{k})$ is complex with the following properties:

$$\left.\begin{aligned} \varepsilon_{ij}\left(\omega,\vec{k}\right) &= \varepsilon_{ij}^{*}\left(-\omega,-\vec{k}\right) \\ \operatorname{Re}\varepsilon_{ij}\left(\omega,\vec{k}\right) &= \operatorname{Re}\varepsilon_{ij}\left(-\omega,-\vec{k}\right) \\ \operatorname{Im}\varepsilon_{ij}\left(\omega,\vec{k}\right) &= -\operatorname{Im}\varepsilon_{ij}\left(-\omega,-\vec{k}\right) \end{aligned}\right\} \tag{5.117}$$

In the case of an uniform isotropic medium, in which the properties are the same at any point of space and in any direction, the tensor is only a function of $\vec{k}$ and does not change when a replacement of $\vec{k} \to -\vec{k}$ takes place. The tensor $\varepsilon_{ij}\left(\omega,\vec{k}\right)$ can only contain expressions with the unit tensor δ_{ij} and the tensor $\left(k_i k_j\right)$. Other tensors of the second rank with components of $\vec{k}$ are impossible. That is why for an isotropic medium the tensor $\varepsilon_{ij}\left(\omega,\vec{k}\right)$ can be represented by [*Furutsu*, 1963]:

$$\varepsilon_{ij}\left(\omega,\vec{k}\right) = \left(\delta_{ij} - \frac{k_i k_j}{k^2}\right)\varepsilon^{tr}\left(\omega,\vec{k}\right) + \frac{k_i k_j}{k^2}\varepsilon^{l}\left(\omega,\vec{k}\right) \tag{5.118}$$

This means that of the 9 components of the tensor $\varepsilon_{ij}\left(\omega,\vec{k}\right)$ only two are independent in an isotropic medium, i.e., $\varepsilon^{tr}\left(\omega,\vec{k}\right)$ and $\varepsilon^{l}\left(\omega,\vec{k}\right)$. These components are called transverse and longitudinal permittivities, respectively.

This can be understood when we multiply the tensor $k_i k_j / k^2$ with E_j. This operation extracts the longitudinal (with respect to the wave vector $\vec{k}$) part of the field $\left(\vec{k}\cdot\vec{E}\right)$. In the same way, the value $\varepsilon^{l}\left(\omega,\vec{k}\right)$ characterizes the electromagnetic properties of the medium with respect to the longitudinal field. After multiplication with $\vec{E}$, the tensor $\left(\delta_{ij} - k_i k_j / k^2\right)$ extracts the transverse part of the field. For this reason, $\varepsilon^{tr}\left(\omega,\vec{k}\right)$ characterizes the electromagnetic properties of the medium with respect to the transverse field.

In general, for an anisotropic medium, an electromagnetic field in the substance is neither absolutely longitudinal nor absolutely transverse. For an anisotropic but spatially homogeneous medium, e.g., an effective medium modelling grass cover (see Fig. 5.2), the wave equation (5.25) can be written as

$$\vec{\nabla}\times\vec{\nabla}\times\vec{E} - k_0^2 \int_{-\infty}^{D} dt' \int d\vec{r}'\,\overline{\overline{\varepsilon}}\left(t-t',\vec{r}-\vec{r}'\right)\cdot\vec{E}\left(t',\vec{r}'\right) = 0 \tag{5.119}$$

Going from a $(t,\vec{r})$ to a $(\omega,\vec{k})$ representation, we obtain

$$\left\{k^2\delta_{ij} - k_i k_j - k_0^2\varepsilon_{ij}\left(\omega,\vec{k}\right)\right\}E_j = 0 \tag{5.120}$$

The condition for solving this system of homogeneous equations is given by

$$\Lambda = \left|k^2\delta_{ij} - k_i k_j - k_0^2\varepsilon_{ij}\left(\omega,\vec{k}\right)\right| = 0 \tag{5.121}$$

where $|\ \cdot\ |$ denotes the matrix determinant. The dispersion relationship (5.121) connects the frequency ω with the wave vector $\vec{k}$ for electromagnetic waves existing in the medium.

In the case of an isotropic medium, for which the tensor has the form as given in Eq. (5.118), the dispersion relation (5.121) splits into two equations, i.e., the determinant Λ is factorized, assuming the following form:

$$\varepsilon^l\left(\omega,\vec{k}\right)\left[k^2 - k_0^2\varepsilon^{tr}\left(\omega,\vec{k}\right)\right] = 0 \tag{5.122}$$

The first multiplicative factor represents the existence of longitudinal waves and the second factor the existence of transverse waves in the medium. Unfortunately, such a factorization cannot be done in the general case of an anisotropic medium. It should be pointed out that a longitudinal field by definition is a potential field (with the property for a plane monochromatic wave)

$$\vec{E}\left(\omega,\vec{k}\right) = -\vec{k}U\left(\omega,\vec{k}\right) \tag{5.123}$$

Using Maxwell`s equation $\vec{\nabla}\cdot\vec{D} = 0$ we find

$$k_i\varepsilon_{ij}\left(\omega,\vec{k}\right)k_j U\left(\omega,\vec{k}\right) = 0 \tag{5.124}$$

or

$$\frac{k_i k_j \varepsilon_{ij}\left(\omega,\vec{k}\right)}{k^2} = 0 \tag{5.125}$$

This represents the dispersion equation for longitudinal or potential waves in an anisotropic medium.

From the dispersion relation (5.121), it is possible to understand the propagation of the field into a medium when an electromagnetic wave is incident on its boundary. However, strictly speaking, the tensor $\varepsilon_{ij}(\omega, \vec{k})$ can only be introduced for an unbounded and spatially homogeneous medium. That is why in Sec. 5.3.5 we considered the boundary value problem. The analysis of waves in a plane layer of finite thickness, with discontinuities at the boundaries, can be carried out reasonably well on the basis of the dispersion relationship if the dimensions of the medium are much larger than the wavelength. In this case, the dispersion equation (5.121) correctly describes the spatial change of electromagnetic waves at distances from the boundary that are large compared to the wavelength. At such distances, a spatial change of the field is determined by the properties of the medium itself and not by the discrete boundary value conditions. Such an approach is widely used when analyzing the propagation of electromagnetic waves in a plasma [*Li,* 1992].

When solving the boundary value problem, the complex projection $\vec{k}(\omega)$ on a specified direction is usually determined under the assumption that ω and two other orthogonal projections of $\vec{k}(\omega)$ have real values. The spatial change of the field in such case is determined by

$$\vec{E}(t, \vec{r}) \approx \Sigma \vec{E}_n e^{-i\omega t + i\vec{k}_n(\omega) \cdot \vec{r}} \tag{5.126}$$

where $\vec{k}_n(\omega)$ obeys the dispersion equation (5.121). In the general case of a complex wave vector $\vec{k}(\omega)$, the wave in (5.126) can be called "conditionally" plane, because in this case the planes of constant phase (i.e., the planes perpendicular to the vector $\mathrm{Re}\,\vec{k}(\omega)$) do not coincide with the planes of constant amplitude (i.e., the planes perpendicular to the vector $\mathrm{Im}\,\vec{k}(\omega)$). Such waves are called inhomogeneous plane waves.

If the wave has a small attenuation factor, i.e., $\left|\mathrm{Im}\,\vec{k}(\omega)\right| << \left|\mathrm{Re}\,\vec{k}(\omega)\right|$ (e.g., in a weakly-absorbing, almost transparent medium), then, with great accuracy, it is permissible to say that phase and group velocities of waves coincide with velocities in the absence of absorption. We focus our attention on the phase velocity of coherent waves in the medium, viz.,

$$\vec{v}_{ph} = \frac{\omega \vec{k}(\omega)}{k^2} \tag{5.127}$$

which determines the phases shift between orthogonal polarized waves.

The above-stated properties of the permittivity tensor of an anisotropic medium are used when we derive the system of equations for the averaged field and its distortions within the framework of the second-order approximation, viz.,

$$\begin{aligned}&\frac{d^2\tilde{\vec{E}}(\vec{q},z)}{dz^2}+\left\{ik_0^2\left(\overline{\overline{I}}+4\pi^2 n\Delta\varepsilon_0(0)\overline{\overline{S}}(\vec{q})\right)-q^2\overline{\overline{I}}\right\}\cdot\tilde{\vec{E}}(\vec{q},z)=\\&=-k_0^2\overline{\overline{S}}(\vec{q})\cdot\int d\vec{\mu}\Delta\varepsilon_0(\vec{q}-\vec{\mu})\delta\tilde{\vec{E}}(\vec{q},z)\end{aligned} \tag{5.128}$$

and

$$\begin{aligned}&\frac{d^2}{dz^2}\left\{\delta\tilde{\vec{E}}(\vec{q},z|\vec{\chi})\right\}+\left\{ik_0^2\left(\overline{\overline{I}}+4\pi^2 n\Delta\varepsilon_0(0)\overline{\overline{S}}(\vec{q})\right)-q^2\overline{\overline{I}}\right\}\cdot\delta\tilde{\vec{E}}(\vec{q},z|\vec{\chi})\\&=-k_0^2\overline{\overline{S}}(\vec{q})\cdot\int d\vec{\mu}\Delta\varepsilon_0(\vec{q}-\vec{\mu})\tilde{\vec{E}}(\vec{\mu},z|\vec{q}-\vec{\mu}+\vec{\chi})\end{aligned} \tag{5.129}$$

where

$$S_{il}(\vec{q})=\frac{1}{2}\left(\delta_{ij}-\frac{k_i k_j}{k^2}\right)d_{jl} \tag{5.130}$$

After Fourier transformation over z in (5.128) and (5.129), the relationship (5.90) is used. As a result, we derive the expressions:

$$\left\{k_0^2\left(\overline{\overline{I}}+4\pi^2 n\Delta\varepsilon_0(0)\overline{\overline{S}}(\vec{q})\right)-k^2\overline{\overline{I}}\right\}\cdot\tilde{\vec{E}}(\vec{k})=-k_0^2\overline{\overline{S}}(\vec{q})\cdot\int d\vec{\mu}\Delta\varepsilon_0(\vec{q}-\vec{\mu})\delta\tilde{\vec{E}}(\vec{\mu},k_z|\vec{q}-\vec{\mu}) \tag{5.131}$$

and

$$\left\{k_0^2\left(\overline{\overline{I}}+4\pi^2 n\Delta\varepsilon_0(0)\overline{\overline{S}}(\vec{q})\right)-k^2\overline{\overline{I}}\right\}\cdot\delta\tilde{\vec{E}}(\vec{q},k_Z|\vec{\chi})=$$
$$-k_0^2\overline{\overline{S}}(\vec{q})\cdot\int d\vec{\mu}\Delta\varepsilon_0(\vec{q}-\vec{\mu})\delta\tilde{\vec{E}}(\vec{\mu},k_Z|\vec{q}-\vec{\mu}+\vec{\chi})- \tag{5.132}$$
$$-k_0^2 4\pi^2\Delta\varepsilon_0(-\vec{\chi})\overline{\overline{S}}(\vec{q})\cdot\tilde{\vec{E}}(\vec{q}+\vec{\chi},k_Z)$$

Now it is necessary to express the right-hand side of (5.131) as a function of $\vec{E}(\vec{k})$. This is done by means of (5.132), where we take advantage of the fact that $\vec{\chi}$ is an arbitrary vector and that a change in notation can be carried out via

$$\vec{q}\rightarrow\vec{v};\vec{\chi}\rightarrow\vec{q}-\vec{v} \tag{5.133}$$

Instead of (5.132), we obtain

$$\delta\tilde{\vec{E}}(\vec{v},k_Z|\vec{q}-\vec{v})=-\left(\overline{\overline{A}}\right)^{-1}(\vec{v})\cdot k_0^2 4\pi^2\Delta\varepsilon_0(\vec{v}-\vec{q})\tilde{\vec{E}}(\vec{q},k_Z) \tag{5.134}$$

under the assumption that the perturbation in the field, $\delta\vec{E}$, is small in comparison to the average field. Here, $\left(\overline{\overline{A}}(\vec{v})\right)^{-1}$ is the inverse of the matrix:

$$\overline{\overline{A}}(\vec{v})=k_0^2\left(\overline{\overline{I}}+4\pi^2 n\Delta\varepsilon_0(0)\overline{\overline{S}}(\vec{v})\right)-k^2\overline{\overline{I}} \tag{5.135}$$

Next, we multiply (5.134) with $\Delta\varepsilon_0(\vec{q}-\vec{v})$ and integrate over $\vec{v}$ to get the expression

$$-4\pi^2 k_0^2\int d\vec{v}\left(\Delta\varepsilon_0(\vec{q}-\vec{v})\right)^2\left(\overline{\overline{A}}(\vec{v})\right)^{-1}\cdot\tilde{\vec{E}}(\vec{q},k_Z) \tag{5.136}$$

Eq. (5.131) can now be rewritten in the form

$$\left\{k_0^2\left[\overline{\overline{I}}+4\pi^2 n\Delta\varepsilon_0(0)\overline{\overline{S}}(\vec{q})-\int d\vec{v}\left[\Delta\varepsilon_0(\vec{q}-\vec{v})\right]^2\left(\overline{\overline{A}}(\vec{v})\right)^{-1}\right]-k^2\overline{\overline{I}}\right\}\cdot\vec{E}(\vec{k})=0 \tag{5.137}$$

The dispersion relationship (5.121) follows unambiguously from this system of linear equations. The permittivity tensor in accordance with (5.137) has the following form:

$$\overline{\overline{\varepsilon}}\left(\omega,\vec{k}\right)=(k_0^2-k^2)\ \overline{\overline{I}}+4\pi^2\overline{\overline{S}}(\vec{q})\left\{n\Delta\varepsilon_0(0)\overline{\overline{I}}-k_0^2\int d\vec{v}\left(\overline{\overline{A}}(\vec{v})\right)^{-1}\left(\Delta\varepsilon_0(\vec{q}-\vec{v})\right)^2\right\} \quad (5.138)$$

For further analysis, it is necessary to calculate the integral in (5.138), but first we need an explicit expression for the elements of the inverse matrix $\left(\overline{\overline{A}}(\vec{v})\right)^{-1}$, with the matrix $\overline{\overline{A}}(\vec{v})$ given by

$$\overline{\overline{A}}=\begin{Vmatrix} \left(k_0^2-k^2\right)+2\pi^2n\Delta\varepsilon_0(0)\left(k_0^2-k_x^2\right) & -2\pi^2n\Delta\varepsilon_0(0)k_xk_y & -\dfrac{2\pi^2n\Delta\varepsilon_0(0)}{\varepsilon}k_xk_y \\ -2\pi^2n\Delta\varepsilon_0(0)k_yk_z & \left(k_0^2-k^2\right)+2\pi^2n\Delta\varepsilon_0(0)\left(k_0^2-k_y^2\right) & -\dfrac{2\pi^2n\Delta\varepsilon_0(0)}{\varepsilon}k_yk_z \\ -2\pi^2n\Delta\varepsilon_0(0)k_zk_y & -2\pi^2n\Delta\varepsilon_0(0)k_zk_y & \dfrac{\left(k_0^2-k^2\right)+2\pi^2n\Delta\varepsilon_0(0)\left(k_0^2-k_z^2\right)}{\varepsilon} \end{Vmatrix} \quad (5.139)$$

Since the calculations are lengthy and complex, we analyze the particular case for horizontally polarized waves. Taking into consideration that the model of the scatterers has axial symmetry with respect to the OZ axis, the wave vector $\vec{k}=\left\{0,k_{Y,}k_Z\right\}$ in the YOZ plane is studied. In this case, the determinant of the matrix $\overline{\overline{A}}$ is equal to

$$\det\overline{\overline{A}}=\frac{1}{\varepsilon}\begin{bmatrix}\left(k_0^2-k^2\right)+\\ +2\pi^2n\Delta\varepsilon_0(0)k_0^2\end{bmatrix}\left\{\begin{matrix}\left(k_0^2-k^2\right)^2+\\ +4\pi^4n^2\left(\Delta\varepsilon_0\right)^2k_0^2\left(k_0^2-2k^2\right)+2\pi^2n\Delta\varepsilon_0\left(k_0^2-k^2\right)\left(2k_0^2-k^2\right)\end{matrix}\right. \quad (5.140)$$

Equating (5.140) to zero leads to the dispersion relationship for a field in an effective medium in the first approximation. It is obvious that $\det\overline{\overline{A}}=0$ when:

$$k^2=k_0^2\left(1+2\pi^2n\Delta\varepsilon_0(0)\right) \quad (5.141)$$

which corresponds to the earlier result [cf. Eq. (5.82)].

When $\det\overline{\overline{A}}\neq 0$, the elements of the inverse matrix $\left(\overline{\overline{A}}\right)^{-1}$ can be written in the form

$$a_{il}^{(-1)} = \frac{A_{li}}{\det A} \tag{5.142}$$

where A_{li} is the co-factor of the element a_{li} in the determinant [*Gantmacher,* 1988].

It is clear that the integral in (5.138) is determined by means of the distinct poles of the integrand at the points defined by the relationship (5.141). The detailed calculation scheme allows us to estimate the accuracy of the derived effective permittivity in the case of radar remote sensing by means of horizontally and vertically polarized waves.

5.4 Conclusions and applications

In this chapter, the process of scattering of radiowaves from a surface is considered. The main direction of the study is the determination of appropriate approaches for the solution of inverse scattering problems, leading, specifically, to the determination of the dielectric permittivity of a surface. General solutions of such problems are impossible due to the large variety of surfaces that can be investigated with the help of remote sensing methods. In this chapter, a specific type of surface is considered, with a grass cover modelled in terms of cylinders perpendicular to the plane of incidence. The full analysis of such a model, especially with full account of multiple scattering, represents a very complicated electrodynamic problem. For this reason, several additional simplifications are introduced. First, it is assumed that the cylinders on a pertinent portion of the surface under consideration are uniformly distributed. Then, single and higher–order scattering approximations are considered sequentially. Three specific models of scattering from cylinders are investigated in detail.

Under the aforementioned simplifications, it is possible to reach certain conclusions regarding the averaged scattered fields, accounting for appropriate boundary conditions, the statistical inhomogeneity and anisotropy of the grass cover model and the polarization properties of the incident field. With respect to the latter, only simple cases of field polarization, e.g., linear polarizations are considered.

The most important outcome is that it is possible to obtain (by an analytical approach) the rather complicated dependence of the averaged scattered field and the effective dielectric permittivity of the surface. The latter depends on the grass cover model itself, as well as the properties of the incident field, e.g., the angle of incidence. For example, the anisotropy of the effective permittivity decreases with an increase of the angle of incidence.

The results obtained in this chapter may be useful in several physical applications. For example, agricultural areas, such as meadows, crops of wheat, corn, etc., can be modelled in terms of uniformly distributed cylinders. In such cases, knowledge of the effective surface permittivity allows one to determine the humidity and biomass of the surface cover.

CHAPTER 6

Review of Vegetation Models

6.1 Introduction

Many papers have been devoted to the problem of radio wave reflection from vegetation. The investigations have been pursued over a wide range of wavelengths (meterwave, microwave, millimeterwave, optical wave) using active methods (radar, scatterometers) and passive methods (radiometry, photography) as well. Ground-based, aircraft and satellite-based measurements and theoretical analyses have been performed. The results have been published in a number of reviews [*Potapov,* 1992; *Yakovlev*, 1994; *Chuhlantzev*,1980] and monographs [*Shutko*, 1986; *Ratchkulik*, 1981; *Kondratiev*, 1984].

The present review is limited to electrodynamic models of vegetation. The frequency band covers meter waves down to millimeter waves. Passive methods (radiometry) are not discussed.

Sections 6.1 and 6.2 contain a brief description of biometrical characteristics and (electro)physical properties of vegetation; in section 6.3 the main models of vegetation are described and attention is devoted in sections 6.4 and 6.5 to determining the biometrical characteristics of vegetation on the basis of radar sensing data.

6.2 Biometrical characteristics of vegetation

The following parameters are used as biometrical characteristics of vegetation [*Shutko*, 1986; *Ross*, 1975; *Ulaby*, 1983]:

1) biomass of vegetation = the weight of green mass per unit area (centner/ha; 1 centner = 100kg, 1 hectare ha = 10000m^2)
2) density of planting = the number of plants per unit area (units/m^2)
3) height of vegetation, cm
4) relative weight moisture content, %
5) moisture storage of vegetation cover = the thickness of deposited water, mm
6) area of foliage surface, m^2/ha
7) sparseness of planting = that part of the area without cultured vegetation
8) utility yield = the weight of utility (valuable organisms of plants)

9) biological yield = the total weight of plant organisms formed during the period of vegetation
10) above-ground vegetation mass
11) evolution - homogeneity of a phenomenological composition of crops
12) dry photomass of plants = the mass of plants taking part in the process of photosynthesis

The above mentioned biometrical characteristics of vegetation are the most widely used.

An example of biometrical characteristics of vegetation is shown in Table 6.1 [*Shutko*, 1986; *Ross*, 1975; *Peake*, 1959; *Chuhlantzev*, 1979, *Finkelshtein*, 1994]:

Crops	Density of planting, units/m²	Height of vegetation, cm	Weight of green mass, centner/ha	Moisture content, %	Moisture storage, mm
Winter wheat	400...1000	60...120	20...40 35...130	70...85 50...60	0.26...1.1 0.60...2.2
Winter rye	300...700	120...150	20...45	75...85 50...60	0.75...1.3
Spring wheat	300...500	80...120	50...150	55...65	1.1...2.5
Spring barley	400...900	40...90		60...70	0.7...2.4
Corn	5...30	120...320		80...85	
Sugar beet	7.4	30...50	125...300	80...90	1.2...2.8
Sorghum		300...400			
Oats		82...100	225...534	73...77	
Wheat		88...96	208	56...71	
Barley		74...77	188...251	77...81	
Rye		84...85	201...207	74...75	
Alfalfa with grass		44.5	236		

Table 6.1 Example biometrical characteristics of vegetation

The characteristics of vegetation given in Table 6.1 only give approximate knowledge of biometrical characteristics of vegetation because they depend on the stage of plant vegetation, weather and climatic conditions, soil conditions (moisture content, fertilizer content, etc.) and many other factors.

Remote measurement of all biometrical characteristics of vegetation is hardly feasible. A more effective way is detecting correlation relationships between different parameters. As an example, the correlation matrix of biometrical characteristics of winter wheat in a phase of heading is shown in Table 6.2 [*Kuusk*, 1982]:

	DC	N_0	N_k	h	Q	M_k	M_s	M_{yl}	M_{gl}	Q_d	S_{fs}	Y
Designed coverage (DC), %	1.00											
Total number of stems (N_0), units/m^2	0.78	1.00										
Number of stems with heads (N_k), units/m²	0.84	0.97	1.00									
Height of plants (h), cm	0.84	0.55	0.52	1.00								
(Wet) biomass (Q), centner/ha	0.80	0.72	0.73	0.75	1.00							
Mass of heads (M_k), centner/ha	0.74	0.72	0.75	0.89	0.79	1.00						
Mass of stems (M_s), centner/ha	0.71	0.78	0.81	0.73	0.91	0.85	1.00					
Mass of yellow leaves (M_{yl}), centner/ha	0.57	0.54	0.53	0.66	0.72	0.67	0.68	1.00				
Mass of green leaves (M_{gl}), centner/ha	0.68	0.65	0.62	0.88	0.87	0.84	0.78	0.62	1.00			
(Dry) biomass (Q_d), centner/ha	0.64	0.51	0.54	0.66	0.95	0.81	0.84	0.55	0.81	1.00		
Area of foliage surface (S_{fs}), m³ ha	0.74	0.68	0.71	0.73	0.92	0.81	0.94	0.36	0.98	0.80	1.00	
Crop-producing power (Y), t/ha	0.71	0.75	0.68	0.80	0.78	0.86	0.54	0.40	0.77	0.73	0.79	1.00

Table 6.2 Correlation matrix of agrometeorological and photometrical characteristics of winter wheat

Analysis of this table shows that there is quite a strong correlation between some parameters, especially between height and mass of heads, the height and crop-producing power (correlation coefficient 0,8), the crop-producing power and biomass (correlation coefficient 0,78). The table also presents the correlation between crop-producing power y and biomass.

The following linear regression equation [*Kondratiev*, 1984, 1986] gives the relationship for winter wheat:

$$y = kQ + b \tag{6.1}$$

Here, y denotes the crop-producing power in tons/ha and Q is the above-ground biomass in centner/ha. For different stages of growing, the parameter k ranges from 0.1 to 0.35 and the range of b is from 0 to 14.

The relationship between above-ground vegetation mass and crop-producing power of raw cotton [*Ratchkulik*, 1981] was found to be

$$y = \frac{aQ}{b + Q} \tag{6.2}$$

where the constants "a" and "b" for different regions of central Asia range from 36 to 50 and 22 to 116, respectively.

The height of plants h is one of the most significant parameters determining the structure, the form and the crop-producing power. According to [*Ratchkulik*, 1981] linear regression equations for winter wheat in the central regions of the Nechernozem zone (NZ, south of Moscow) and in the south of Ukraine (SU) become

$$y = 0.15h + 9.3 \text{ for NZ} \tag{6.3}$$

$$y = 0.41h - 3.19 \text{ for SU} \tag{6.4}$$

The measurement of the height and (or) biomass by remote sensing methods allows us to carry out calculations of the crop-producing power.

6.3 Electrophysical characteristics of vegetation

Vegetation is a multi-component structure consisting of free water and the actual vegetation itself, which is a mixture of bounded water and air. That is why the dielectric permittivity of vegetation $\dot{\varepsilon} = \varepsilon' - j\varepsilon''$ must be calculated as the dielectric permittivity of a mixture. When the permittivity ε' of air and dry vegetation mass are equal and ε'' is zero, the water content in vegetation is a parameter that determines the permittivity $\dot{\varepsilon}_{ve}$ of a vegetation element. According to [*Shutko*, 1986] the dielectric permittivity of a vegetation element is given by

$$\dot{\varepsilon}_{ve} \approx \rho_v \dot{\varepsilon}_w \tag{6.5}$$

where $\dot{\varepsilon}_{\upsilon}$ is the dielectric permittivity of water and ρ_{υ} is the volume moisture content of a vegetation element. A different formula was proposed in [*Ulaby*, 1984] for leaves, viz.,

$$\dot{\varepsilon}_{ve} \approx (\varepsilon'_w - 3)\rho_v^2 + 3 - j\varepsilon''_w \rho_v^2 \tag{6.6}$$

where $\varepsilon'_w = \mathrm{Re}\,\dot{\varepsilon}_w$ and $\varepsilon''_w = \mathrm{Im}\,\dot{\varepsilon}_w$. Data for $\varepsilon'_{\upsilon e}$ and $\varepsilon''_{\upsilon e}$ of some vegetation elements are presented in Table 6.3 [*Shutko*, 1986; *Ulaby,* pp. 714-725, 1987; *Ulaby*, pp. 541-557, 1987].

Knowing the complex permittivity $\dot{\varepsilon}_{ve}$ of vegetation elements, we can calculate $\dot{\varepsilon}_{\upsilon}$ of vegetation, according to [*Chuhlantzev,* 1979] as follows:

$$\dot{\varepsilon}_{\upsilon} \approx 1 + \frac{u}{3}\dot{\varepsilon}_{\upsilon e} V \tag{6.7}$$

Here, V is the relative volume of the vegetation component, u=1 for blades and u=2 for prickles. According to [*Redkin*, 1973], $\dot{\varepsilon}_{\upsilon}$ of vegetation can be evaluated using the theory for two-component mixtures; specifically,

$$\dot{\varepsilon}_v = \frac{\dot{\varepsilon}_{ve}(1 + pu) + u(1 - p)}{\dot{\varepsilon}_{ve}(1 - p) + p + u} \tag{6.8}$$

where u is an index characterizing the form (cylindrical to spherical), with a range 10 ≤ u ≤ 20 and p is a filling coefficient, ranging from 10^{-2} to 10^{-3}.

Vegetation	f, GHz	ρ_v	ε'_{ve}	ε''_{ve}
Branches of Indian poplar	2 6 12...22	0.166 0.166 0.166	7.0 5.0 4.0	2.0 1.8 1.0
Leaves of asp	1.5...8.5	0.60 0.20 0.04	3.0 18.0 3.0	9.5 6.0 0.2
Leaves of grains	2...14 2...20 2...20 2...11	0.413 0.137 0.268 0.600	25...20 10...7 16...10 27	7...9 3 5 6
Stems of grains	2...11	0.47	15	4
Grains and grass	15 15 3.0 3.0 1.5 1.5 1.0 1.0	dry wet dry wet dry wet dry wet		37.2 37.9 13.0 16.6 6.7 13.9 4.2 15.0

Table 6.3 Dielectric permittivity of vegetation elements

Laboratory measurements and model calculations are of little use in radar remote sensing of vegetation. We have to rely to a large extent upon representative ground tests. According to [*Ulaby*, 1984] ε' for forest is almost independent of frequency in the range from 0.1 to 10 GHz. Characteristic values are given as follows:

$$\left.\begin{array}{l} 48 \le \varepsilon' \le \;\; 68 \text{ - for soft forest} \\ 30 \le \varepsilon' \le \;\; 53 \text{ - for solid forest} \end{array}\right\} \text{if the field } \vec{E} \text{ is parallel to the grains } (E_{11})$$

$$\left.\begin{array}{l} 37 \le \varepsilon' \le \;\; 54 \text{ - for soft forest} \\ 13 \le \varepsilon' \le \;\; 28 \text{ - for solid forest} \end{array}\right\} \text{if the field } \vec{E} \text{ is perpendicular to the grains } (E_{\perp})$$

For frequencies between 0.6 and 10 GHz, the imaginary part of $\dot{\varepsilon}_v$ for forest becomes

$$\varepsilon'' = \frac{A}{f^{0.96}} + \frac{B(f/f_c)}{1 + f/f_c}, \tag{6.9}$$

where f denotes the frequency (GHz) ; $f_c \approx$ 20 GHz (summer); $f_c \approx 10$ GHz (winter); $A \approx 1.5 \cdot 10^9 (E_{11})$; $A \approx 4 \cdot 10^8 (E_{\perp})$; $B \approx 0$ to 80 depending on the percentage of water content.

According to data over the microwave band, $\varepsilon' \approx 1.1$ for grains and

$$\varepsilon'' \approx 2.9 \cdot 10^{-5} \frac{mQ}{h} \tag{6.10}$$

where m denotes moisture content, Q is the biomass (centner/ha) and h is the height of vegetation (m). Formula (6.10) is derived from linear regression after processing the experimental results. That explains the numerical value in front of the parameter mQ/h in this formula.

Experimental results yield $\varepsilon'' \approx 0.0005$ to 0.01. For instance, at f = 35 GHz, $\dot{\varepsilon}_{grass} \approx$ 1,1 + j0.01 and $\dot{\varepsilon}_{tree} \approx$ 1.015 + j0.002 [*Potapov*, 1992]. Measurements of the permittivity of grains of some crops are presented in [*Radio Engineering, VINITI*, 1977] at wavelengths λ = 3 and 10 cm. Dielectric characteristics of crops derived from calculations and measurements over the frequency range from 8.6GHz to 18GHz are presented in [*Ulaby*, 1978]. Values of the specific attenuation Γ (dB/m) were obtained after evaluation of the signal intensity reflected from a single scatterer placed within the vegetation. It was found that Γ = 12 dB/m for oats; Γ = 15 dB/m for potatoes (Γ in dB/m = $54.6\left(\mathrm{Im}\sqrt{\dot{\varepsilon}}\right) / \lambda$, where λ is the wavelength in m).

For a two-component mixture, of water and dry vegetation, the following relationships are given in [*Peake*, 1959; *Chuhlantzev*, 1979; *Finkelshtein*, 1994]:

$$\varepsilon_v' = 1 + m \frac{\varepsilon_w' Q}{2\rho_v h} 10^{-0.5} \tag{6.11}$$

$$\varepsilon_v'' = m \frac{\varepsilon_w'' Q}{3\rho_v h} 10^{-0.5} \tag{6.12}$$

Using these two expressions, we obtain

$$\dot{\varepsilon}_v = \varepsilon_v' - j\varepsilon_v'' \tag{6.13}$$

$$\dot{\varepsilon}_w = \varepsilon_w' - j\varepsilon_w'' \tag{6.14}$$

where $\dot{\varepsilon}_v$ is the dielectric permittivity of vegetation, $\dot{\varepsilon}_w$ is the dielectric permittivity of water, Q is vegetation biomass (centner/ha), h is the height of the vegetation cover VC (m), m denotes the weight moisture capacity and ρ_v is the density of plants (g/cm²). For most agricultural crops, m = 0.5-0.9, h = 0.2-4m, Q = 20-500 centner/ha and ρ_v = 0.5-0.8g/cm².

6.4 Electrodynamical model of vegetation

Most, if not all, of the recent publications [*Potapov*, 1992; *Ulaby*, 1984; *Ulaby* pp. 714-725, 1987; *Ulaby*, pp. 541-557, 1987; *Redkin*, 1973; *Karam*, 1982, 1988, 1989, 1992; *Ferrazzoli*, 1989; *Lasinski*, 1989, 1990; *Chauhan*, 1989; *Durden*, 1989, 1990; *Pitts*, 1988; *Ulaby*, 1990; *Ulaby*, pp. 83-92, 1987; *Toan*, 1989; *Shwering*, 1986, 1988; *Mo*, 1987; *Richardson*, 1987; *Chuhlanzev*, pp. 256-264, 1979, 1986, 1989; *Redkin*, 1977; *Redkin*, 1973; *Chuhlanzev*, 1988; *Armand*, 1977; *Chuhlanzev*, pp. 2269-2278, 1989; *Lang*, 1983, 1985] on electromagnetic sensing of vegetation deal with the analysis of radio-wave reflection from vegetation. Moreover, in these publications, the analyses are largely restricted to extreme cases (wavelength $\lambda \gg$ characteristic size of the vegetation or wavelength $\lambda \ll$ characteristic size of the vegetation). The vegetation is viewed as a layer consisting either of elements of a specified shape (disks, cylinders, ellipsoids) or of randomly distributed elements with a specified number of elements in a unit of volume.

In this section we shall limit the discussion to wavelengths ranging from millimeters to meters.

6.4.1 Homogeneous and cylindrical model

The simplest vegetation model is a homogeneous layer with effective permittivity $\dot{\varepsilon}$. In this case, the reflection coefficients R_u and R_l from the upper and lower boundaries are given by [*Finkelshtein*, 1977, 1994]

$$R_u \approx m\frac{\varepsilon'_w Q}{8\rho_v h}; \; R_l \approx \left|\frac{1-\sqrt{\dot{\varepsilon}_s}}{1+\sqrt{\dot{\varepsilon}_s}}\right| e^{-0.23\Gamma_v h} \tag{6.15}$$

Here, $\Gamma_v = 0.3\left(m\varepsilon''_w Q/(\rho_v h)\right) f \cdot 10^{-7}$ is the specific attenuation of radiowaves in vegetation, $\dot{\varepsilon}_s$ is the dielectric permittivity of soil, $\varepsilon'_s \approx 3.2 + \varepsilon'_w W_s$; $\varepsilon''_s \approx 10^{-4} + \varepsilon''_w W_s$, W_s is the moisture content of soil and $\dot{\varepsilon}_w$, ρ_v, Q, m and h are defined in previous sections of this chapter.

The dependence of R_l, R_u *and* $r = {R_u}/{R_l}$ on frequency is shown in Fig. 6.1, for the parameters $h = 1m$, $W_s = 10\%$, $\xi = 0.001 - 0.005$, where the relative volume ξ is proportional to $1/(\rho_v h)$. The data were taken from [*Finkelshtein*, 1994] and [*Karpuhin*, 1989].

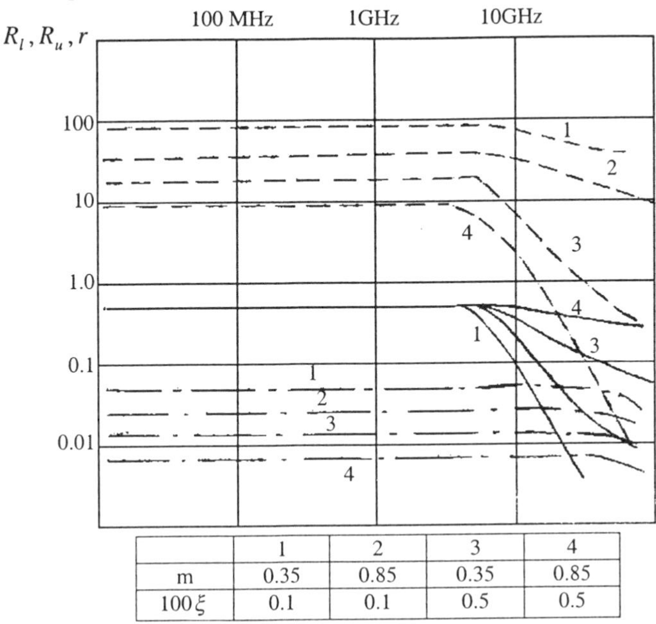

	1	2	3	4
m	0.35	0.85	0.35	0.85
100ξ	0.1	0.1	0.5	0.5

Fig. 6.1 Dependence of the reflection coefficients on frequency.
R_u ———————, R_l -·-·-·-·-·-·-·-, r - - - - - - - - - - - -

As can be seen from Fig. 6.1, the moisture content m ranges from 0.35 to 0.85 and the relative volume ξ from 10^{-3} to 5.10^{-3} (characteristic for all agricultural crops). R_l, R_u *and* $r = {R_u}/{R_l}$ are practically constant within the frequency range 10 to $4 \cdot 10^3$ MHz. Calculations show that R_u and r vary less with soil moisture content (ranging from 10 to 30%) than with vegetation moisture content. (The decrease of R_u and r for $f > 4 \cdot 10^3 MHz$ is caused by an increase in the attenuation of radio waves in the vegetation cover.

In another model the (vegetation) elements are simulated by means of cylinders. Such a model is appropriate for grain crops, trunks of trees, thorns, vertical branches of trees and thin grass layers.

Referring to [*Ulaby*, pp. 714-725, 1987; pp. 83-92, 1987] and [*Chuhlantzev*, 1986], we now consider the case where a plane wave $\vec{E}_{inc}$ is incident on an infinitely long cylinder and the scattered wave $\vec{E}_{ref}$ in the (x, y) plane makes an angle ψ with the x axis (see Fig. 6.2).

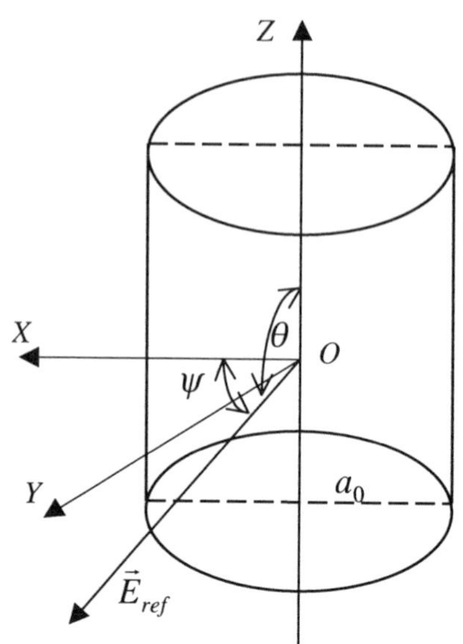

Fig. 6.2 Cylindrical model

The field at large distances will then be

$$\vec{U}_p(r,\psi) = \vec{U}_{p,in} \cdot \left(\frac{2}{\pi kr}\right)^{0.5} \exp\left\{jkr - j\frac{\pi}{4}\right\} T_p(\psi) \tag{6.16}$$

where $p = h$ for horizontal and $p = v$ for vertical polarization, $k = 2\pi/\lambda_0$, λ_0 is the wavelength in vacuum, U_p is the amplitude of the field, $\vec{U} = \vec{E}$ or $\vec{U} = \vec{H}$ and r is the distance to the point of the observer. $T_p(\psi)$ is determined according to the following formulas:

$$T_v(\psi) = \sum_{n=-\infty}^{n=\infty} (-1)^n C_n^{TM} e^{jn\psi} \tag{6.17}$$

$$T_h(\psi) = \sum_{n=-\infty}^{n=\infty} (-1)^n C_n^{TE} e^{jn\psi} \tag{6.18}$$

$C_n^{TM}(\varphi)$, $C_n^{TE}(\varphi)$ are functions of the angle $\varphi = \frac{\pi}{2} - \theta$. (In Fig. 6.2, $\varphi = 0$ and for this case the functions are calculated in [*Ruck*, 1970] and [*Chuhlantzev*, 1986]). The scattering cross-section σ_r is given by

$$\sigma_{r3} = 4\pi \lim_{r\to\infty} r^2 \frac{\vec{E}_{ref}\vec{E}^*_{ref}}{\vec{E}_{in}\vec{E}^*_{in}} = 4\pi \lim_{r\to\infty} r^2 \frac{\vec{H}_{ref}\vec{H}^*_{ref}}{\vec{H}_{in}\vec{H}^*_{in}} \tag{6.19}$$

$$\sigma_{r2} = 2\pi \lim_{r\to\infty} r \frac{\vec{E}_{ref}\vec{E}^*_{ref}}{\vec{E}_{in}\vec{E}^*_{in}} = 2\pi \lim_{r\to\infty} r \frac{\vec{H}_{ref}\vec{H}^*_{ref}}{\vec{H}_{in}\vec{H}^*_{in}} \tag{6.20}$$

where the symbol * denotes complex conjugation. Eqs (6.19) and (6.20) are applicable to three-dimensional and two-dimensional cases, respectively. Substituting Eq. (6.16) into (6.20) and integrating over ψ, we obtain

$$\sigma_r = \int_0^{2\pi} \sigma_{r2} d\psi = \frac{\lambda_0}{\pi^2} \int_o^{2\pi} \left| T_p(\psi) \right|^2 d\psi \tag{6.21}$$

It should be noted that the cross-section of absorption equals

$$\sigma_{n1} = \sigma_{r2}(\pi) = \frac{2\lambda_0}{\pi} \left| \mathrm{Re}\left(T_p(\pi) \right) \right| \tag{6.22}$$

In the following we consider a number of cylinders which are distributed in a layer of thickness d (cf. Fig. 6.3).

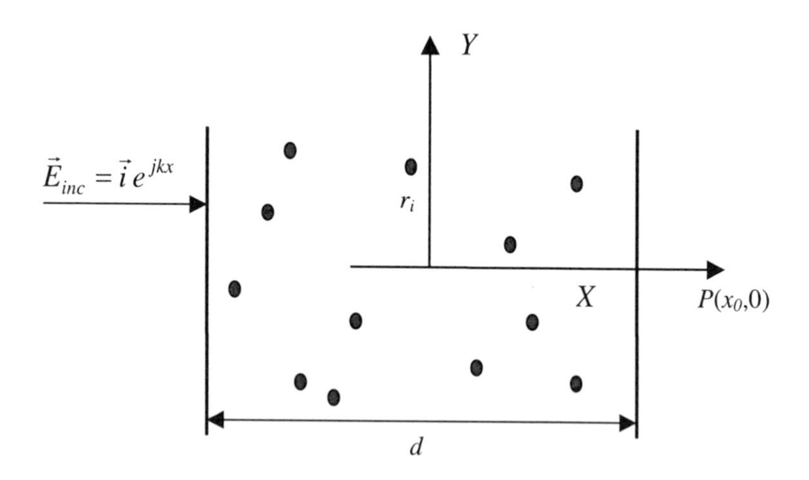

Fig. 6.3 Distribution of cylinders in a layer

The field at point P due to cylinder i, becomes

$$U_{ref} = \left(\frac{2}{\pi k r_i}\right)^{0.5} \exp\left\{ j\left(kr_i - \frac{\pi}{4}\right)\right\} T(\psi_i) \tag{6.23}$$

where r_i is the distance between cylinder i and point P, $T(\psi_i)$ is the amplitude of the field at a large distance and $U_{in} = U_0 e^{jkx}$.
Assuming that $-0.5b \le y \le 0.5b;\ x_0 >> 0.5d$, we derive the expression

$$r_i = \sqrt{(x_i - x_0)^2 + y_i^2} \approx (x_0 - x_i) + \frac{y_i^2}{2x_0} \tag{6.24}$$

If N cylinders are randomly distributed over the area

$$-0.5b \le y \le 0.5b;\ -0.5d \le x \le 0.5d$$

we find

$$U_{ref}(x_0,0) = \left(\frac{2}{\pi x_0 k}\right)^{0.5} \exp\left(j\left(kx_0 - \frac{\pi}{4}\right)\right) T(\pi)$$

$$\times N \int_{-\frac{d}{2}}^{\frac{d}{2}} dx \int_{-\frac{b}{2}}^{\frac{b}{2}} \exp\left(\frac{jky^2}{2x_0}\right) dy \sim \frac{2Nd}{k} T(\pi) U_{in} \tag{6.25}$$

In case $2b \le x_0 \le \frac{\pi b^2}{2\lambda}$ and the integration bound in (6.25) has been set to infinity.

The total field becomes

$$U = U_{in} + U_{ref} = U_{in}\left(1 + \frac{2Nd}{k} \cdot T(\pi)\right) \tag{6.26}$$

If we assume that an equivalent homogenous layer with index of refraction $\dot{n} = \sqrt{\varepsilon}$ can model a layer of cylinders and we take into account that $U = U_{in} \exp\{jk(\dot{n} - 1)d\}$, then for $kd(\dot{n} - 1) << 1$, we obtain

$$U \approx U_{in}\left(1 + jkd(\dot{n} - 1)\right) \tag{6.27}$$

Comparison of Eqs (6.26) and (6.27) shows that the equivalent index of refraction "$\dot{n}$" of the layer is given by

$$\dot{n} = 1 - j\frac{2N}{k^2} T(\pi) \tag{6.28}$$

When the incidence angle of the plane wave equals θ, a multiplier $1/\sin\theta$ appears in front of the sums in Eqs (6.17) and (6.18).

For determining the backscatter characteristics we refer to [*Toan*, 1989]. It is shown in this reference that the backscatter coefficient can be represented in the form

$$\delta^\circ = \delta_s^\circ T + \delta_{veg}^\circ + \delta_{in}^\circ \tag{6.29}$$

where $\delta_s^0, \delta_{veg}^0, \delta_{in}^0$ are reflection contributions from the boundary surface vegetation to land, from the vegetation layer itself and from a cross term, respectively. The quantities entering into Eq. (6.29) are defined as follows:

$$\delta^{\circ}_{veg} = \frac{2\pi \cos\theta_i}{\langle k_e \rangle} \cdot \langle f(-\tau,\tau) \rangle^2 \cdot \left(1 - \exp\frac{-2\langle k_e \rangle h}{\cos\theta_i} \right) \tag{6.30}$$

$$\delta^{\circ}_{\text{int}}{}^{\bullet} = \frac{2\pi \cos^3\theta_i}{\langle k_e \rangle} \left| \langle f(-\tau,\tau) \rangle \right|^2 \cdot \left(1 - T^2\right) T^2 R^2 \times \left(1 + \frac{2N_0 h \langle k_e \rangle \left(\left| \langle f(\tau,\tau) \rangle \right|^2 + \left| \langle f(-\tau,\tau) \rangle \right|^2 \right)}{\cos^2\theta_i \left| \langle f(-\tau,\tau) \rangle \right|^2 \left(1 - T^2\right)} \right) \tag{6.31}$$

$$\delta^{\circ}_s = \exp\left(\frac{-2\langle k_e \rangle h}{\cos\theta_i} \right) \tag{6.32}$$

$$\langle k_e \rangle = \frac{4\pi N_0}{k_0} \mathrm{Im} \langle f(\tau,\tau) \rangle \tag{6.33}$$

N_0 = cylinder density (number of cylinders per unit area)
h = vegetation height
θ_i = incidence angle (with respect to normal)

$$T = \exp\{ (\langle k_e \rangle h) / \cos\theta_i \} \tag{6.34}$$

R = (1-T) is the Fresnel reflection coefficient from the boundary surface of vegetation to land

$\langle \rangle$ indicates ensemble averaging

$\langle f(-\tau,\tau) \rangle$ is the average value of the backscattered field amplitude

$$\langle f(\bar{0},\tau) \rangle = N_0 \int\int\int\int \rho(2l,a,\theta,\varphi) \cdot f(\bar{0},\tau) dl \cdot da \cdot d\theta \cdot d\varphi \tag{6.35}$$

$2l$ = cylinder length
a = cylinder radius
θ, φ are the angles of orientation of a cylinder in spherical coordinates
$\rho(2l,a,\theta,\varphi)$ is the density distribution of the cylinders

$f(\overline{0},\tau)$ is the reflection amplitude field from an equivalent infinite cylinder [cf. Eq. (6.16)].

6.4.2 Disk model

Another vegetation model is represented by a set of round disks. This model is suitable for the description of radiowave reflection from foliage and from crops (corn, tomatoes, potatoes, etc.)

Following [*Karam,* 1989; *Lasinski,* 1989; *Lang,* 1983, 1985; *Ishimaru;* 1981] we consider a plane wave as the incident field, viz.

$$\vec{E}_{in}(\vec{x}) = \vec{\alpha} e^{jk_0 \vec{x}\cdot\vec{i}} \tag{6.36}$$

where

$$\vec{\alpha} \in \{\vec{h}, \vec{v}\} \tag{6.37}$$

This wave is incident in a direction $\vec{i}$ on a disk with radius r and thickness T (see Fig. 6.4). The wave can be polarized horizontally ($\vec{h} = \dfrac{\vec{i}\times\vec{z}}{|\vec{i}\times\vec{z}|}$) or vertically ($\vec{v} = \vec{h}\times\vec{i}$).

The normal $\vec{n}$ to the disk makes an angle θ with the z-axis. The canting angle of the disk in relation to the z-axis is ψ and the permittivity $\dot{\varepsilon}$ of the disk is denoted by $\dot{\varepsilon}_d$.

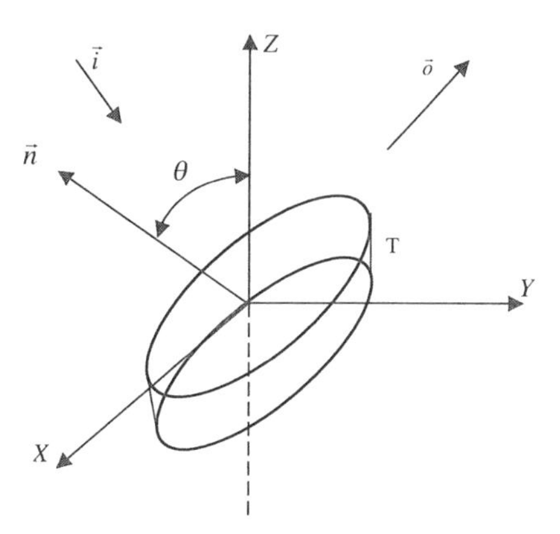

Fig. 6.4 Disk model

The reflected field in a direction $\vec{o}$ becomes [*Le Vine*, 1983]

$$\vec{f}\left(\vec{o},\vec{i},\vec{\alpha}\right)=\frac{k^2\left(\dot{\varepsilon}_d-1\right)}{4\pi}\int_V \vec{E}_{in}\left(\vec{x}',\vec{\alpha}\right)e^{-jk\vec{o}\cdot\vec{x}}\cdot\left(I-\vec{o}\vec{o}\right)dV \tag{6.38}$$

where $\vec{E}_{in}$ is the field inside a disk, I is the unit dyadic and V denotes the volume of the disk. If the disk radius greatly exceeds its thickness $\left(r>>T\right)$, and if the boundary conditions on the disk are used, it can be shown that

$$\vec{E}_{in}\left(\vec{x},\vec{\alpha}\right)=\left(\frac{1}{\dot{\varepsilon}_d}\left(\vec{n}\cdot\vec{\alpha}\right)\vec{n}+\vec{\alpha}-\left(\vec{n}\cdot\vec{\alpha}\right)\vec{n}\right)e^{jk\vec{i}\cdot\vec{x}} \tag{6.39}$$

where $\vec{x}\in V$. Assuming that $\vec{E}_{in}$ in the direction $\vec{n}$ is constant, Eq. (6.38) results into

$$\vec{f}\left(\vec{o},\vec{i},\vec{\alpha}\right)=\frac{k^2 rT\left(\dot{\varepsilon}_d-1\right)}{2}\left(I-\vec{o}\vec{o}\right)\cdot\left[\vec{\alpha}-\frac{\dot{\varepsilon}_d-1}{\dot{\varepsilon}_d}\left(\vec{n}\cdot\vec{\alpha}\right)\vec{\alpha}\right]\frac{J_1\left(vr\right)}{v} \tag{6.40}$$

where

$$\vec{v}=k\vec{n}\times\left(\vec{i}-\vec{o}\right)\times\vec{n} \tag{6.41}$$

$$v=\left|\vec{v}\right| \tag{6.42}$$

and $J_1(\rho)$ is the first-order ordinary Bessel function. By combining results for $\vec{\alpha}=\vec{h}$ and $\vec{\alpha}=\vec{\upsilon}$, we obtain

$$f\left(\vec{o},\vec{i}\right)=\vec{f}\left(\vec{o},\vec{i},\vec{h}\right)\cdot\vec{h}+\vec{f}\left(\vec{o},\vec{i},\vec{v}\right)\cdot\vec{v} \tag{6.43}$$

We now consider a plane wave incident on a layer of round disks, with a layer thickness *d*, and assume that the distribution of the disks within the layer is uniform (cf. Fig. 6.5). The electric field vector is directed along $\vec{h}$.

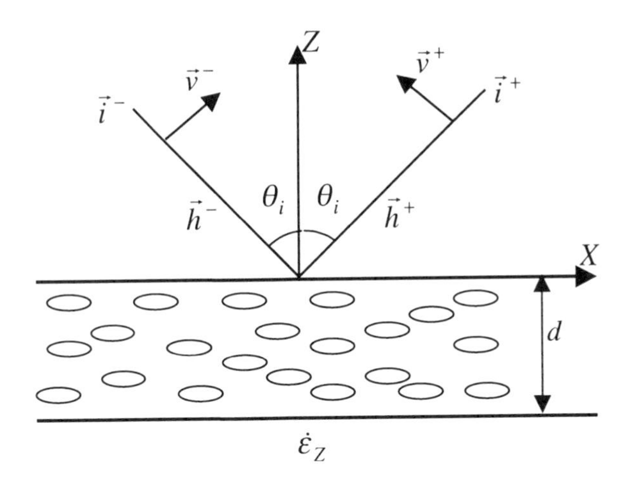

Fig. 6.5 Disks within a layer

The average electric field $\langle \vec{E}(x) \rangle$ includes the field scattered from the layer of thickness d [*Lang*, 1983; *Brehovskih*, 1957] and is expressed by

$$\left\langle \vec{E}\,(x,z) \right\rangle = \left(e^{-jk_0 z} + \mathrm{R} e^{jk_0 z} \right) e^{jk_0 x} \vec{h} \tag{6.44}$$

where $k_0 = 2\pi / \lambda_0$ and R is the reflection coefficient from the layer, given by

$$R = \frac{R_{12} + R_{23} e^{2 jk_h d}}{1 + R_{12} R_{23} e^{2 jk_h d}} \tag{6.45}$$

where R_{12} and R_{23} are the Fresnel reflection coefficients, given respectively by

$$R_{12} \cong \frac{\cos\theta_i - \left(\dot{\varepsilon}_d - \sin^2 \theta_i \right)^{0,5}}{\cos\theta_i + \left(\dot{\varepsilon}_d - \sin^2 \theta_i \right)^{0,5}} \tag{6.46}$$

and

$$R_{23} \cong \frac{\cos\theta_i - \left(\dot{\varepsilon}_d / \dot{\varepsilon}_g - \sin^2\theta_i\right)^{0,5}}{\cos\theta_i + \left(\dot{\varepsilon}_d / \dot{\varepsilon}_g - \sin^2\theta_i\right)^{0,5}} \tag{6.47}$$

$\dot{\varepsilon}_g$ being the soil permittivity. In Eq. (6.45),

$$k_h = k\cos\theta_i + \frac{2\pi p}{k\cos\theta_i}\overline{f_{hh}} \tag{6.48}$$

Here, p is the disk density distribution as function of angle θ and radius r,

$$\overline{f_{hh}} = \int p(r,\theta) f_{hh}(r,\theta) dr d\theta \tag{6.49}$$

$$f_{hh}(r,\theta) = \frac{1}{2\pi}\int_0^{2\pi} \overline{h}\cdot\overline{f}\left(\vec{i}^{\pm},\vec{i}^{\mp}\right)d\psi \tag{6.50}$$

Where $\vec{i}$ is the direction vector and $\vec{h}$ the unit polarization vector (see Fig. 6.5). In Eqs (6.46) and (6.47) it is assumed that $\dot{\varepsilon}_l \approx \dot{\varepsilon}_d$, where $\dot{\varepsilon}_l$ is the dielectric permittivity of the layer. If this is not true, then $\dot{\varepsilon}_d$ has to be substituted by $\dot{\varepsilon}_l$; the latter can be calculated from the formula of a two-component mixture [cf. Eq. (6.8)].

If the total volume of disks is small compared to the volume of the layer and, furthermore, if the albedo of individual scatterers is also small, then the backscatter coefficient σ^0_{hh} for horizontal polarization and disk density ρ (number of disks per unit area) can be derived using the "distorted-wave Born approximation" specifically,

$$\sigma^0_{hh} = \sigma^0_{hhd} + \sigma^0_{hhr} + \sigma^0_{hhdr} \tag{6.51}$$

where

$$\sigma^0_{hhd} = \rho\sigma_{hhd}\frac{1 - e^{-4\,\mathrm{Im}\,k_h d}}{4\,\mathrm{Im}\,k_h} \tag{6.52}$$

$$\sigma^0_{hhr} = \rho\sigma_{hhr}|R|^4\frac{e^{-4\,\mathrm{Im}\,k_h d} - e^{-8\,\mathrm{Im}\,k_h d}}{4\,\mathrm{Im}\,k_h} \tag{6.53}$$

$$\sigma^0_{hhdr} = \rho\sigma_{hhdr} d \left|R\right|^2 e^{-4\,\mathrm{Im}\,k_h d} \tag{6.54}$$

The backscatter coefficients σ^0_{hhd}, σ^0_{hhr}, σ^0_{hhdr} characterize the direct reflection of the incident wave, in which the reflection from the vegetation ground boundary plays a role as well. The coefficient σ^0_{hh} is related to σ^0_{hhd}, σ^0_{hhr}, σ^0_{hhdr} and the field $\overline{f}$ (for direction vectors τ^-, τ^+) as follows:

$$\sigma_{hhd} = 4\pi \overline{\left|\vec{h}\left(\overline{f}\left(-\tau^-,\tau^-\right)\right)\vec{h}\right|^2} \tag{6.55}$$

$$\sigma_{hhr} = 4\pi \overline{\left|\vec{h}\left(\overline{f}\left(-\tau^+,\tau^+\right)\right)\vec{h}\right|^2} \tag{6.56}$$

$$\sigma_{hhdr} = 4\pi \overline{\left|\vec{h}\left(\overline{f}\left(-\tau^+,\tau^-\right)+\overline{f}\left(-\tau^-,\tau^+\right)\right)\vec{h}\right|^2} \tag{6.57}$$

The averaging has been carried out over the disk angle orientations and radii [see Eqs (6.49) and (6.50)]. Similar formulas can be derived for the vertical wave polarization (vectors v^+, v^- as shown in Fig. 6.5).

6.4.3 Three-dimensional model

In this model, vegetation is represented by a volume of scatterers. This model satisfactorily describes scattering from dense vegetation when the reflection from the underlying surface is not taken into account.

We assume that there is a set of scatterers with a packing density N and average cross-sections for absorption σ_t, scattering σ_s and bistatic scattering σ_{bi} [*Potapov*, 1992; *Ishimaru*, 1981; *Ulaby*, 1988]. We write

$$\sigma_t = \frac{1}{N}\int \sigma_t(D)p(D)dD \tag{6.58}$$

where $p(D)$ is the particle size distribution and $\sigma_t(D)$ is the absorption cross-section of a particle with size *D*.

When the bi-static scattering cross-section is azimuthally symmetric with respect to scattering in the forward direction ($\psi = 0$), we can write

$$\sigma_{bi}(\psi) = \sigma_s g(\psi) \tag{6.59}$$

where $g(\psi)$ is the angular scattering pattern, satisfying

$$\iint_{4\pi} g(\psi) d\Omega = 4\pi \tag{6.60}$$

as normalization condition and

$$\sigma_s = \frac{1}{4\pi} \iint_{4\pi} \sigma_{bi}(\psi) d\Omega \tag{6.61}$$

In Eqs (6.60) and (6.61), the integration is over the entire solid angle Ω.

In first approximation of repeated scattering with constant coefficients of the antenna gain, the total received power is determined by

$$P(\psi) = N \frac{P_1 G_1 G_2 \lambda^2}{(4\pi)^3} \iiint_V \sigma_{bi}(\vec{r}, \psi) \frac{\exp\{-\tau_1(\vec{r}) - \tau_2(\vec{r})\}}{L_1^2(\vec{r}) L_2^2(\vec{r})} dV \tag{6.62}$$

where P_1 is the power of the transmitter, $\vec{r}$ is the radius vector, $\tau_1(\vec{r})$, $\tau_2(\vec{r})$ are respectively the optical paths from the transmitter and receiver to an element of volume dV, L_1, L_2 are respectively the distances from the transmitter and receiver to an element of volume dV and V denotes the volume occupied by the scatterers.

Dividing Eq. (6.62) by the power P_0 in free space at $\psi = 0$, i.e.,

$$P_0 = P_1 G_1 G_2 \left(\frac{\lambda L}{8\pi} \right)^2 \tag{6.63}$$

and assuming that $\sigma_{bi}(\vec{r}, \psi) = \sigma_{bi}(\psi)$, we may derive from the quotient $P(\psi)/P_0$, the coefficient of bi-static scattering

$$k_{bi} = N\sigma_{bi}(\psi) \frac{P(\psi)}{P_0} \cdot \frac{\pi}{L^2 I(\psi)} \tag{6.64}$$

where $I(\psi)$ is the illumination function in Eq. (6.62). Taking Eq. (6.59) into account, we derive from the volume coefficient of scattering $k_s = N\sigma_s$

$$k_s g(\psi) = \frac{P(\psi)}{P_0} \cdot \frac{\pi}{L^2 I(\psi)} \tag{6.65}$$

For an elemental volume with a concentration of N particles, the bi-static (cross-polar) scattering coefficient becomes:

$$k_s = N \langle \sigma_{vh} \rangle = 4\pi N \langle s_{vh} s_{vh}^* \rangle \tag{6.66}$$

where $\langle s_{vh} s_{vh}^* \rangle = \langle |s_{vh}|^2 \rangle = p_{21}$ is an element of the Muller matrix (see Appendix B).

From remote sensing experiments of forests in the millimeter-wave band [*Potapov*, 1992], it follows that

$$k_{hh}(\psi) \approx k_{vv}(\psi) = k_1(\psi) \tag{6.67}$$

$$k_{hv}(\psi) \approx k_{vh}(\psi) = k_2(\psi) \tag{6.68}$$

and the volume scattering coefficient of k_s is expressed by

$$k_s^h = \frac{1}{4\pi} \iint\limits_{4\pi} \left(k_{hh}(\vec{k}_s) + k_{vh}(\vec{k}_s) \right) d\Omega \tag{6.69}$$

$$k_s^v = \frac{1}{4\pi} \iint\limits_{4\pi} \left(k_{vv}(\vec{k}_s) + k_{hv}(\vec{k}_s) \right) d\Omega \tag{6.70}$$

for horizontally and vertically polarized waves, respectively.

It was proposed in [*Shwering*, 1986] that for remote sensing of vegetation by millimeter waves the following equation for $g(\psi)$ can be used:

$$g(\psi) = \alpha f(\psi) + (1 - \alpha) \tag{6.71}$$

Here, $f(\psi)$ is the Gaussian forward lobe and α is the ratio of scattering power in the forward direction relative to the total power.

Experimental evidence [*Shwering*, 1986; *Ulaby*, 1988] indicates that

$$f(\psi)=2\left(1+\frac{1}{\beta_s^2}\right)e^{-\frac{|\psi|}{\beta_s}} \tag{6.72}$$

where β_s is half-power beam width.

6.4.4 Model using transport theory

In this section the field intensity is calculated using the emission transport equation.

Following [*Potapov*, 1992] and [*Shwering*, 1986, 1988] we consider a plane wave incident at a planar boundary separating air from a vegetation half-space characterized by a set of parameters $\Delta\gamma_0, \alpha, \sigma_a$ and σ_s. $\Delta\gamma_0$ is the width of the forward lobe of the scattering pattern

$$g(\gamma_0)=\alpha\left(\frac{2}{\Delta\gamma_0}\right)^2 e^{-\left(\frac{\gamma_0}{\Delta\gamma_0}\right)^2}+(1-\alpha) \tag{6.73}$$

where $\gamma_0=\arccos\left(\vec{k}_s,\vec{k}_i\right)$ is the angle between the vectors describing the scattering $\vec{k}_s(\theta_s,\varphi_s)$ and incident $\vec{k}_i(\theta_s,\varphi_s)$ directions; σ_a, σ_s are the absorption and scattering cross-sections, respectively.

The transport equation, governing the field intensity $I(z,\theta)$ in vegetation, is given by

$$\frac{\partial I(z,\theta)}{\partial z}\cos\theta+(\sigma_a+\sigma_s)I(z,\theta)=\frac{\sigma_s}{4\pi}\int_0^{2\pi}\int_0^{\pi} g(\gamma_0)I(z,\theta')\sin\theta' d\theta' d\varphi', \quad for \ z>0 \tag{6.74}$$

Here,

$$\cos\gamma_0=\vec{k}_i\cdot\vec{k}_s=\cos\theta\cos\theta'+\sin\theta\sin\theta'\cos(\varphi-\varphi') \tag{6.75}$$

The boundary conditions for Eq. (6.74) are:

$$\left.\begin{array}{l} I(0,\theta) = S_P \delta(\theta) \dfrac{1}{2\pi \sin\theta} \\ \quad I(z,\theta) \to 0, \quad z \to \infty \end{array}\right\} \tag{6.76}$$

$$0 \le \theta \le \frac{\pi}{2}$$

where $\vec{S}_P$ is the Pointing vector and $\delta(\theta)$ denotes the Dirac delta function.

The solution of Eq. (6.74) is presented in the form [*Shwering,* 1986]:

$$I = I_{ri} + I_d = I_{ri} + I_1 + I_2 \tag{6.77}$$

where I_{ri} and I_d are the coherent and diffused components of the field intensity, respectively. I_{ri} is given as follows:

$$I_{ri}(z,\theta) = S_P \frac{\delta(\theta)}{2\pi \sin\theta} e^{-\tau} \qquad (z > 0) \tag{6.78}$$

On the other hand, the two components of the diffused intensity assume the forms

$$I_1(z,\theta) = \frac{1}{4\pi} S_P\, e^{-\tau} \sum_{m=1}^{M} \frac{1}{m!} (\alpha\Lambda\tau)^m q_m(\theta) \qquad (z > 0) \tag{6.79}$$

$$I_2(z,\theta) = \frac{1}{2\pi} S_P\, A \frac{s-1}{s-\cos\theta} e^{-\tau'} \qquad (z \to \infty) \tag{6.80}$$

The following definitions are used for the quantities appearing in Eqs (6.78)-(6.80):

$$\tau = (\sigma_a + \sigma_s) z \tag{6.81}$$

$$\tau' = [\sigma_a + (1-\alpha)\sigma_s] \frac{z}{s} \tag{6.82}$$

$$q_m(\theta) = \frac{1}{m} \left(\frac{2}{\Delta\gamma_0} \right)^2 \exp\left\{ -\frac{1}{m} \left(\frac{\theta}{\Delta\gamma_0} \right)^2 \right\} \tag{6.83}$$

$$\Lambda = \frac{\sigma_s}{\sigma_s + \sigma_a} \tag{6.84}$$

We know that the quantity M appearing in the summation in Eq. (6.79) is a large integral number and that proper values for the amplitude factor A and s are functions of $\Lambda' = \frac{(1-\alpha)\Lambda}{1-\alpha\Lambda}$.

As an example, values of s are presented in Table 6.4:

Λ	0	0.25	0.50	0.60	0.70	0.80	0.85	0.90	0.95	0.99	1.00
s	1.00	1.01	1.04	1.10	1.21	1.41	1.59	1.90	2.64	5.80	$\propto$

Table 6.4 Dependence of s on Λ

It should be mentioned that the expression for I_{ri} is exact; however, I_1 given in Eq. (6.80) is an approximation. The approximation becomes more accurate when $\Delta\gamma_0 << \pi$. Expressions for I_2 are valid in asymptotic cases, but for short distances I_2 is calculated numerically.

Analysis of the solution to the transport equation indicates that the coherent component I_{ri} is dominant at short distances and gives a precise defined direction. It can be considered as an extension of the incident wave in a random medium with an exponential attenuation due to absorption and scattering. The incoherent component $I_d = I_1 + I_2$ is formed as a result of scattering of the coherent component. It consists of a great number of waves propagating in different directions.

The parameters $\Delta\gamma_0, \alpha, \sigma_a, \sigma_s$, defined earlier in this section, are determined by the micro- and macrostructure of vegetation. It is quite difficult to obtain accurate values for these parameters. Therefore, it may be more suitable to select them from experimental data.

Experimental remote sensing of forests at frequencies f = 9.6; 28.8 and 57.6 GHz [*Shwering*, 1988] has shown a satisfactory agreement with theoretical calculations.

6.5 Determination of biometrical characteristics of vegetation from radar remote sensing data

A field reflected from vegetation actually depends on the biometrical characteristics of vegetation. In order to derive information about these characteristics from radar data it is necessary to solve an inverse problem.

Sometimes it is possible to connect the biometrical characteristics of vegetation with remote sensing data in a direct way. From experiments [*Finkelshtein*, 1994], it has been found that the reflection coefficient (R_l) of the air-to-vegetation boundary is connected with the specific moisture content of agricultural crops WQ/h (W = moisture, Q = biomass, h = height) by the formula

$$R_l = 2.4 \cdot 10^{-5} \frac{WQ}{h} \tag{6.85}$$

where a normal angle of incidence has been assumed. The specific attenuation $\Gamma(dB)$ equals

$$\Gamma(dB) = 0.02Q(c/ha) \tag{6.86}$$

This specific attenuation is determined by

$$\Gamma = 20\log\frac{A_1}{hA_2} \tag{6.87}$$

where A_1 is the amplitude of a pulse reflected from a corner reflector placed above the vegetation cover and A_2 is the same quantity in the absence of vegetation.

The dependence of the vegetation height h and the biomass Q on the time difference Δt (in ns) between the time lag of a pulse reflected from an air-vegetation boundary surface and a land-vegetation boundary is given as follows:

$$h(m) = 20.02\Delta t \tag{6.88}$$

$$Q(c/ha) = 60.9\Delta t \tag{6.89}$$

The measurement of pulse amplitudes reflected from an air-vegetation boundary surface allows us to define a degree of sparseness in vegetation.

In the review article [*Yakovlev,* 1994], the following relationships are given for the dependence of the scattering cross-section σ on the volume moisture content M_V of different vegetation environments:

$\sigma = 35.73M_v + 16.66$ for maize

$\sigma = 32.78M_v - 20.78$ for gathered maize and plowed land

$\sigma = 30.2M_v + 25.7$ for orchards

6.6 Classification of vegetation

Observations carried out by means of airplane- or space-based radars are used for classification of agricultural crops.

Methods of identification are based on the analysis of hysteresis average values and dispersions obtained during measurements. Methods of classification assume a period of training (or evaluation of histogram parameters and measurement results) first and after that verification using other data. The capability of identification/recognition is characterized by an error matrix derived as a result of the verification process.

As a criterion of distinction between two classes "a" and "b", it has been proposed to use the expression [*Yakovlev*, 1994]:

$$s = \frac{M_a - M_b}{D_a - D_b} \tag{6.90}$$

where M_a, M_b are average values and D_a, D_b are variances in the criteria of recognition of classes "a" and "b", respectively. It is assumed that the recognition is unsatisfactory for $s < 0.5$, the recognition is good for $0.5 < s < 1.0$, the recognition is very good when $1.0 < s < 1.5$ and the recognition is excellent if $s > 1.5$. The effectiveness of these criteria was verified on the basis of measurements carried out at $\lambda = 19cm$ and $\lambda = 6cm$ with two polarizations (HV and HH), and at $\lambda = 2cm$ with a VV polarization.

Detailed data concerning the results of classification of agricultural crops performed on the basis of ground, airplane and space observations on testing grounds of USA,

Canada and European countries have been presented in the review article [*Yakovlev*, 1994].

6.7 Conclusions and applications

This chapter deals with survey-type material on the electrodynamic characteristics of terrestrial surfaces with various types of vegetation covers. Primary attention is given to agricultural covers (wheat, rye, barley, corn, sugar-beet, oats, sorghum, lucerne); also to grassy covers. Among the large number of biometric characteristics of vegetation covers, we mention the biomass of vegetation, the density of crops, the relative humidity content, etc. These, as well as other, characteristics are clearly interconnected. It is expedient, then, to determine appropriate correlation connections. For example, those between the height of crops and their mass, between the biomass and yield amount of grain, etc. In this chapter, we have not provided detailed descriptions of experiments that have been carried out, nor have we included complicated mathematical calculations. Our purpose has been to provide references to the pertinent literature for the interested reader.

The majority of specific results, taken from appropriate sources, have an approximate, rough character and are based on considerations involving broad limits of correction and normalization factors. These limits depend on the stage of maturing of an appropriate structure, season, moisture content in vegetation, etc. Therefore, the derivation of more precise relations for concrete agricultural covers is hardly possible.

For example, the productivity of winter wheat is a linear function of the ground biomass, with a constant of proportionality varying in limits from 0.1 up to 0.35 (i.e., more than 3 times) and with a correction factor lying in limits from 0 up to 14. A similar picture applies to other agricultural covers. For cotton, the constant of proportionality varies from 36 up to 50 and the correction factor lies in limits from 22 up to 116.

From the point of view of scattering of radiowaves from vegetation covers, it is necessary to consider an effective composite structure consisting of water, air and vegetation. Therefore, the permittivity of vegetation is considered as that of a mixture and may have a complex character. Its real and imaginary parts are indicators of the biological properties of vegetation. The latter may depend significantly on frequency. The necessity of application of multi-frequency sensing of the same object is clearly evident.

Except for sensing agricultural covers, the remote sensing of forest regions is very common. Of very important significance, in this case, is the polarization of the sensing wave, as a forest region is a rather well structured environment. Results based on vertical and horizontal wave polarizations may differ substantially, depending on the type of trees, their density, etc.

The scattering of radiowaves from vegetation earth covers requires appropriate modeling. Such canonical models may involve disks, cylinders or ellipsoids randomly distributed within a specified region. These models depend significantly on the wavelength of the sensing electromagnetic wave.

Two widely used models of vegetation are a homogeneous layer with an effective dielectric permittivity and a region filled with an appropriate distribution of cylinders. The latter can describe well the reflection of electromagnetic waves from trunks of trees, vertically located branches of trees, laminas of a grassy cover, etc. A third model involving a distribution of disks, can be used to describe the scattering of radiowaves from a deciduous cover, from stalks of tomatoes, potatoes, corn, etc. It is usually assumed that the disks are uniformly distributed within the layer.

Under certain conditions, the interaction of electromagnetic wave with a vegetation layer can be modeled as volumetric scattering, especially if the vegetation is dense and reflections of radiowaves from the underlying surface can be disregarded.

The main application of results discussed in this chapter is in the possibility for determining biometric characteristics of various types of vegetation covers (within the framework of accepted models) by means of remote measurements. For example, the reflection coefficient of radiowaves is directly proportional to the humidity content in agricultural crops and the biomass volume and is inversely proportional to the height of the vegetation. The derivation of the biometric characteristic properties of vegetation layers (at least some of them) from the analysis of reflected and scattered electromagnetic waves is the solution of an inverse problem.

Parameters, such as the moisture content in vegetation and the volume of biomass, allow to predict the crop. The determination of the height of crops allows to determine the degree of their maturity and to give a reasonable prognosis for the yield and an appropriate time for collecting the crop. Any modification of the aforementioned parameters corresponds to respective alterations of the effective surface dielectric permittivity, which, in turn, modifies the characteristics of the reflected radiowaves for different kinds of polarization and different frequencies of sensing.

CHAPTER 7

Electrodynamic and Physical Characteristics of the Earth's Surfaces

7.1 Introduction

In this chapter, the interrelations between electrodynamic and physical characteristics of earth surfaces are considered. The electrodynamics characteristics are measured by the permittivity that reflects the polarizability of the medium under the perturbation of an external electromagnetic field. The physical characteristics are measured by the thermodynamic (e.g., temperature, pressure) or chemical parameters (e.g., moisture, salinity) of the medium. The electrodynamic characteristics depend on the frequency of the external electromagnetic field, the natural frequencies of the medium, as well as the physical properties indicated above. Since the reflectivity properties of the medium depend on its electrodynamic properties (permittivity), we see how important it is to derive the electrodynamic properties from the physical characteristics of the medium.

Earlier in the monograph, we discussed the main problems of classification and identification of radar objects using remote sensing. The necessity was shown for having a certain set of attributes of the sensed objects in order to derive (with the required degree of accuracy) the desired solutions. It is possible to derive such a set of attributes only if the information on characteristics of radio waves reflected from the surface is available. That is why the interrelation between electrodynamic and physical characteristics of the sensed objects are considered in this chapter. It is evident that these physical characteristics of the radar sensed objects form precisely the aforementioned set of object attributes that allow us to solve the problems of classification and identification.

These investigations lay the foundation for further approaches to solving inverse problems. Specifically, we can consider the interrelation between empirical characteristics of surfaces and characteristics determining the radar polarization status. This means that we can derive relations that indirectly connect the physical characteristics of the radar-sensed objects and the polarization characteristics of the received radio waves, i.e., those characteristics that are directly determined by the radar. In other words, the data received by the radar may be reprocessed into the set of attributes of the radar-sensed object for solving the problems of classification and identification.

This chapter contains the analysis of the interrelation between electrodynamic and electrophysical characteristics of layered media. Specifically, the functional and empirical relationships between the complex dielectric permittivity of various earth covers, such as water, ice, snow, ground, vegetation and their humidity, density, salinity, temperature and other physical characteristics are established. A significant amount of work has been carried out in this field. The authors have analyzed the available literature and drawn a number of conclusions that are presented here.

7.2 Complex permittivity

The electrical properties of a non-magnetic medium are determined by its complex dielectric permittivity $\varepsilon = \varepsilon' - i\varepsilon''$, which may be represented in the form

$$\left.\begin{aligned} &\varepsilon = \varepsilon'(1 - i\tan\delta), \\ &\tan\delta = \frac{\varepsilon''}{\varepsilon'} = \frac{60\sigma_{eff}\lambda}{\varepsilon'} \end{aligned}\right\} \tag{7.1}$$

in terms of the effective conductivity σ_{eff}. The complex index of refraction n is often used:

$$n = \sqrt{\varepsilon} \tag{7.2}$$

The magnitude of its real part determines the phase speed v of the electromagnetic wave propagation in the medium, viz.,

$$v = \frac{c}{\mathrm{Re}\sqrt{\varepsilon}} \tag{7.3}$$

The magnitude of its imaginary part determines the electromagnetic wave attenuation Γ; specifically,

$$\Gamma(dB/m) = \frac{54.6}{\lambda}\mathrm{Im}\sqrt{\varepsilon} = 0.182\,\mathrm{Im}\sqrt{\varepsilon}\cdot f(MHz) \tag{7.4}$$

It follows from Eq. (7.1) that

$$\left.\begin{aligned} \operatorname{Im}\sqrt{\varepsilon} &= \sqrt{\frac{\varepsilon'}{2}}\sqrt{\sqrt{1+\tan^2\delta}-1} \\ \operatorname{Re}\sqrt{\varepsilon} &= \sqrt{\frac{\varepsilon'}{2}}\sqrt{\sqrt{1+\tan^2\delta}+1} \end{aligned}\right\} \quad (7.5)$$

In particular, if $\tan\delta << 1$, then

$$\left.\begin{aligned} \operatorname{Re}\sqrt{\varepsilon} &\approx \sqrt{\varepsilon'}\left(1+\frac{\tan^2\delta}{8}\right) \\ \operatorname{Im}\sqrt{\varepsilon'} &\approx \frac{\sqrt{\varepsilon'}}{2}\tan\delta \\ \Gamma(dB/m) &\approx 9.1\cdot 10^{-2}\sqrt{\varepsilon'}\cdot f\,(MHz)\cdot\tan\delta \end{aligned}\right\} \quad (7.6)$$

7.3 Dielectric and physical parameters

The complex dielectric permittivity (DP) depends on many parameters, such as frequency, temperature, moisture, salinity, density, porosity, etc. To derive the DP dependence upon only one parameter is very difficult (especially under full-scale conditions).

In this section, attention is paid to the dependence of the permittivity on the physical parameters of a substance for centimeter waves; specifically, wavelengths of 1.8 cm and 3.2 cm are used. However, results for other values of wavelengths in the SHF band are also presented, primarily because the main relationships differ slightly with a change in wavelength in the SHF band; at least they do not differ qualitatively and they change little quantitatively.

7.3.1 Dielectric permittivity and moisture

In the radio-frequency band, the permittivity of dry rocks is much smaller than the permittivity of water. The permittivity dependence upon moisture (in volumetric percentages or in volumetric part), frequency f and temperature (t in ^{0}C, T in K) can be derived with the use of Debye's formulas [*Wang,* 1980]:

$$\left.\begin{aligned}\varepsilon' &= \varepsilon'_\infty - \frac{\varepsilon'_s - \varepsilon'_\infty}{2\ln\xi_0}\ln\frac{\xi_0^{-1} + \omega^2\tau_m^2}{\xi_0 + \omega^2\tau_m^2}\\ \varepsilon'' &= -\frac{\varepsilon'_s - \varepsilon'_\infty}{\ln\xi_0}\left(\tan\frac{(1-\xi_0)\omega\tau_m}{\left(1+\omega^2\tau_m^2\right)\sqrt{\xi_0}}\right)^{-1} = 60\lambda\sigma_{eff}\\ \varepsilon'_s &= \lim_{\omega\to 0}\varepsilon';\quad \varepsilon'_\infty = \lim_{\omega\to\infty}\varepsilon';\quad \tau_m = \frac{1}{2\pi f_m};\quad \xi_0 = \exp\left\{-\frac{\Delta\alpha}{kT}\right\} = e^{-\beta}\end{aligned}\right\} \tag{7.7}$$

where the term $60\lambda\sigma_{eff}$ is equal to [cf. Eq. (7.1)]

$$60\lambda\sigma_{eff} = \frac{\sigma}{\omega\varepsilon_0} \tag{7.7a}$$

Furthermore, τ_m is relaxation time, f_m is relaxation frequency and ξ_0 is a parameter dependent on the temperature T (in Kelvin), the activation energy band $\Delta\alpha$ and the Boltzmann's constant k.

[*Ulaby et al.*, 1987] point out that the dependence of the reflection coefficient $|R|$ on the moisture for soil containing 49% clay, 16% sand and 35% silt is practically linear within the frequency band $f = 2.75 - 7.25 GHz$.

For frozen rocks, the complex dielectric permittivity depends on moisture and temperature. This phenomenon takes place due to the fact that the complex dielectric permittivity decreases when passing through the ice freezing point $T \approx 271.54\ K$.

The ε' dependence upon moisture for frozen sand shows a maximum when the relative humidity W, defined by the ratio between the vapour pressure and the saturated vapour pressure, is about 13% (see Fig. 7.1 for $f = 20GHz$ and sand-grain sizes between $0.3mm$ and $0.5mm$) [*Ilijn*, 1993, 1994, 1995]. The authors explained this behaviour in the following way. Ice formed from bounded water has ferro-electric properties. This results in an increase of ε' and, as a consequence, in the occurrence of peaks in the $\varepsilon'(W)$ relationship.

[*Vasiliev*, 1995] has proved that the temperature and moisture of frozen rocks are connected by $T - T_0 = \alpha(W - W_0)$, where α depends on the physical characteristics of soil (density, specific heat capacity, etc.).

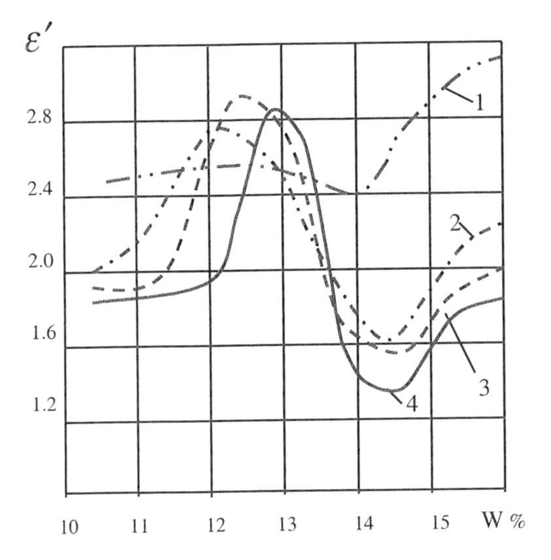

Fig. 7.1 ε' of frozen sand as function of relative humidity in %
(1)T=273.14 K; (2)T=270.14 K; (3)T=264.14 K; (4)T=259.14 K

Peat consists of water, gas (in volumetric sense not more than 5%) and dry material. It can be shown that the dielectric permittivity ε' of peat depends on moisture according a linear relationship [*Tiuri*, 1982, 1983].

The calculated results are shown in Fig. 7.2. Such a simple relationship allows us to determine peat reserves with the use of radar remote sensing.

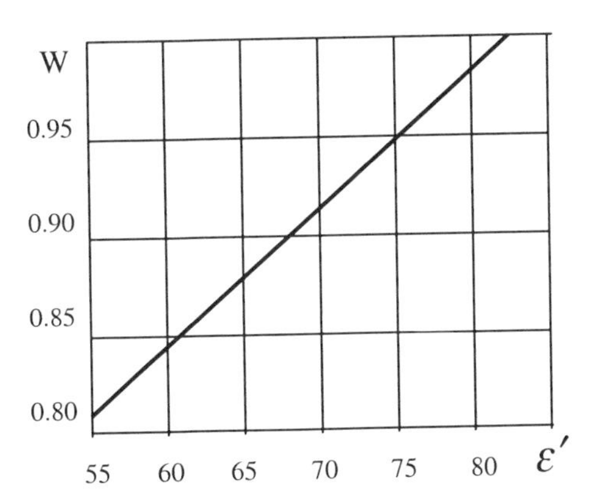

Fig. 7.2 ε' of peat as a function of relative humidity W in cm^3/cm^3

The dielectric permittivity of vegetation cover also depends on moisture. The relationship can be given in the following form [*Finkelshtein*, 1994]:

$$\left.\begin{aligned} \varepsilon_v' &= 1 + \frac{\varepsilon_w' QW}{2h\rho_v} \cdot 10^{-5} \\ \varepsilon_v'' &= \frac{\varepsilon_w'' QW}{3h\rho_v} \cdot 10^{-5} \end{aligned}\right\} \tag{7.8}$$

Here, ε_v, ε_w are respectively the dielectric permittivities of vegetation and water, W is the vegetation moisture (weight of volumetric part, cm^3/cm^3), ρ_v is the vegetation density (g/cm^3), h is the vegetation height (m) and Q denotes the vegetation biomass $\left(c/ha,\ c = 100Kg\right)$.

For most agricultural crops, $W = 0.5 - 0.9$; $h = 0.2m - 0.4m$; $\mathrm{Q} = 20 - 50\ c/ha$; $\rho_v = 0.5 - 0.8\ g/cm^3$. Eq. (7.8) can sometimes be simplified in the SHF band: $\varepsilon_v' \approx 1.1$; $\varepsilon_v'' = 2.9 \cdot \dfrac{QW}{h} \cdot 10^{-5}$ [*Finkelshtein*, 1994].

A linear dielectric permittivity dependence on moisture has been pointed out by [*Dobson*, 1986] and [*Ulaby*, 1987] in the frequency range $f = 1 - 18GHz$.

The complex dielectric permittivity of trees depends on moisture as given by [*Yakovlev*, 1994]:

$$\left.\begin{aligned} \varepsilon_v' &= 5 + 5\left(51.56W - 0.5\right)\left(1 + 0.185W\right)^2 \\ \varepsilon_v'' &= 0.185\left(51.56W - 0.5\right)\left(1 + 0.185W\right)^2 \end{aligned}\right\} \tag{7.9}$$

Although the linear dependence of permittivity on moisture can describe different media, e.g., peat and trees, the media can be distinguished by radar remote sensing from the slope of the line that is previously calibrated for different values of temperature.

7.3.2 Dielectric permittivity and medium density

In the case of surface remote sensing, the medium density profile is of interest. However, if the medium depth is large (glaciers, very dry rocks, planetary surfaces), density variations with depth may have a substantial effect on the complex dielectric permittivity. Design formulas have been proposed for densities in the range

$\rho \approx 0-2\ g/cm^3$. In this case, the dependence of the real part of the permittivity and the loss tangent on density is given as

$$\left.\begin{aligned} \varepsilon' &= (1+a\rho)^2 \\ \tan\delta &= b\rho \end{aligned}\right\} \tag{7.10}$$

where a and b are constants. These formulas have been used by [*Basharinov*, 1989] for a two-component mixture (air and solid materials). The relationship between the parameters a and b for some materials (in particular, for quartz, sand with $a = 0.41$ [*Schmulevitch*, 1971; *Basharinov*, 1989]) is given by

$$\varepsilon'_s = \varepsilon'_w W + (1+0.5\rho_s)^2 \tag{7.11}$$

where ε'_s and ε'_w are the permittivities for soil and water, respectively, ρ_s is the soil density (g/cm^3) and W is the weight of the volumetric part of water in soil.

As an example we give the relationship between the dielectric permittivity of moon soil $\left(f > 10^5 Hz\right)$ and the density ρ $\left(g/cm^3\right)$ [*Olhoeft*, 1975]:

$$\varepsilon' = (1.93 \pm 0.17\rho)^2 \tag{7.12}$$

A similar relationship was derived by [*Hanna*, 1982]. Thirty-two samples of talc, barite and dolomite were investigated at $f = 10GHz$. For 28 samples of clay and kaolin, the coefficient 1.93 in Eq. (7.12) should be replaced by the coefficient 2.13. According to the authors` opinion, this is connected with the non-negligible contribution of AL_2O_3, Fe_2O, TiO_2. A relationship of the form (7.12) is also valid for rocks e.g. gabbro (a granular igneous rock), silicates, etc. [*Olhoeft*, 1975].

Glaciers (especially Antarctic glaciers) and snow are media for which effect of density on permittivity is substantial. The permittivity of glaciers strongly depends on the density due to the high pressures. This is shown in Fig. 7.3 [*Bogorodsky*, 1975] in the absence of moisture.

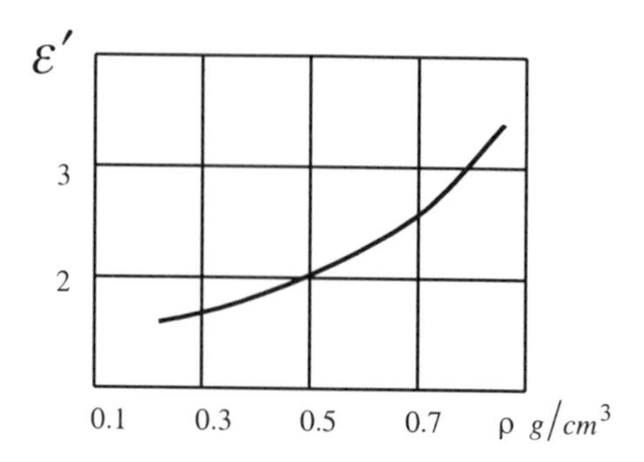

Fig. 7.3 ε' of glacier as a function of density ρ (g/cm^3).

The permittivity dependence on density indicates in Fig. 7.3 a non-linear (quadratic) relationship.

The dielectric permittivity of snow depends also on density. This is due to snow consolidation with age. The formulas for the permittivity of dry snow are given as follows [*Tiuri*, 1984]:

$$\left.\begin{aligned}\varepsilon' &= 1+1.17\rho+0.7\rho^2 \\ \varepsilon'' &= 0.52\varepsilon_i''\rho\left(1+1.19\rho\right)\end{aligned}\right\} \tag{7.13}$$

Measurements have been carried out at frequencies of 5.6 GHz and 12.6 GHz. The ε' measurements were done at $T = 253.14\ K\left(-20^0 C\right)$ and $\varepsilon_i'' = 8\cdot 10^{-4}$ [(marine ice at $T = 253.14\ K\left(-20^0 C\right)$)]. The ε' dependence with temperature and age has not been reported. The loss tangent relationship is approximated by

$$\tan\delta = 0.83\cdot 10^6\cdot\rho\frac{1+1.19\rho}{1+1.7\rho+0.7\rho^2}\left(\frac{1}{f}+1.23\cdot 10^{-14}\cdot f\right)\cdot e^{0.036T} \tag{7.14}$$

with the temperature T expressed in ^0C.

The permittivity of wet snow depends on snow density and moisture. For f = 500 to 1000 MHz it was found that [*Tiuri*, 1984]

$$\left.\begin{array}{r}\varepsilon' = 1 + 1.7\rho + 0.7\rho^2 + 8.7W + 70W^2 \\ \varepsilon'' = 0.9 fW\left(1 + 8.3W\right)\cdot 10^{-9}\end{array}\right\} \quad (7.15)$$

The monographic chart shown in Fig. 7.4 is based on Eq. (7.15). It interconnects W (humidity), ρ (density) and ε (dielectric permittivity) for $f = 1\ GHz$.

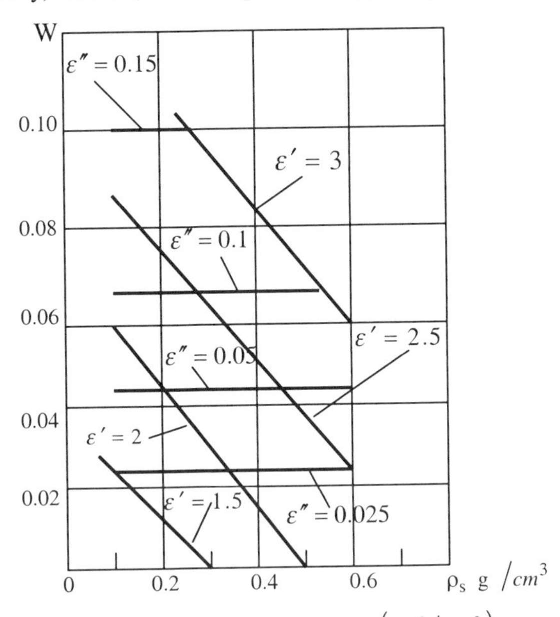

Fig. 7.4 Dielectric permittivity for wet snow: humidity $\left(cm^3/cm^3\right)$, density (g/cm^3)

The following formulas can be used for other frequencies:

$$\left.\begin{array}{r}\varepsilon'_w = 4.9 + \dfrac{82.8}{1 + \left(f / f_0\right)^2} \\ \varepsilon''_w = \dfrac{82.8\left(f / f_0\right)}{1 + \left(f / f_0\right)^2} \\ \varepsilon'' = 0.1W\left(1 + 8W^2\right)\varepsilon''_w \\ \varepsilon' = \varepsilon'_d + 0.1W\left(1 + 8W^2\right)\varepsilon'_w\end{array}\right\} \quad (7.16)$$

Here, ε'_w and ε''_w are respectively the real and the imaginary parts of the water complex dielectric permittivity, $f_0 = 8.84\,GHz$ and $\rho_s = \rho + W$; ρ_s, ρ being the densities of wet and dry snow, respectively.

The chart shown in Fig. 7.4 allows us to determine the permittivity of snow for known W and ρ. The same monographic chart allows determining magnitudes of W and ρ according to the known permittivity of snow. Numerical comparison of the results derived from formulas (7.15) (respectively, (7.14)) and the results derived from formulas (7.16) show that going from a frequency of $1\,GHz$ to higher frequencies (up to $10\,GHz$) the result in practice does not cause considerable changes.

7.3.3 Dielectric permittivity and salinity

Unlike other media, the dielectric permittivity of water and sea ice substantially depend on salinity. The connection between salinity s(g/l) and conductivity σ(mS/cm) is determined by [*Williams*, 1986]

$$s = 0.4665\sigma^{1.0878} \tag{7.17}$$

This relationship was derived from 109 samples of salt lakes in Australia. However, Eq. (7.17) cannot be used when $s < 3(g/l)$ and $s > 70(g/l)$. The dielectric permittivity dependence upon chloride concentration may be represented by a linear function for sea water and for a *NaCl* solution (see Figs 7.5 and 7.6).

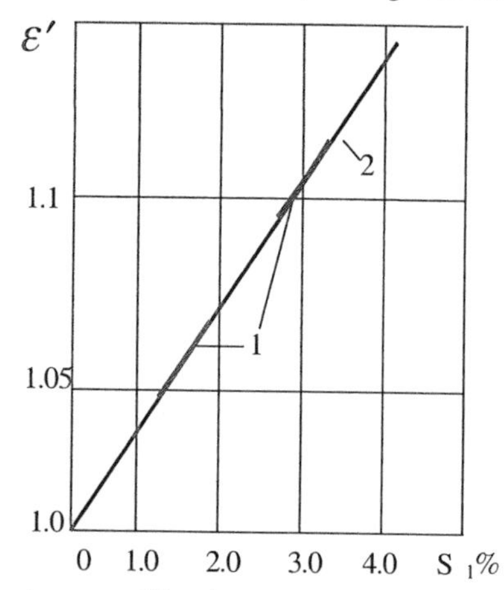

Fig. 7.5 ε' of marine water (1) and a water- NaCl mixture (2) as function of concentration of chlorides s_1 (%) at T = 24.5°C

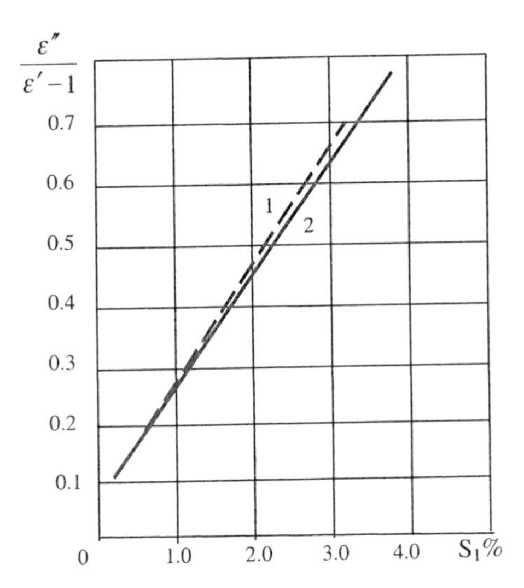

Fig. 7.6 $\frac{\varepsilon''}{\varepsilon'-1}$ of marine water (1) and a water-NaCl mixture (2) as function of concentration of chlorides $s_1(\%)$ at $T = 24.5^0$C

The dielectric permittivity of sea ice depends on salinity; also, on the fact that the sea water dielectric permittivity depends on salinity. The dielectric permittivity and the conductivity of sea ice as functions of salinity s(‰) (g/1000g)at $f = 3\ MHz$ are shown in Figs. 7.7 and 7.8 [*Wentworth*, 1964].

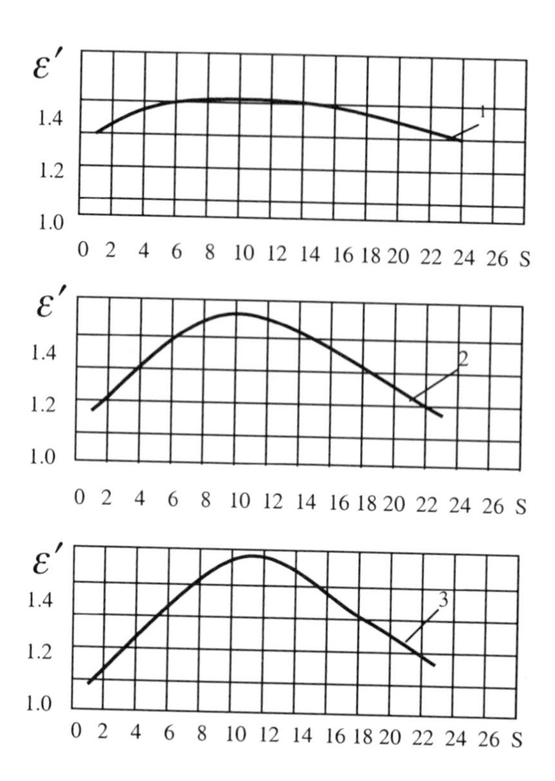

Fig. 7.7 Dependence of ε' of marine ice on salinity s(‰)
1) T= -30 ^{0}C 2) T= -20 ^{0}C 3) T= -10 ^{0}C

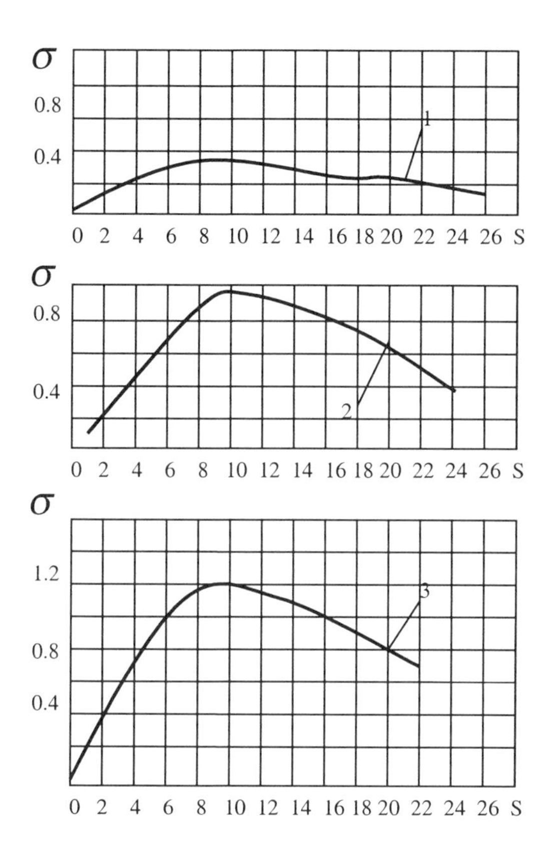

Fig. 7.8 Dependence of conductivity σ (mS/m) of marine ice on salinity s(‰)
1) $T=-30^{\circ}C$ 2) $T=-20^{\circ}C$ 3) $T=-10^{\circ}C$

The $\tan\delta$ (loss tangent) dependence on concentration of free ions H^{+} (pH) is shown in Fig. 7.9 [*Tiuri*, 1984] and can be approximated by

$$\tan\delta = \tan\delta_{d_0} + Ae^{-pH}\ ; \quad A = 6.37 \tag{7.18}$$

where $\tan\delta_{d_0}$ is the loss tangent of dry snow under normal conditions and

$$pH = -Log\left[H^+\right] \tag{7.18a}$$

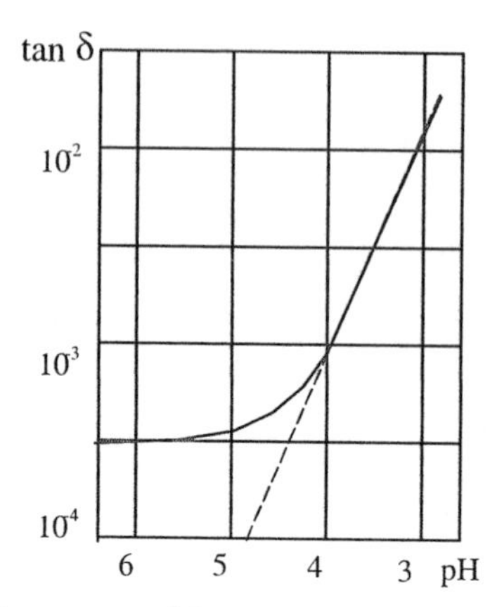

Fig. 7.9 The loss tangent of dry snow as a function of pH for f = 2 GHz

The effect of salinity on the dielectric permittivity of sand at $f = 20\ GHz$ is analyzed in [*Ilijn*, 1994]. Distilled water and different concentrations of NaCl (salting concentrations) were used for humidification of the samples. The dependence of the dielectric permittivity on humidity and temperature is shown in Fig. 7.10 [*Ilijn*, 1994].

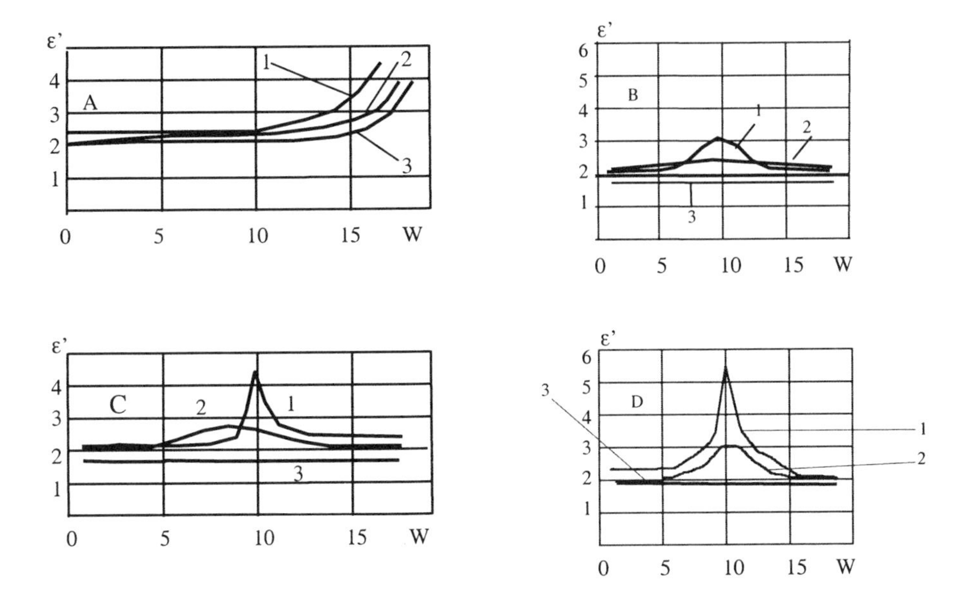

Fig. 7.10 ε' of sand as a function of humidity W(%), different salinities and temperatures at f=20GHz
1) s=0% 2) s=1.33% 3) s=8.5% A) T=5^0 C B) T = -5^{0}C C) T = -10^0 C D) T = -18^{0}C

ε' decreases when salinity increases independently of the temperature. The lower the temperature, the greater the decrease in the complex dielectric permittivity. The curves shown reach a maximum which is reached for humidity $W \approx 10\%$.

It has been observed [*Olhoeft*, 1975] that there exists a dependence of the complex dielectric permittivity upon the chemical composition of rocks. In [*Shuji*, 1992] the complex dielectric permittivity measurements were carried out for ice with acid impurities (*HCl*, HNO_3, H_2SO_4) at the frequency $f = 9.7\ GHz$. It appeared that ε' and ε'' are connected with the acid concentration by a linear relationship. The higher the acid concentration, the higher ε' and ε''.

7.3.4 Dielectric permittivity and temperature

The dielectric permittivity at the surface of most media depends slightly on temperature, except for water, ice, snow and frozen soils. Fresh water is a dipole-type dielectric and its dielectric permittivity can be calculated according Debye's formula [*Hallikainen*, 1977], viz.,

$$\varepsilon = \varepsilon'_\infty + \frac{\varepsilon'_s - \varepsilon'_\infty}{1 + i\omega\tau} \tag{7.19}$$

where $\varepsilon'_\infty = \lim_{\omega\to\infty} \varepsilon' \approx 5$; $\varepsilon_s = \lim_{\omega\to 0} \varepsilon'$ and τ is the relaxation time.

The dependence of ε'_s on temperature (within the range -8°C to +50°C) is given by [*Malmberg*, 1956]:

$$\varepsilon'_s = 87.74 - 0.4008\,T + 9.398 \cdot 10^{-4}\,T^2 + 1.410 \cdot 10^{-4}\,T^3 \tag{7.20}$$

(Sub-zero temperatures correspond to supercooled water).

The dependence of τ on temperature is shown in Fig. 7.14 [*Hallikainen,* 1977]. The dependence of the complex permittivity of fresh water on temperature can be plotted from Eqs (7.19) and (7.20) and using Fig. 7.11. Examples of such relationships are given in [*Hallikainen*, 1977] for $f = 4.7\ GHz$ and $10\ GHz$.

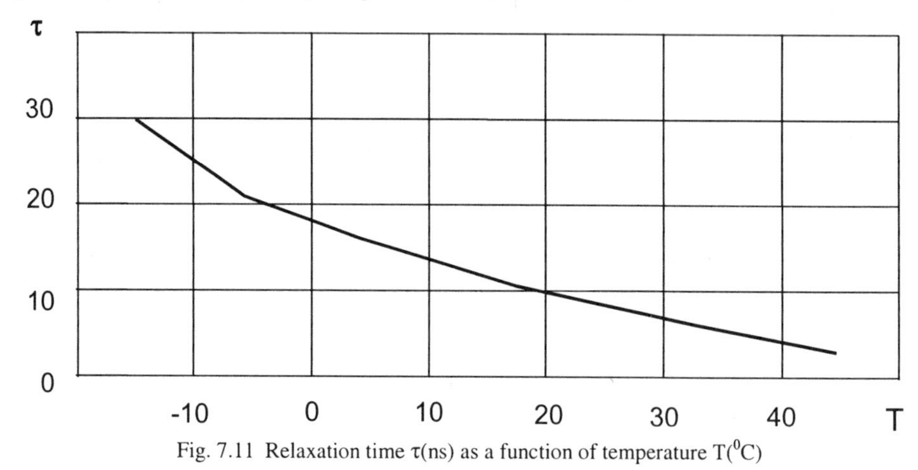

Fig. 7.11 Relaxation time τ(ns) as a function of temperature T(°C)

Salts increase the dielectric loss in water due to an increase of free charge carriers. For such materials, Eq. (7.19) is transformed into

$$\varepsilon = \varepsilon'_\infty + \frac{\varepsilon'_s - \varepsilon'_\infty}{1 + i\omega\tau_1} - i\frac{\sigma}{\varepsilon_0\omega} \tag{7.21}$$

where ε_0, σ and τ_1 are the vacuum permittivity, sea water conductivity and sea water relaxation time, respectively.

The proportion of liquid in ice decreases with decreasing temperatures. This results in a decrease of the dielectric permittivity. Complicated relationships between the permittivity and temperature take place for sea ice. This is due to the fact that parts of salts precipitate. This results in a change in salinity within the chemical composition.

The permittivity of fresh water ice for $f > 1\,MHz$ hardly depends on temperature ($\varepsilon' \approx 3.1884 + 0.00091T$ for $f = 2-10\,GHz$) [*Matzler*, 1987]. Furthermore, for $f > 10^6\,Hz$, ε' and $\tan\delta$ do not depend on frequency.

7.4 Interrelations between dielectric and physical characteristics

Our aim in this section is to discuss quantitatively the dependence of the electric permittivity on physical characteristics, such as frequency, temperature, moisture, salinity, etc. The discussion is carried out for different media.

7.4.1 Water

For fresh water, the dependence of permittivity on frequency and temperature is described in Eqs (7.19) and (7.20).

For sea water, the dependence of permittivity on frequency and temperature is described in Eqs (7.20) and (7.21). It should be noted that the relaxation time for sea water depends on temperature and salinity. In this case, Eq. (7.21) can be replaced by [*Stogryn*, 1971]

$$\varepsilon = a(s)\varepsilon(T) \tag{7.22}$$

where $a(s)$ is a coefficient that takes the salinity into account.

The relationship between the sea water relaxation time τ_1 and salinity becomes

$$\tau_1 = \tau(T)b_N(s) \tag{7.23}$$

where τ is the fresh water relaxation time and $b_N(s)$ depends on salinity.

Typical relationships of ε' and τ_1 as functions of the percentage of salting concentration s are shown in Fig. 7.11 [*Lane*, 1953]. ε'_m and ε'_f are the dielectric permittivities of sea (marine) and fresh water, respectively.

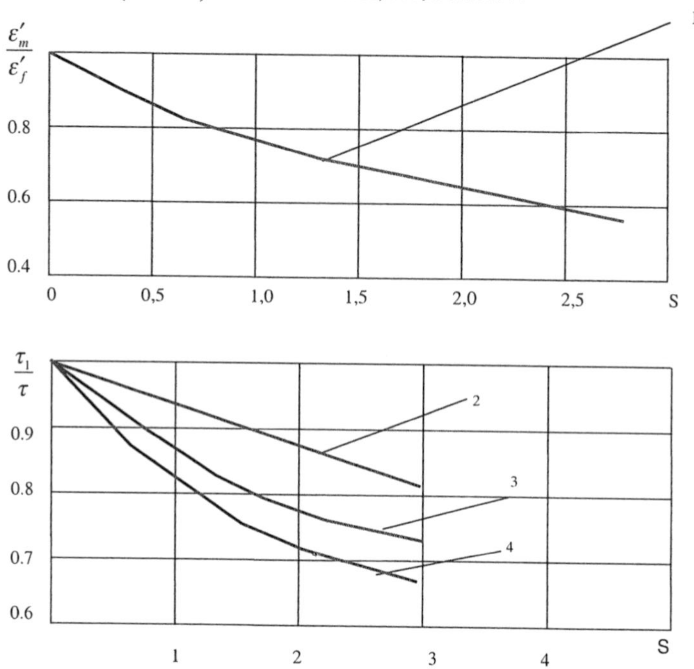

Fig. 7.12 $\varepsilon'_m / \varepsilon'_f$ and τ_1 / τ of NaCl solution as a function of salinity s(%) for different temperatures
1) T = 0^0C - 40^0C 2) T = 30^0C 3) T= 20^0C; 4. T = 0^0C

Approximations of the relationships shown in Fig. 7.12 yield

$$\left. \begin{aligned} a(s) &= 1.255^{-N} \\ b_N(s)\big|_{T=0^0 C} &= 1.14^{-N} \end{aligned} \right\} \tag{7.23a}$$

where

$$N = s \cdot \left(1.707 \cdot 10^{-2} + 1.205 \cdot 10^{-5} \cdot s + 4.058 \cdot 10^{-9} \cdot s^2\right) \tag{7.24}$$

and the salinity s is given in ‰.

In this way, the dependence of the sea water dielectric permittivity upon frequency, temperature and salinity can be determined from Eqs (7.21) to (7.24).

7.4.2 Ice

Sea ice has a complicated structure. It consists of fresh water ice crystals, a solution of frozen and non-frozen cells, supercooled water, air bubbles and other impurities. Sea water contains various salts. The freezing temperatures of the constituent mixtures are different. Therefore, the chemical composition varies with temperature.

The following four different models of the sea water permittivity are based on the work of [*Taylor*, 1965]:

Model 1

Electric field $\vec{E}$ is parallel to the principal axis of impurities:

$$\varepsilon_{mi} = \varepsilon_{fi}\left[1 - V_b\left(1 - \frac{\varepsilon_b}{\varepsilon_{fi}}\right)\right] \tag{7.25}$$

Model 2

Electric field $\vec{E}$ is perpendicular to the principal axis of impurities:

$$\varepsilon_{mi} = \varepsilon_{fi}\left[1 - 2V_b\left(1 - \frac{\varepsilon_b}{\varepsilon_{fi}}\right)\frac{\varepsilon_{mi}}{\varepsilon_{mi} + \varepsilon_b}\right] \tag{7.26}$$

Model 3

Random orientation of impurities:

$$\varepsilon_{mi} = \varepsilon_{fi}\left[1 - \frac{1}{3}V_b\left(1 - \frac{\varepsilon_b}{\varepsilon_{fi}}\right)\left(1 + \frac{4\varepsilon_{mi}}{\varepsilon_{mi} + \varepsilon_b}\right)\right] \tag{7.27}$$

Model 4

Impurities have a spherical form:

$$\varepsilon_{mi} = \varepsilon_{fi}\left(1 - 3V_b\frac{\varepsilon_{mi}}{\varepsilon_{fi}}\cdot\frac{\varepsilon_{fi} - \varepsilon_b}{2\varepsilon_{mi} + \varepsilon_b}\right) \tag{7.28}$$

In these expressions, ε_{mi}, ε_{fi} and ε_b are the dielectric permittivities of sea ice, fresh water ice and chemical composition, respectively. V_b is the ratio of the composition volume to the total volume of sea ice. Its magnitude has been calculated by [*Assur*, 1960] and [*Frankestein*, 1965]:

$$V_b = \begin{cases} s\left(-\dfrac{43.795}{T}+1.189\right), & if \quad -22.9^0 C \le T \le -8.2^0 C \\ s\left(-\dfrac{45.917}{T}+0.930\right), & if \quad -8.2^0 C \le T \le -2.6^0 C \\ s\left(-\dfrac{52.560}{T}-2.280\right), & if \quad -2.6^0 C \le T \le -0.5^0 C \\ s\left(\dfrac{1.003}{T^2}-\dfrac{48.684}{T}+4.092\right), & if \quad -0.5^0 C \le T \le -0.1^0 C \end{cases} \tag{7.29}$$

A plot of V_b as a function of temperature for different salinities is shown in Fig. 7.13, which is based on the work by [*Hallikainen*, 1977].

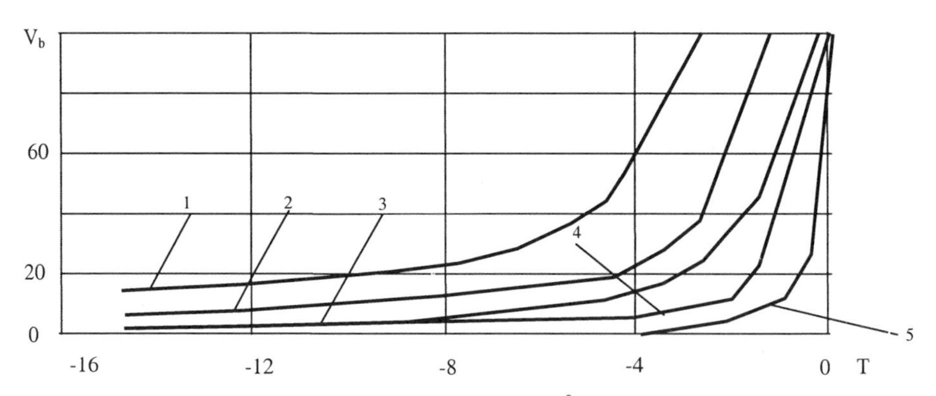

Fig. 7.13 V_b(%) as a function of temperature (T^0C) and different salinities.
1. $s = 4‰$; 2. $s = 2‰$; 3. $s = 1‰$; 4. $s = 0.5‰$; 5. $s = 0.25‰$.

Calculations of the dielectric permittivity of sea ice are carried out in [*Hallikainen*, 1977]. The results are in a good agreement with experiment [*Vant*, 1978; *Hallikainen*, 1977; *Hoekstra,* 1971].

The dielectric permittivity of sea ice depends on temperature in a complicated way. This is due to the fact that the freezing of different components of sea water takes place at different temperatures. The phase diagram of sea ice (s = 1‰) is shown in Fig. 7.14 [*Assur*, 1960], where r is the weight ratio of salts in the composition.

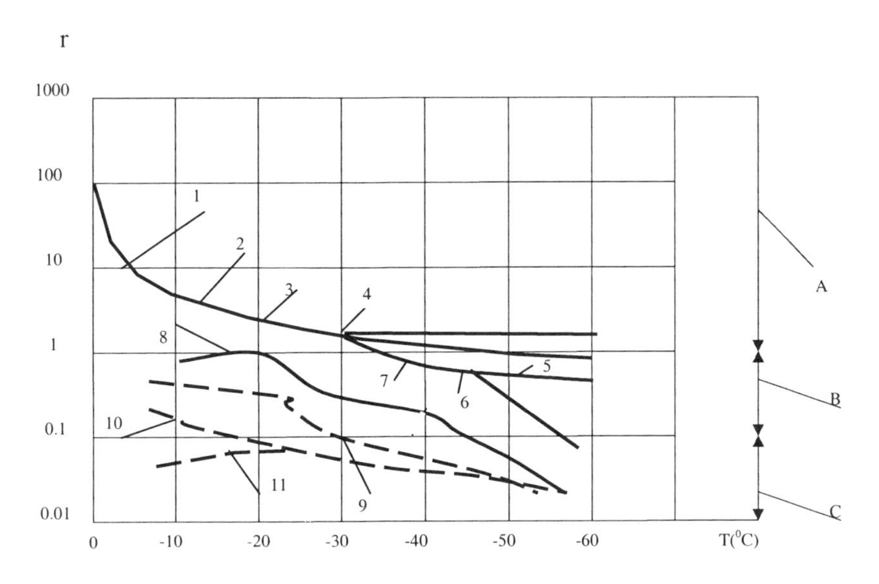

Fig. 7.14 Phase diagram r(g/Kg) of marine ice sample for s = 1‰. A. Ice; B. Salt; C. Brine 1.$CaCO_3$; 2. $Na_2SO_4 \cdot 10H_2O$; 3. $MgCl_2 \cdot 6H_2O$; 4. $NaCl \cdot 2H_2O$; 5. $MgCl_2$; 6. $MgCl_2 \cdot 2H_2O$; 7. KCl; 8. H_2O; 9. Cl^-; 10. Na^+; 11. SO^{-+} $Mg^{++} + Ca^{++} + K^+$

The relations in Fig. 7.14 are well approximated by [*Stogryn*, 1971; *Frankestein*, 1965]

$$r = 508.18 + 14.535T + 0.2018T^2 \quad if \quad -43.2^0 C \le T \le -36.8^0 C \tag{7.30}$$

$$r = 242.94 + 1.5299T + 0.0429T^2 \quad if \quad -36.8^0 C \le T \le -22.9^0 C \tag{7.31}$$

$$r = 57.041 - 9.929T - 0.16204T^2 - 0.002396T^3 \quad if \quad -22.9^0 C \le T \le -8.2^0 C \tag{7.32}$$

$$r = 1.725 - 18.756T - 0.3964T^2 \quad if \quad -8.2^0 C \le T \le -2.0^0 C \tag{7.33}$$

7.4.3 Snow

The permittivity of dry ice can be determined by using equations for a two-component mixture. The mixture consists of ice particles (impurities) and air (the main medium). The Maxwell-Garnet formula [*Bojarsky*, 1991], leading to the permittivity

$$\varepsilon_m = \varepsilon + \frac{3\mu(\varepsilon_{ins} - \varepsilon)\cdot \dfrac{\varepsilon}{\varepsilon_{ins} + 2\varepsilon}}{1 - \mu \dfrac{\varepsilon_{ins} - \varepsilon}{\varepsilon_{ins} + 2\varepsilon}} \tag{7.34}$$

shows a good agreement with the experiment. Here, ε_m is the dielectric permittivity of a mixture, ε is the complex dielectric permittivity of the main medium, ε_{ins} is the permittivity of the impurities of the volume part taken by the impurities.

For snow ($\varepsilon = 1$, $\varepsilon_{ins} = \varepsilon_i$, $\varepsilon_m = \varepsilon_s$), Eq. (7.34) takes on the following form:

$$\varepsilon_s = 1 + \frac{3\mu\left(\varepsilon_{fi} + 2\right)}{\left(\varepsilon_{fi} + 2\right) - \mu\left(\varepsilon_{fi} - 1\right)} \tag{7.35}$$

Where ε_{fi} is the permittivity of fresh water. The parameter μ is determined by the densities of snow, ρ_s and ice $\rho_i \approx 9.17$; specifically,

$$\mu = \frac{\rho_s}{\rho_i} \tag{7.36}$$

The permittivity of snow can then be calculated from Eqs (7.19), (7.35) and (7.36).

Another expression for the permittivity of snow was proposed by [*Wobxhall*, 1977] and [*Bojarsky*, 1991]:

$$\varepsilon_s = \frac{2\varepsilon_s + \varepsilon_i}{2\varepsilon_s + \varepsilon_i - v\left(\varepsilon_i + 2\right)\left(\dfrac{\lambda}{2\pi}\right)^2} \tag{7.37}$$

where ε_i is the permittivity of ice. Here, the parameter v depends on snow density and on parameters characterizing the ice grains. This formula is in a good agreement with experiments carried out for frequencies up to $20\ GHz$.

The permittivity of sea snow can be determined by using Eqs (7.16) and (7.19) [*Hallikainen*, 1986]. The result is in a good agreement with experiments carried out for frequencies 3 to 37 GHz.

The permittivity of snow can also be determined with the use of equations for a three-component mixture (water, air, ice). Comparison of calculated results with experiments does not show a satisfactory agreement so far [*Sihlova*, 1988].

7.4.4 Soil

The permittivity of soil can be determined with the use of Eq. (7.7) including the dependence upon temperature, moisture and salinity.

The permittivity of soil can be written as [*Dmitriev*, 1990; *Hanai*, 1961; *Sherman*, 1968]

$$\frac{\varepsilon_m - \varepsilon_d}{\varepsilon - \varepsilon_d}\left(\frac{\varepsilon}{\varepsilon_m}\right)^{\frac{1}{3}} = 1 - \Phi \tag{7.38}$$

where ε_m, ε_d and ε are respectively the permittivities of the mixture, the dispersed phase and the substance in which the dispersed phase is distributed; Φ is the volume part of the dispersed phase. The parameters ε_d, ε and Φ are considered to be known.

The equations for the permittivity of the mixture can be derived from the expressions

$$\frac{(\varepsilon'_m - \varepsilon'_d)^2 + (\varepsilon''_m - \varepsilon''_d)^2}{(\varepsilon' - \varepsilon'_d)^2 + (\varepsilon'' - \varepsilon''_d)^2} \cdot \frac{\varepsilon'^2 + \varepsilon''^2}{\varepsilon'^2_m + \varepsilon''^2_m} = (1 - \Phi)^2 \tag{7.39}$$

$$\frac{\dfrac{\varepsilon''_m}{\varepsilon'_m} - \dfrac{\varepsilon''}{\varepsilon'}}{1 + \dfrac{\varepsilon''_m}{\varepsilon'_m} \cdot \dfrac{\varepsilon''}{\varepsilon'}} = \frac{X(3 - X^2)}{1 - 3X^2} \tag{7.40}$$

where

$$X = \frac{\dfrac{\varepsilon''_m - \varepsilon''_d}{\varepsilon'_m - \varepsilon'_d} - \dfrac{\varepsilon'' - \varepsilon''_d}{\varepsilon' - \varepsilon'_d}}{1 + \dfrac{\varepsilon''_m - \varepsilon''}{\varepsilon'_m - \varepsilon'_d} \cdot \dfrac{\varepsilon'' - \varepsilon''_d}{\varepsilon' - \varepsilon'_d}} \tag{7.41}$$

Dielectric permittivity calculations have been carried out for soil with the parameters $\varepsilon'_d = 3.5$; $\varepsilon'_{H_2O} = 79$; $\varepsilon''_d \approx 0$; $\sigma_{H_2O} = 0.01 - 50 mS$ [*Dmitriev*, 1990].

Wet soil can be considered as a mixture. This mixture consists of an air medium and quartz impurities covered with a film. In this case, we can use Eq. (7.37). In this formula, the following parameters should be replaced: permittivity of snow by permittivity of soil ($\varepsilon_s \Rightarrow \varepsilon_g$); permittivity of ice by permittivity of a water-quartz mixture ($\varepsilon_i \Rightarrow \varepsilon_{wq}$). The mixture complex dielectric permittivity is calculated with the use of the formulas for a two-phase medium [*Sihlova*, 1988; *Tinga*, 1973; *Taylor*, 1965].

Frozen soil can also be considered as a mixture. This mixture consists of air medium and of spherical particles of quartz. The air medium contains also spherical particles of ice. Quartz particles are covered with a water film. The film thickness decreases with decreasing soil temperature [*Hallikainen*, 1977]. The permittivity of soil can be calculated from

$$\varepsilon_g = \frac{\varepsilon_1 \varepsilon_2}{\varepsilon_1 \varepsilon_2 - \mu \varepsilon_2 (\varepsilon_2 + 2) \left(\dfrac{\lambda}{2\pi} \right)^2 - \nu \varepsilon_1 (\varepsilon_1 + 2) \left(\dfrac{\lambda}{2\pi} \right)^2} \tag{7.42}$$

where $\varepsilon_1 = 2\varepsilon_g + \varepsilon_{wq}$; $\varepsilon_2 = 2\varepsilon_g + \varepsilon_i$; μ, ν are parameters dependent on the density and moisture of the soil and its components [*Bojarsky*, 1995].

The dependence of permittivity upon frequency is taken into account in the formula for the permittivity of the mixture. The permittivities in Eq. (7.42) are considered to be independent of frequency [*Bojarsky*, 1995].

A method for calculating earth and soil permittivities in the microwave range is shown in [*Podkovkov*, 1990]. That method takes the effect of moisture and soil salinity into account. The dependence of permittivity upon frequency can be calculated with the use of the expression

$$\varepsilon(\omega) = \varepsilon_\infty + \frac{\varepsilon_0 - \varepsilon_\infty}{1 + (i\omega\tau)^{1-\alpha}} \tag{7.43}$$

where $\varepsilon_\infty = \lim\limits_{\omega\to\infty} \varepsilon$; $\varepsilon_0 = \lim\limits_{\omega\to 0} \varepsilon$, τ is relaxation time and α is a distribution parameter.

For frozen clay, Eq. (7.43) is reduced to the Debye's formula with $\alpha = 0$ and is in good agreement with experimental data.

7.4.5 Vegetation

The permittivity of vegetation, ε_v, can be calculated by using the equations for a two-component mixture i.e. cellulose (chief part of the cell walls of plants) and air:

$$\varepsilon_v = \frac{(1+qp)\varepsilon_{ve} + (1-p)q}{\varepsilon_{ve}(1-p) + q + p} \tag{7.44}$$

Here, ε_{ve} is the permittivity of the vegetation part (excluding air), q is the shape index and p the coefficient of filling.

The shape index and coefficient of filling (for most plants) have the following ranges: $10 \le q \le 20$, $p \approx 10^{-2} - 10^{-3}$. The complex dielectric permittivity is calculated by

$$\varepsilon_{ve} = 3 + (\varepsilon'_w - 3)\cdot W_v^2 + i\varepsilon''_w W_v^2 \tag{7.45}$$

where $\varepsilon'_w = \mathrm{Re}\,\varepsilon_{H_2O}$; $\varepsilon''_w = \mathrm{Im}\,\varepsilon_{H_2O}$ and W_v is the volume moisture content.

Values for vegetation elements are given in [*Ulaby*, 1987] and [*Shutko*, 1986].The permittivity of trees (spruces) can be calculated by the formula for a three-component mixture (cellulose, air, water) [*Tinga*, 1973]. The calculation results are shown in Fig. 7.15. As an example, we mention that $W = 0.3$ corresponds to 10% of the tree volume.

The calculation has been carried out for $f = 2450\ MHz$, $T = 20\ ^0C$ and a relative water density equal to 0.4. The permittivity of water was reported in [*Windle J.J.*, 1954] for 3 GHz and equals $\varepsilon = 77.2 - i\cdot 13.1$. (We note that though the calculation has been carried out for $f = 2450\ MHz$, it is valid for frequencies up to 10000 MHz .)

Similar relations as shown in Fig. 7.15 have been derived for leaves of grains [*Ulaby*, 1987].

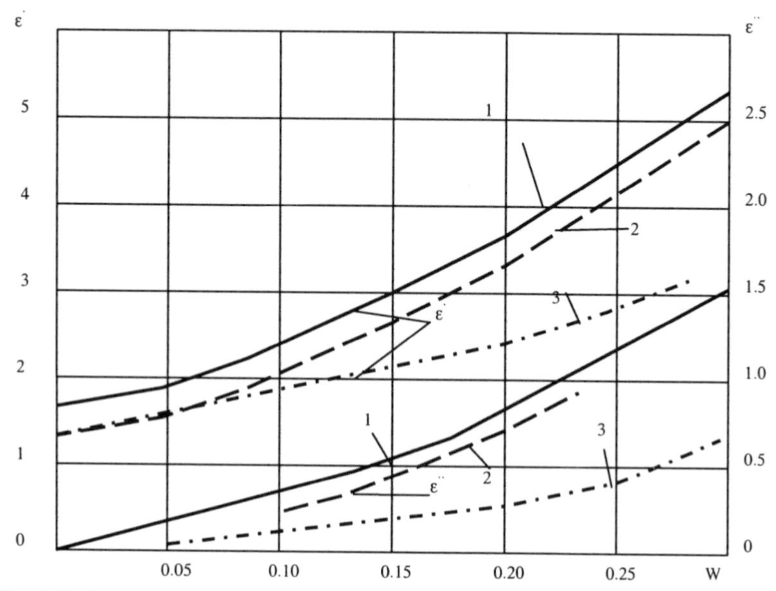

Fig. 7.15 Dielectric permittivity of trees (spurces) as a function of moisture content W
f = 2450 MHz, T = 20^{0}C, • — • – • = experiment.
1) theory (thin con-focal ellipsoidal small water drops and ellipsoidal air holes);
2) experiment and 3) theory (spherical inclusions of water and ellipsoidal air holes)

7.5 Conclusions and applications

In this chapter, we have discussed the relations between electrodynamic and electrophysical characteristics of layered media. The electrodynamic characteristics are determined by the conditions of scattering of sensing radar signals. However, we know that the electrical and physical properties are those associated with the medium as a physical object. The number of electrodynamic characteristic properties available may be limited (e.g., they may include the scattering cross-section of an object, reflection coefficients, the scattering matrix elements, etc.), while the number of electrical and physical characteristic properties may be very large and, furthermore, dependent on the type of a sensed surface. Therefore, in practice, a certain set of the main physical and chemical characteristics is to be chosen for each surface and then this set is to be processed. For bare soil, for example, these characteristics are humidity, density, salinity, etc., and for sea ice the salinity, content of liquid phase, elasticity, etc.

Taking into account that the electrical properties of a non-magnetic medium are determined first of all by its complex permittivity, primary attention is paid to the relationship of this quantity to the physical characteristics of this medium. Electromagnetic characteristics of a sensed object are also determined to a great extent by the permittivity of this object.

As was shown, in some cases the problem of determination of the aforementioned relations was solved sufficiently easily and was confirmed by experimental investigations. For example, the radio wave reflection coefficient depends linearly on the soil humidity within the frequency range f = 2.75–7.25 GHz. A similar linear relation is observed between the real part of the permittivity and the humidity of peat. However, simple relations are rather particularities than a general rule. The same relation for frozen sand is very complicated in character and, furthermore, it depends on the frost penetration temperature. In a general case, it is impossible to derive analytic dependencies between the permittivity of the surface and its moisture content. As a consequence, empirical relations are used for vegetation covers, agricultural crops, trees and other objects. But any empirical relation is determined to a great extent by particular conditions of measurements. This makes its application difficult and allows us to use such relations only for "order of magnitude" approximations.

Similar electrical dependencies are used for the determination of the relation between the permittivity and the medium density, especially if the medium density changes with depth. This relates to solid media, such as mountains, rocks, glaciers and also to snow. Our investigations show that, for snow, this dependence is determined also by the impact of the temperature and by the age of deposited dry snow. For wet snow, the permittivity is determined both by the snow density and its moisture content.

For the investigation of the properties of sea water and sea ice, salinity plays an important role. In some cases, a linear relation between the permittivity and the surface salinity may be observed. But in most cases, this relation is much more complicated and usually is empirical in character. Furthermore, for sea ice, this relation to a great extent depends on temperature.

The permittivity of most surfaces depends weakly on temperature, except for water, ice, snow and frozen soils. For fresh water, the permittivity depends on frequency and temperature; for sea water, it depends, in addition, on salinity. All these relations are generally empirical.

Sea ice is a complicated medium composed of inclusions that are different in their structure. Therefore, its properties are to a great extent determined by temperature. There are four models of sea ice, dictated by the orientation of its inclusions relative to the direction of the incident electric field vector, i.e., parallel orientation, perpendicular orientation, random orientation and orientation in the form of spheres. For each model, there are available analytic dependencies of sea ice permittivity on the volumes of the inclusions, which, in turn, are determined by temperature. In general, dependence of sea ice permittivity on temperature is very complicated in character.

Snow permittivity may be determined by means of application of the equations for a two-component mixture consisting of ice particles and air. These equations include the permittivities of the mixture components and parameters depending on the snow density and the structure of ice particles. Application of these equations show a good agreement with experiment. However, consideration of snow as a three-component mixture and derivation of the respective equations does not result in such an agreement.

Wet soil may be considered as a two-component mixture with the use of the above-mentioned equations. For soil covered with vegetation, an analytic determination of the permittivity may be established using the equations describing a two-component mixture consisting of cellulose and air. But these equations include coefficients that may be determined only empirically. In addition, the permittivity depends on the moisture content in vegetation; this dependence is close to a linear one, as was mentioned above.

Thus, summarizing the results in this chapter, we may state that the determination of relations between physical, chemical, mechanical and other properties of sensed objects is based first of all on experimental results and that their dependencies are expressed in the form of empirical relations. If, in a number of cases, it is possible to derive analytic expressions for the aforementioned relations, then nearly in all these cases these analytic expressions include coefficients that can only be determined empirically.

CHAPTER 8

Reflection of Electromagnetic Waves from Non-Uniform Layered Structures

8.1 Introduction

Reflection of electromagnetic waves from layered structures under different polarization conditions is studied. The main medium electrodynamic characteristic property taken into account is the electric permittivity. The analysis is performed using either a deterministic or a probabilistic (stochastic) approach. Various permittivity profiles are chosen: linear, exponential and polynomial. In the case that the permittivity has a random fluctuating part, a stochastic approach leading to an integral equation is used to determine the ensemble-averaged reflection coefficient and the average power.

8.2 Deterministic approach

8.2.1 Multi-layered structure with an exponential permittivity profile

- **Arbitrary angle of incidence**

Scattering of a plane electromagnetic wave from a medium with two dielectric layers (media II and III in Fig. 8.1) is studied here for various relative permittivity profiles.

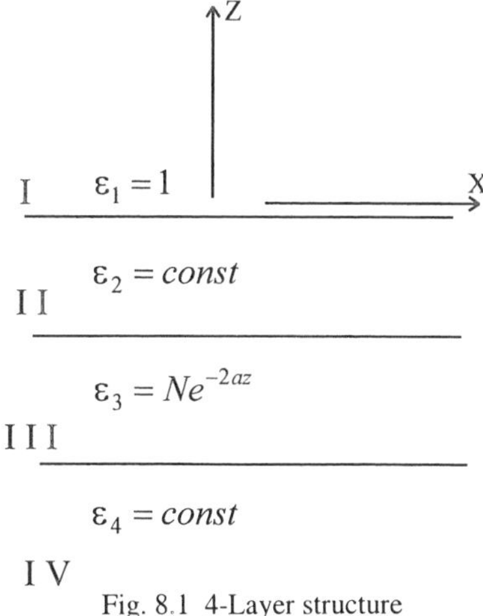

Fig. 8.1 4-Layer structure

Medium II is homogeneous (its dielectric permittivity is constant and equal to ε_2). In medium III the permittivity changes exponentially according to

$$\varepsilon_3 = Ne^{-2az} \tag{8.1}$$

At the top border of the homogeneous layer $\varepsilon_3 = \varepsilon_2$ and at the bottom border, $\varepsilon_3 = \varepsilon_4$. Above layer II there is free space (medium I) with $\varepsilon_1 = 1$ and beneath layer III a homogeneous dielectric (medium IV) with $\varepsilon = \varepsilon_4$, is located. (The origin of the z coordinate is on the border between media II and III). The change in the relative permittivity with depth is shown in Fig. 8.2.

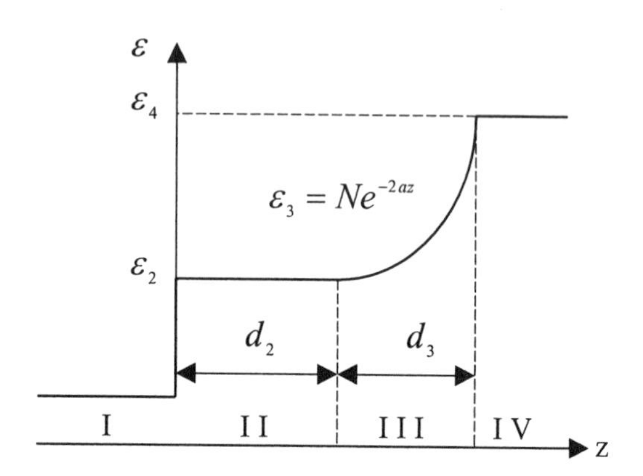

Fig. 8.2 Relative permittivity profile

The parameters N and a, included in Eq. (8.1), satisfy the relations

$$\left.\begin{aligned} N &= \varepsilon_2\left(\frac{\varepsilon_2}{\varepsilon_4}\right)^{\frac{d_2}{d_3}} \\ a &= \frac{1}{2d_3}\ln\left(\frac{\varepsilon_4}{\varepsilon_2}\right) \end{aligned}\right\} \tag{8.2}$$

The reflection coefficient of a plane electromagnetic wave incident from medium I at an angle β will be determined later on.

Within the limits of this model various types of grounds (permeated with moisture or covered by vegetation), sea surfaces (fresh and/or covered with continental ice) and also clouds (with drops of different types and sizes) are being considered.

- **Horizontal polarization**

In the case of horizontal polarization and monochromatic fields, Maxwell's equations in all media give rise to the Helmholtz equation

$$\Delta E_s + k_0^2 \varepsilon_s E_s = 0 \tag{8.3}$$

where E_s is the y-component of the electric field intensity $\vec{E}$ in medium s (s=1 (I), 2 (II), 3 (III), 4 (IV)) (The coordinate axes are shown in Fig. 8.1). The condition of validity of Eq. (8.3) is derived from Maxwell's equations [*Jackson*, 1975]:

$$-\vec{\nabla}\left\{\frac{\vec{\nabla}\varepsilon_0\varepsilon_s(z)\cdot\vec{E}}{\varepsilon_0\varepsilon_s(z)}\right\} \cong 0 \tag{8.3a}$$

The incident plane wave is written as

$$E_{inc} = e^{ik_0(x\sin\beta - z\cos\beta)} \tag{8.4}$$

The solution to the Helmholtz Eq. (8.3), E_s is assumed to have the form

$$E_s = U_s(z)e^{ik_0 x\sin\beta} \tag{8.5}$$

where $U_s(z)$ is the solution to the ordinary differential equation

$$\frac{d^2U}{dz^2} + k_0^2\left(\varepsilon_s - \sin^2\beta\right) = 0 \tag{8.6}$$

For homogeneous media I, II and IV, Eq. (8.6) leads to the following expressions:

In medium I:

$$E_1 = e^{ik_0 x\sin\beta}\left(e^{-ik_0 z\cos\beta} + \mathrm{R}e^{+ik_0 z\cos\beta}\right) \tag{8.7}$$

In medium II:

$$E_2 = e^{ik_0 x \sin\beta}\left(C_1 e^{-ik_0 z\sqrt{\varepsilon_2 - \sin^2\beta}} + C_2 e^{+ik_0 z\sqrt{\varepsilon_2 - \sin^2\beta}} \right) \tag{8.8}$$

In medium IV:`

$$E_4 = e^{ik_0 x \sin\beta} T\, e^{-ik_0 z\sqrt{\varepsilon_4 - \sin^2\beta}} \tag{8.9}$$

In medium III Eq. (8.6) becomes

$$\frac{d^2 U_3}{dz^2} + k_0^2\left(Ne^{-2az} - \sin^2\beta\right)U_3 = 0 \tag{8.10}$$

Introducing the change of variable $\eta = e^{-az}$, Eq. (8.10) is transformed into the Bessel equation

$$\eta^2 \frac{d^2 U_3}{d\eta^2} + \eta \frac{dU_3}{d\eta} + k_0^2\left(\frac{N}{a^2}\eta^2 - \frac{\sin^2\beta}{a^2} \right)U_3 = 0 \tag{8.11}$$

which has the following general solution:

$$U_3(z) = C_3 J_\nu\left(\frac{k_0\sqrt{N}}{a} e^{-az} \right) + C_4 N_\nu\left(\frac{k_0\sqrt{N}}{a} e^{-az} \right) \tag{8.12}$$

Here, $\nu = (k_0 \sin\beta)/a$ and $J_\nu(\chi)$, $N_\nu(\chi)$ are the Bessel and Neumann functions of the order ν, respectively.

For E_3 we write

$$E_3 = e^{ik_0 x \sin\beta}\left[C_3 J_\nu\left(\frac{k_0\sqrt{N}}{a} e^{-az} \right) + C_4 N_\nu\left(\frac{k_0\sqrt{N}}{a} e^{-az} \right) \right] \tag{8.13}$$

The continuity of the tangential components of the electric and magnetic fields at the boundaries, leads to the following relationship for the coefficients R, C_1, C_2, C_3, C_4 and T:

$$\begin{Vmatrix} -1 & 1 & 1 & 0 & 0 & 0 \\ 1 & \frac{\sqrt{\varepsilon_2 - \sin^2\beta}}{\cos\beta} & -\frac{\sqrt{\varepsilon_2 - \sin^2\beta}}{\cos\beta} & 0 & 0 & 0 \\ 0 & \frac{1}{\alpha} & \alpha & -\delta_1 & -\delta_2 & 0 \\ 0 & \frac{\sqrt{\varepsilon_2 - \sin^2\beta}}{\alpha} & -\alpha\sqrt{\varepsilon_2 - \sin^2\beta} & \delta_3 & \delta_4 & 0 \\ 0 & 0 & 0 & \delta_5 & \delta_6 & -g \\ 0 & 0 & 0 & \delta_7 & \delta_8 & g\sqrt{\varepsilon_4 - \sin^2\beta} \end{Vmatrix} \cdot \begin{Vmatrix} R \\ C_1 \\ C_2 \\ C_3 \\ C_4 \\ T \end{Vmatrix} = \begin{Vmatrix} 1 \\ 1 \\ 0 \\ 0 \\ 0 \\ 0 \end{Vmatrix} \tag{8.14}$$

The parameters in Eq. (8.14) are explained in Appendix C.

For the reflection coefficient *R*, the following expression is obtained, for $\varepsilon_2 = \varepsilon_4$ (Eq. 8.14).

$$R_{21} = \frac{\sqrt{\varepsilon_2 - \sin^2\beta} - \cos\beta}{\sqrt{\varepsilon_2 - \sin^2\beta} + \cos\beta} \tag{8.15}$$

Here,

$$R = \frac{R_{21} + We^{-2ik_0 d_2 \sqrt{\varepsilon_2 - \sin^2\beta}}}{1 + R_{21} We^{-2ik_0 d_2 \sqrt{\varepsilon_2 - \sin^2\beta}}} \tag{8.16}$$

is the Fresnel reflection coefficient. *W* in Eq. (8.16) is given by

$$W = \frac{\begin{vmatrix} 1 & -\delta_1 & -\delta & 0 \\ \sqrt{\varepsilon_2 - \sin^2\beta} & \delta_3 & \delta_4 & 0 \\ 0 & \delta_5 & \delta_6 & -1 \\ 0 & \delta_7 & \delta_8 & \sqrt{\varepsilon_4 - \sin^2\beta} \end{vmatrix}}{\begin{vmatrix} 1 & -\delta_1 & -\delta & 0 \\ -\sqrt{\varepsilon_2 - \sin^2\beta} & \delta_3 & \delta_4 & 0 \\ 0 & \delta_5 & \delta_6 & -1 \\ 0 & \delta_7 & \delta_8 & \sqrt{\varepsilon_4 - \sin^2\beta} \end{vmatrix}} \tag{8.17}$$

The value of W characterizes internal reflections from the boundaries of medium III. Eq. (8.17) cannot be simplified further and requires, therefore, a numerical approach for completion of a direct analysis. It should be pointed out that Eq. (8.15) must also hold when reflections from a homogeneous layer are described.

- **Vertical incidence**

At vertical incidence $(\beta = 0)$, the expression for W is simplified to

$$W = \frac{\left[J_0(\tilde{X}) + iJ_1(\tilde{X})\right]\left[N_0(\tilde{Y}) + iN_1(\tilde{Y})\right] - \left[J_0(\tilde{Y}) + iJ_1(\tilde{Y})\right]\left[N_0(\tilde{X}) + iN_1(\tilde{X})\right]}{\left[J_0(\tilde{X}) - iJ_1(\tilde{X})\right]\left[N_0(\tilde{Y}) + iN_1(\tilde{Y})\right] - \left[J_0(\tilde{Y}) + iJ_1(\tilde{Y})\right]\left[N_0(\tilde{X}) - iN_1(\tilde{X})\right]} \tag{8.18}$$

where

$$\left.\begin{aligned} \tilde{X} &= \frac{4\pi\sqrt{\varepsilon_2}}{\ln\dfrac{\varepsilon_4}{\varepsilon_2}} \cdot \frac{d_3}{\lambda} \\ \tilde{Y} &= \frac{4\pi\sqrt{\varepsilon_4}}{\ln\dfrac{\varepsilon_4}{\varepsilon_2}} \cdot \frac{d_3}{\lambda} \end{aligned}\right\} \tag{8.19}$$

For a "thick" transition layer $(d_3 >> \lambda)$, with

$$\left.\begin{aligned}\tilde{X} &= \frac{4\pi\sqrt{\varepsilon_2}}{\ln\frac{\varepsilon_4}{\varepsilon_2}} \cdot \frac{d_3}{\lambda} >> 1 \\ \tilde{Y} &= \frac{4\pi\sqrt{\varepsilon_4}}{\ln\frac{\varepsilon_4}{\varepsilon_2}} \cdot \frac{d_3}{\lambda} >> 1\end{aligned}\right\} \tag{8.20}$$

the expression for W can be approximated by

$$W = \frac{i}{4} e^{i\left(\tilde{X}-\tilde{Y}\right)} \left[\frac{e^{-i\left(\tilde{X}-\tilde{Y}\right)}}{X} - \frac{e^{-i\left(\tilde{X}-\tilde{Y}\right)}}{Y} \right] + O\left(\frac{1}{\left(\frac{d_3}{\lambda}\right)^2} \right) = O\left(\frac{1}{\frac{d_3}{\lambda}} \right) \tag{8.21}$$

It is seen, then, that for large layer thickness relative to wavelength, W tends to zero; as a consequence, $R \to R_{21}$ which is the reflection coefficient associated with the transition from medium 1 (I) to medium 2 (II). The fulfillment of condition (Eq. 8.20) means that the reflection from the interface between media III and IV practically disappears.

The fact that a transition layer of large thickness acts itself as a matching device is due to the assumptions in the model. As the change in permittivity with depth occurs more smoothly (corresponding to a large thickness of the transition layer), the reflection effects of medium II reduce and disappear in the limit.

Using (8.15) and (8.21), R is approximated by

$$R \cong R_{21} + i \frac{\ln\frac{\varepsilon_4}{\varepsilon_2}}{16\pi\frac{d_3}{\lambda}} \left[\frac{e^{-i\ (\tilde{X}-\tilde{Y})}}{\sqrt{\varepsilon_2}} - \frac{e^{-i\ (\tilde{X}-\tilde{Y})}}{\sqrt{\varepsilon_4}} \right] \frac{1-R_{21}}{WR_{21} + e^{2ik_0 d_2\sqrt{\varepsilon_2}}} \tag{8.22}$$

The weak dependence of the reflection coefficient R on the thickness of the transition layer is seen from Eq. (8.22) in cases where the layer is sufficiently thick. With

$$\frac{d_3}{\lambda} << \frac{\ln\frac{\varepsilon_4}{\varepsilon_2}}{4\pi\sqrt{\varepsilon_4}} \tag{8.23}$$

Eq. (8.22) leads to

$$R = R' + \frac{i\tilde{X}}{2(1+\rho)} \frac{\left(1 - R_{12}^2\right)\left(1 + 2\rho\ln\rho - \rho^2\right)}{WR_{12} + e^{2ik_0 d_2\sqrt{\varepsilon_2}}} + O\left(\frac{d_3^2}{\lambda^2}\right) \tag{8.24}$$

where

$$\rho = \sqrt{\frac{\varepsilon_4}{\varepsilon_2}}$$

$$R' = \frac{R_{12} + R_{24}e^{-2ik_0 d_2\sqrt{\varepsilon_2}}}{1 + R_{12}R_{24}e^{-2ik_0 d_2\sqrt{\varepsilon_2}}} \tag{8.25}$$

$$R_{24} = \frac{\sqrt{\varepsilon_2} - \sqrt{\varepsilon_4}}{\sqrt{\varepsilon_2} + \sqrt{\varepsilon_4}} \tag{8.25a}$$

Where R_{24} is the reflection coefficient with no transition layer.

Expressions (8.23) and (8.24) show that for thin transition layers ($\tilde{X}, \tilde{Y} << 1$), R differs significantly from the situation in which the transition layer is absent.
In other words, R varies significantly from R' for small $\frac{d_3}{\lambda}$, up to R_{21} for large $\frac{d_3}{\lambda}$, as function of thickness d_3.

- **A numerical example**

The layered structure under consideration may describe an air (medium I) to marine ice (medium II) to a transition layer (medium III) to water (medium IV) layered structure. The formation of the transition layer is explained by the variation of temperature, salinity and density of ice with depth. The thickness of this layer depends upon the type and age of ice. For example, two-meter thick marine ice (one-year-old ice) makes transition layers of 30-40 centimeters thickness. For a fresh ice layer of 2 meters thickness, the transition thickness becomes 10-12 centimeters. Specific results dealing with the magnitude of the reflection coefficient R for such a structure are drawn in the figures below.

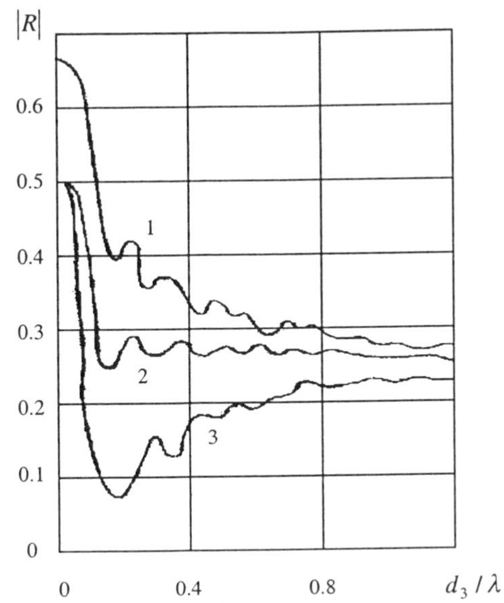

Fig. 8.3 The dependence of $|R|$ on d_3/λ. $\varepsilon_2 = 3.06$; $\varepsilon_4 = 40$

1.$d_2/\lambda = 0.2$; 2.$d_2/\lambda = 0.15$; 3.$d_2/\lambda = 0.1$

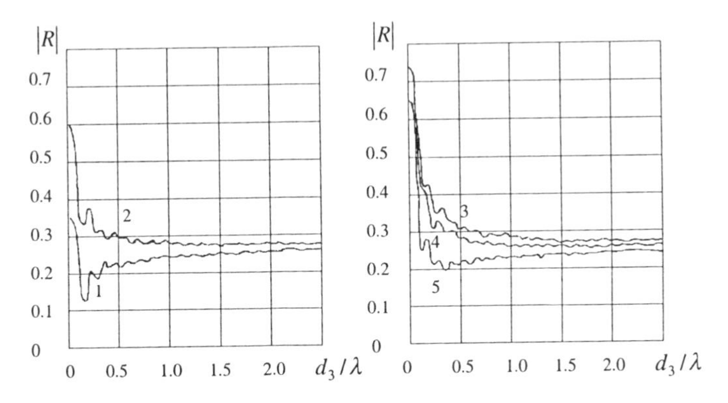

Fig. 8.4 The dependence of $|R|$ on d_3/λ. $\varepsilon_2 = 3.06$; $\varepsilon_4 = 40$

1. $d_2/\lambda = 0.7$; 2. $d_2/\lambda = 0.75$; 3. $d_2/\lambda = 0.8$; 4. $d_2/\lambda = 0.85$; 5. $d_2/\lambda = 0.9$

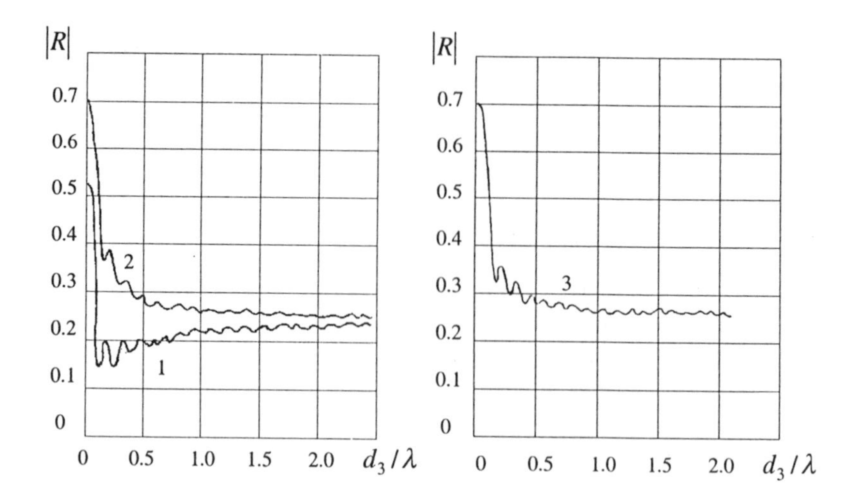

Fig. 8.5 The dependence of $|R|$ on d_3 / λ. $\varepsilon_2 = 3.06$; $\varepsilon_4 = 40$
1. $d_2 / \lambda = 9.0$; 2. $d_2 / \lambda = 9.5$; 3. $d_2 / \lambda = 10$

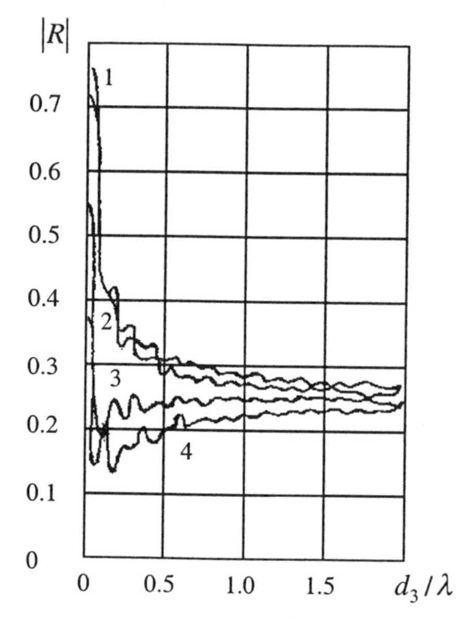

Fig. 8.6 The dependence of $|R|$ on d_3 / λ. $\varepsilon_2 = 3.06$; $\varepsilon_4 = 56$
1.$d_2 / \lambda = 0.5$; 2.$d_2 / \lambda = 1.0$; 3.$d_2 / \lambda = 1.5$; 4.$d_2 / \lambda = 2.0$

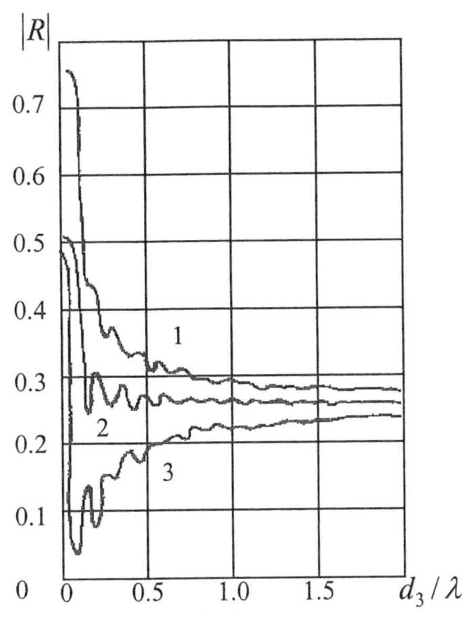

Fig. 8.7 The dependence of $|R|$ on d_3/λ. $\varepsilon_2 = 3.06$; $\varepsilon_4 = 56$
1.$d_2/\lambda = 6.0$; 2.$d_2/\lambda = 6.5$; 3.$d_2/\lambda = 7.0$

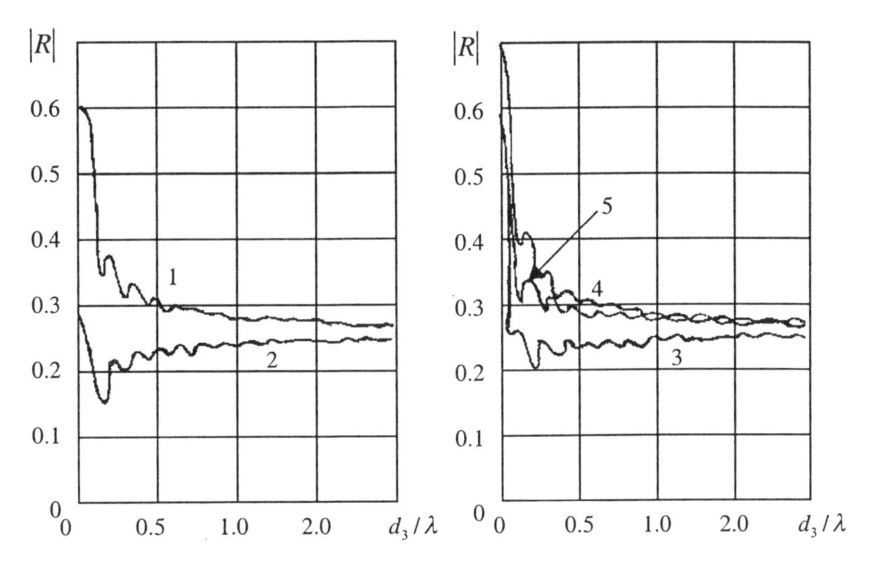

Fig. 8.8 The dependence of $|R|$ on d_3/λ. $\varepsilon_2 = 3.06$; $\varepsilon_4 = 40$
1.$d_2/\lambda = 2.5$; 2.$d_2/\lambda = 3.0$; 3.$d_2/\lambda = 3.5$; 4.$d_2/\lambda = 4.0$; 5.$d_2/\lambda = 40$

Figs 8.3 to 8.8 are based on the exact Eq. (8.15) and neglecting the attenuation in ice. From these figures we learn that ice layers without absorption and with a transition layer may give similar reflections as homogeneous ice layers with absorption. For $d_3/\lambda \approx 1.5-2.0$, the results agree well with the asymptotic representation of R.

Most effects occur for values of d_3/λ between 0.1 and 1.0. The strong interference from borders I-II and II-III is responsible for this. Figs 8.3 to 8.8 also show that for thick transition layers information concerning the internal structure is lost.

Eqs (8.15) to (8.18) permit the determination of the R dependence on angle β, the complex dielectric permittivities $\varepsilon_2, \varepsilon_3, \varepsilon_4$, the thickness of the layers d_2 and d_3, the polarization, etc. As an illustration, results for vertical incidence $(\beta = 0)$ and for marine and fresh ice at wavelengths ranging from centimeters, decimeters to meters are shown.

If $\beta = 0$, we get $|R_h| = |R_v| = |R|$. The expression for R can then be derived using Eqs (8.15) and (8.17):

$$R = \frac{\left[J_0(\tilde{X}) + iJ_1(\tilde{X})\right]\left[N_0(\tilde{Y}) + itN_1(\tilde{Y})\right] - \left[J_0(\tilde{Y}) + itJ_1(\tilde{Y})\right]\left[N_0(\tilde{X}) + iN_1(\tilde{X})\right]}{\left[J_0(\tilde{X}) - iJ_1(\tilde{X})\right]\left[N_0(\tilde{Y}) + itN_1(\tilde{Y})\right] - \left[J_0(\tilde{Y}) + itJ_1(\tilde{Y})\right]\left[N_0(\tilde{X}) - iN_1(\tilde{X})\right]} \quad (8.26)$$

where

$$t = \sqrt{\frac{\varepsilon_3}{\varepsilon_4}} \quad (8.27)$$

R was computed at different wavelengths because the electrophysical parameters of ice and water depend strongly on frequency. In the calculations, the assumptions were made that permittivities $\varepsilon_2 = 3.06$ and $\varepsilon_4 = 56$, respectively, media II and IV have $\varepsilon_1 = 0$.

In Figs 8.9 to 8.11, $|R|^2$ is shown for marine ice at centimeter, decimeter and meter wavelengths.

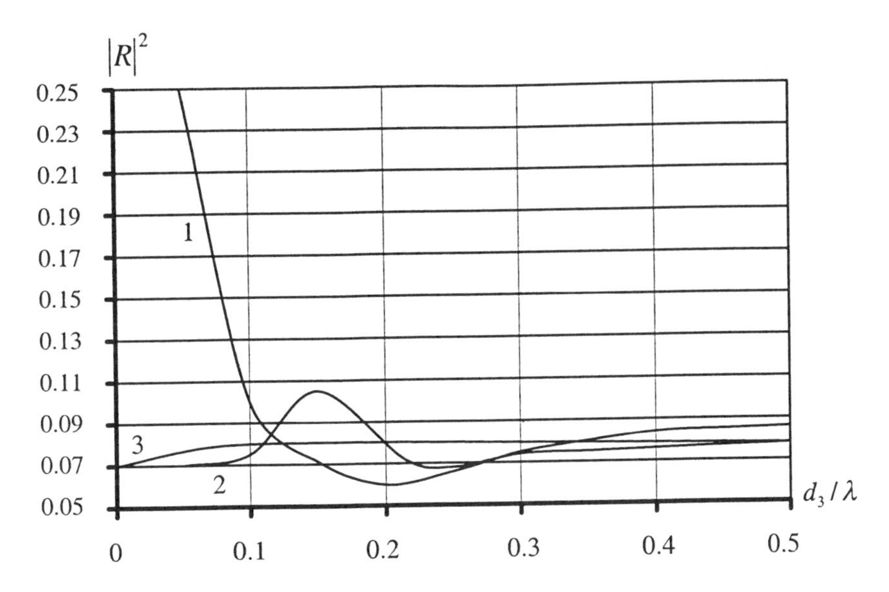

Fig. 8.9 The dependence of $|R|^2$ on d_3/λ

$1.d_2/\lambda = 5$; $2.d_2/\lambda = 10$; $3.d_2/\lambda = 20$

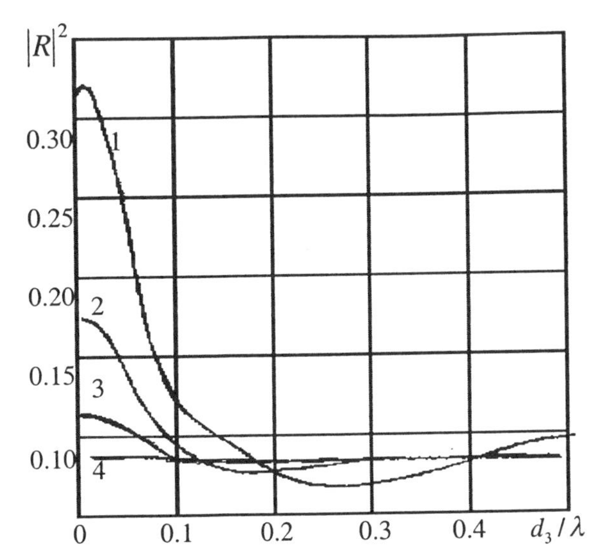

Fig. 8.10 The dependence of $|R|^2$ and d_3/λ

$1.d_2/\lambda = 2.5$; $2.d_2/\lambda = 5$; $3.d_2/\lambda = 7.5$; $4.d_2/\lambda = 10$

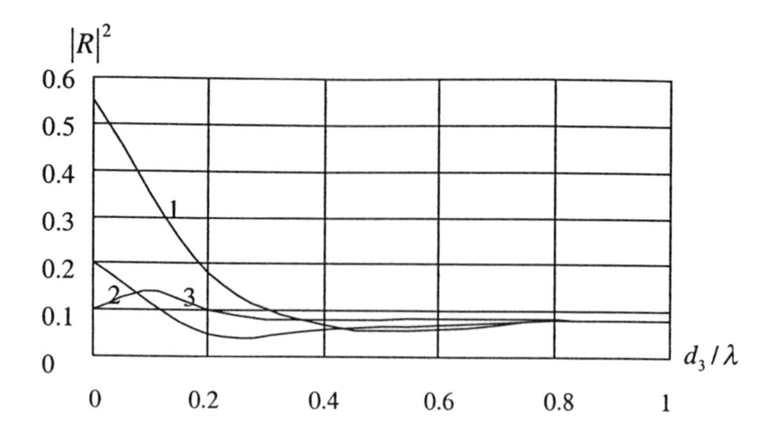

Fig. 8.11 The dependence of $|R|^2$ on d_3/λ
1. $d_2/\lambda = 0.25$; 2. $d_2/\lambda = 1.0$; 3. $d_2/\lambda = 25$

Results for fresh ice using centimeter-waves are given in Figs 8.12 and 8.13.

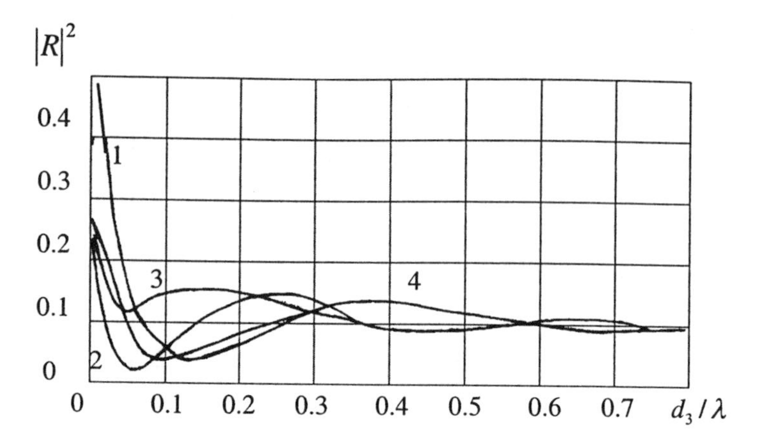

Fig. 8.12 The dependence of $|R|^2$ on d_3/λ
1. $d_2/\lambda = 5$; 2. $d_2/\lambda = 25$; 3. $d_2/\lambda = 75$; 4. $d_2/\lambda = 50$

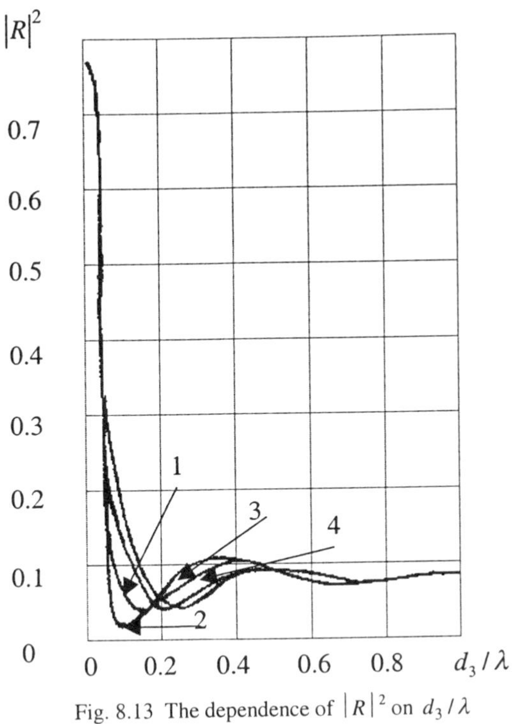

Fig. 8.13 The dependence of $|R|^2$ on d_3/λ
1. $d_2/\lambda = 10$; 2. $d_2/\lambda = 7.5$; 3. $d_2/\lambda = 5$; 4. $d_2/\lambda = 2.5$

An examination of Figs 8.9 to 8.13 enables us to draw the following conclusions:

- A transition layer with thickness in the order of 0.3λ to 0.5λ in decimeter and centimeter wavelengths results in the reduction of $|R|$ up to a value, corresponding to a thick ice layer. Essentially, the transition layer acts as a matching device.

- For marine ice, the thickness of the transition layer appears to be more than a wavelength and $|R|$ approaches its asymptotic value for an infinitely thick marine ice layer.

- For a thicker homogeneous ice layer, the asymptotic value of $|R|$ is attained at smaller thicknesses. For instance, for marine ice and wavelengths in the centimeter range, this value is reduced for thicknesses larger than 20 wavelengths regardless the thickness of the transition layer.

- Stronger attenuation of a radiowave is found in marine ice in comparison with fresh ice. This means that the asymptotic behavior for marine ice is reached at smaller thicknesses.

- **Vertical polarization**

In the case of vertical polarization of the incident wave, analogous calculations for the reflection coefficient, with reference to the magnetic field $\vec{H}$, result in an expression similar to Eq. (8.15). Thus, the reflection coefficient R_{21} is determined by Eq. (8.16) and the parameter W is determined by Eq. (8.17). The significance of the parameters δ_i entering into the expression for W is explained in Appendix D. The order of the cylindrical functions in this case is determined by

$$\nu = \frac{\sqrt{a + k_0^2 \sin^2 \beta}}{a} \tag{8.28}$$

8.2.2 Layer with exponential permittivity profile

In this subsection, expressions are derived for reflections from a single layer, in which the permittivity changes exponentially. Such a structure may represent ice lying on water or on earth. The permittivity of ice does not remain constant with thickness. Furthermore, due to an increase in density, the permittivity increases monotonically with depth of the underlying surface. This change is described by an exponential or a polynomial expression (cf. Figs 8.14 and 8.15).

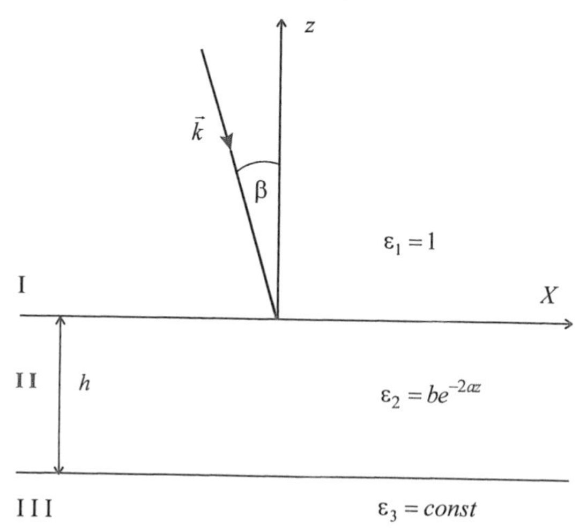

Fig. 8.14 Layer with exponential permittivity profile

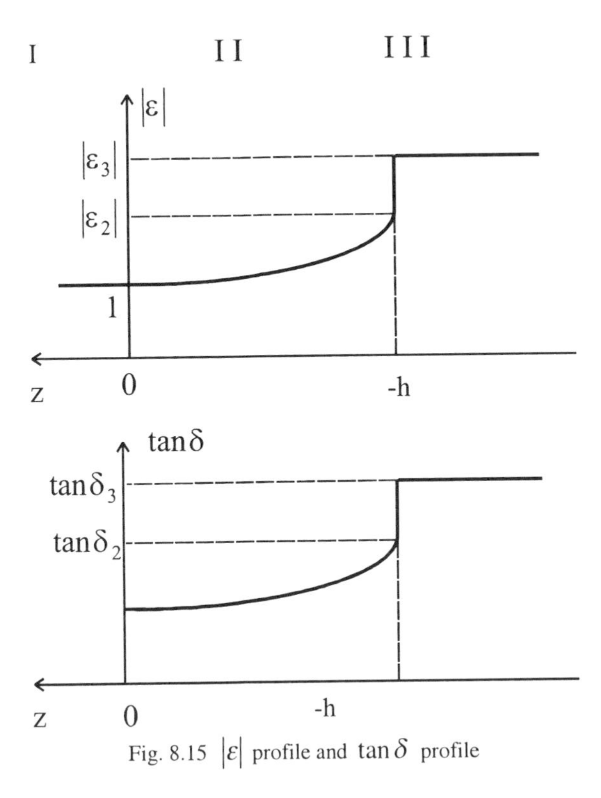

Fig. 8.15 $|\varepsilon|$ profile and $\tan\delta$ profile

We solve the problem following a procedure analogous to that used in Sec. 8.2.1. In the case of horizontal polarization, the field representations in media I and III are given by

$$\left.\begin{aligned} E_1 &= e^{ik_0 x\sin\beta}\left(e^{ik_0 z\cos\beta} + R_h e^{+ik_0 z\cos\beta}\right) \\ E_3 &= e^{ik_0 x\sin\beta}\cdot C_1 e^{ikz\sqrt{\varepsilon_3 - \sin^2\beta}} \end{aligned}\right\} \tag{8.29}$$

In the second medium, the electric field intensity $\vec{E}_2$ satisfies the equation

$$\Delta\vec{E}_2 + \vec{\nabla}(\vec{\nabla}\cdot\vec{E}_2) = 0 \tag{8.30}$$

For the exponential permittivity profile, the solution of this equation can be written as

$$E_2 = e^{ikx\sin\beta}\left[A_1 J_s\left(\frac{k\sqrt{N}}{a}e^{-az}\right) + B_1 N_s\left(\frac{k\sqrt{N}}{a}e^{-az}\right)\right] \tag{8.31}$$

where

$$s = \frac{k\sin\beta}{a} \tag{8.32}$$

and $J_s(u), N_s(u)$ are Bessel and Neumann functions of order s, respectively.

The requirement of continuity of the tangential components of the electric and magnetic fields at the boundaries between media I-II and II-III gives rise to the following system of equations for the unknown coefficients A_1, B_1, C_1, R_h:

$$\left.\begin{aligned} 1 - R_h &= A_1 J_s(\tilde{X}) + B_1 N_s(\tilde{X}) \\ 1 - R_h &= \frac{1}{ik\cos\beta}\left(A_1 \frac{dJ_s(\tilde{X})}{dz} + B_1 \frac{dN_s(\tilde{X})}{dz}\right) \\ A_1 J_s(\tilde{Y}) + B_1 N_s(\tilde{Y}) &= C_1 e^{-ik\sqrt{\varepsilon_3 - \sin^2\beta}} \\ A_1 \frac{dJ_s(\tilde{Y})}{dz} + B_1 \frac{dN_s(\tilde{Y})}{dz} &= ikC_1\sqrt{\varepsilon_3 - \sin^2\beta}\, e^{-ik\sqrt{\varepsilon_3 - \sin^2\beta}} \end{aligned}\right\} \tag{8.33}$$

Here,

$$\left.\begin{aligned} \tilde{X} &= \frac{4\pi\sqrt{\varepsilon_2}}{\ln\frac{\varepsilon_3}{\varepsilon_2}} \cdot \frac{h}{\lambda}, \quad if \quad (z = 0) \\ \tilde{Y} &= \frac{4\pi\sqrt{\varepsilon_3}}{\ln\frac{\varepsilon_3}{\varepsilon_2}} \cdot \frac{h}{\lambda}, \quad if \quad (z = -h) \end{aligned}\right\} \tag{8.34}$$

From these equations we obtain

$$R_h = -\frac{\left[\frac{1}{ik\cos\beta}\frac{dJ_s(\tilde{X})}{dz} - J_s(\tilde{X})\right]\left[\frac{1}{ik\sqrt{\varepsilon_3 - \sin^2\beta}}\frac{dN_s(\tilde{X})}{dz} - N_s(\tilde{Y})\right] - \rightarrow}{\left[\frac{1}{ik\cos\beta}\frac{dJ_s(\tilde{X})}{dz} + J_s(\tilde{X})\right]\left[\frac{1}{ik\sqrt{\varepsilon_3 - \sin^2\beta}}\frac{dN_s(\tilde{X})}{dz} - N_s(\tilde{Y})\right] - \rightarrow}$$

$$\frac{\rightarrow - \left[\frac{1}{ik\sqrt{\varepsilon_3 - \sin^2\beta}}\frac{dJ_s(\tilde{X})}{dz} - J_s(\tilde{Y})\right]\left[\frac{1}{ik\cos\beta}\frac{dN_s(\tilde{X})}{dz} - N_s(\tilde{Y})\right]}{\rightarrow - \left[\frac{1}{ik\sqrt{\varepsilon_3 - \sin^2\beta}}\frac{dJ_s(\tilde{X})}{dz} - J_s(\tilde{Y})\right]\left[\frac{1}{ik\cos\beta}\frac{dN_s(\tilde{X})}{dz} + N_s(\tilde{Y})\right]} \quad (8.35)$$

In the case of vertical polarization, the magnetic fields are given by:

$$\left.\begin{array}{c} H_1 = e^{ikx\sin\beta}\left(e^{ikz\cos\beta} + R_v e^{-ikz\cos\beta}\right) \\ H_2 = e^{ikx\sin\beta} \cdot e^{-az}\left[A_2 J_p\left(\frac{k\sqrt{N}}{a}e^{-az}\right) + B_2 N_p\left(\frac{k\sqrt{N}}{a}e^{-az}\right)\right] \\ H_3 = e^{ikx\sin\beta} \cdot C_2 e^{ikz\sqrt{\varepsilon_3 - \sin^2\beta}} \end{array}\right\} \quad (8.36)$$

The continuity requirement for the tangential components leads to:

$$\left.\begin{array}{c} 1 - R_v = A_2 J_p(\tilde{X}) + B_2 N_p(\tilde{X}) \\ 1 - R_v = A_2 Q_1 + B_2 Q_2 \\ A_2 J_p(\tilde{Y})e^{ah} + B_2 N_p(\tilde{Y})e^{ah} = C_2 e^{-ikh\sqrt{\varepsilon_3 - \sin^2\beta}} \\ A_2 Q_3 + B_2 Q_4 = C_2 e^{-ikh\sqrt{\varepsilon_3 - \sin^2\beta}} \end{array}\right\} \quad (8.37)$$

The coefficients Q_1, Q_2, Q_3, Q_4, the parameter p and the reflection coefficient R_v are given as follows:

$$\left.\begin{aligned}
Q_1 &= \frac{1}{ik\varepsilon_2 \cos\beta} \cdot \frac{d}{dz}\left[e^{-az} J_p\left(\frac{k\sqrt{N}}{a} e^{-az} \right) \right]\Bigg|_{z=0} \\
Q_2 &= \frac{1}{ik\varepsilon_2 \cos\beta} \cdot \frac{d}{dz}\left[e^{-az} N_p\left(\frac{k\sqrt{N}}{a} e^{-aZ} \right) \right]\Bigg|_{z=0} \\
Q_3 &= \frac{1}{ik\varepsilon_2 \cos\beta} \cdot \frac{d}{dz}\left[e^{-az} J_p\left(\frac{k\sqrt{N}}{a} e^{-az} \right) \right]\Bigg|_{z=-h} \\
Q_4 &= \frac{1}{ik_2\sqrt{\varepsilon_3 - \sin^2\beta}} \cdot \frac{d}{dz}\left[e^{-az} N_p\left(\frac{k\sqrt{N}}{a} e^{-az} \right) \right]\Bigg|_{z=-h} \\
p &= \frac{1}{a}\sqrt{a^2 + k^2 \sin^2\beta}
\end{aligned}\right\} \tag{8.38}$$

$$R_v = -\frac{\left[Q_1 - J_p(\tilde{X})\right]\left[Q_4 - e^{ah} N_p(\tilde{Y})\right] - \left[Q_2 - N_p(\tilde{X})\right]\left[Q_3 - e^{ah} J_p(\tilde{Y})\right]}{\left[Q_1 + J_p(\tilde{X})\right]\left[Q_4 - e^{ah} N_p(\tilde{Y})\right] - \left[Q_2 + N_p(\tilde{X})\right]\left[Q_3 - e^{ah} J_p(\tilde{Y})\right]} \tag{8.39}$$

Eqs (8.35) and (8.39) enable us to find the required reflection coefficients as functions of the incidence angle β, the thickness of layer II and the type of polarization. As an example, specific reflection results are shown for vertical incidence ($\beta = 0$) for certain types of marine ice, snow on earth and for wavelengths in the centimeter and meter ranges. If $\beta = 0$, obviously $|R_h| = |R_v| = |R|$, where

$$R = \frac{\left[J_0(\tilde{X}) + i\sqrt{\varepsilon_2}\, J_1(\tilde{X})\right]\left[N_0(\tilde{Y}) + iN_1(\tilde{Y})\right] - \left[N_0(\tilde{X}) + i\sqrt{\varepsilon_2}\, N_1(\tilde{X})\right]\left[J_0(\tilde{Y}) + iJ_1(\tilde{Y})\right]}{\left[J_0(\tilde{X}) - i\sqrt{\varepsilon_2}\, J_1(\tilde{X})\right]\left[N_0(\tilde{Y}) + iN_1(\tilde{Y})\right] - \left[N_0(\tilde{X}) - i\sqrt{\varepsilon_2}\, N_1(\tilde{X})\right]\left[J_0(\tilde{Y}) + iJ_1(\tilde{Y})\right]} \tag{8.40}$$

The results are shown in Figs 8.16 – 8.18. In these figures, the reflection from ice lying on water is given as function of the ice thickness for centimeter, decimeter and meter wavelengths. Figs 8.19 and 8.20 illustrate results for snow permeated with water and lying on a ground surface or on ice, respectively.

Examining these figures we note the following:

- The layer acts as a good matching device for thickness even less than 0.5λ (e.g., for the ice - water layer). In this case, $|R|^2$ reduces to less than 0.1.

- For a layer thickness, h_2 larger than $\lambda/2$, the underlying structure can no longer be detected and interference effects (characteristic of multi-layer cases) disappear.

- For $h_2 < 0.5\lambda$, no significant dependence on $\tan\delta_2$ and $\tan\delta_3$ is found.

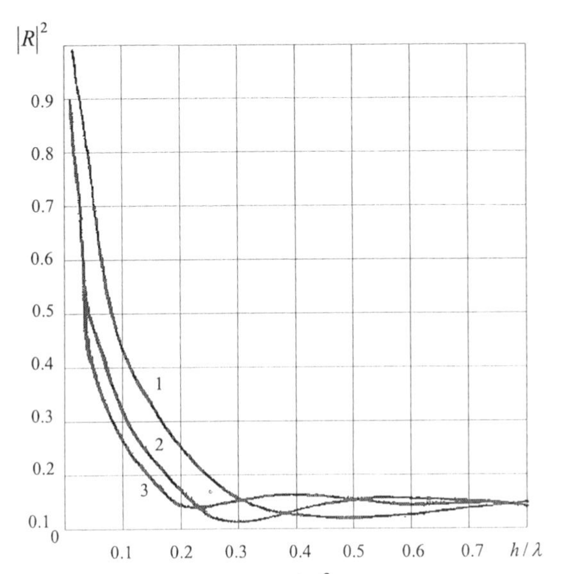

Fig. 8.16 The dependence of $|R|^2$ on h/λ. $\varepsilon_3 = 3.2$

1) $\lambda \sim$ cm, $\varepsilon_2 = 40(1-i)$; 2) $\lambda \sim$ dm, $\varepsilon_2 = 80(1-i)$; 3) $\lambda \sim$ m, $\varepsilon_2 = 80(1-10i)$

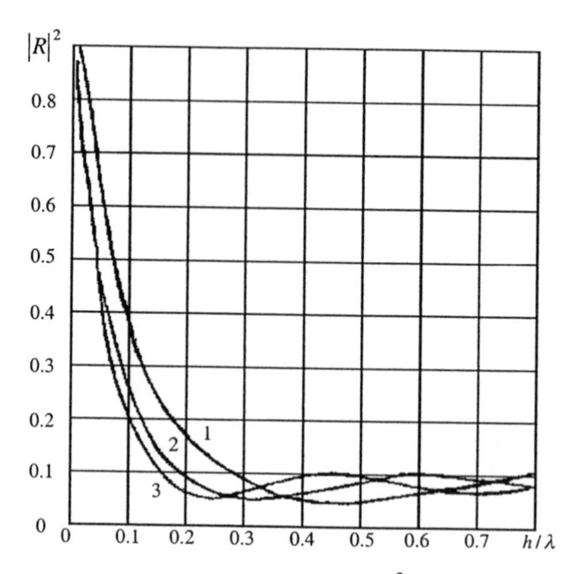

Fig. 8.17 The dependence of $|R|^2$ on h/λ

1) $\lambda \sim$ cm, $\varepsilon_2 = 40(1-i)$, $\varepsilon_3 = 3.2(1-0.05i)$

2) $\lambda \sim$ dm, $\varepsilon_2 = 80(1-i)$, $\varepsilon_3 = 3.2(1-0.05i)$

3) $\lambda \sim$ m, $\varepsilon_2 = 80(1-10i)$, $\varepsilon_3 = 3.2(1-0.1i)$

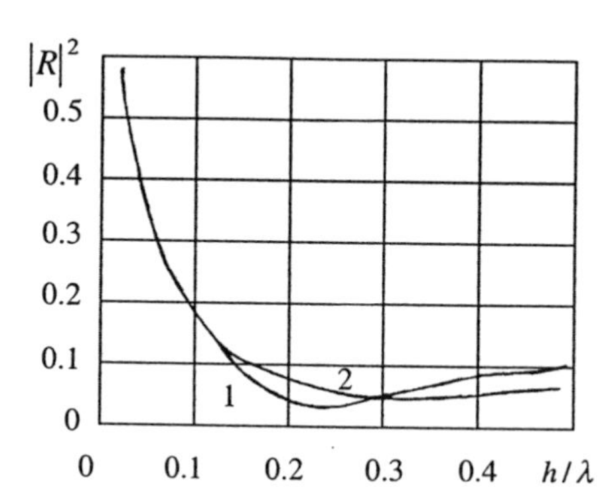

Fig. 8.18 The dependence of $|R|^2$ on h/λ

a) $\lambda \sim$ cm, $\varepsilon_2 = 40(1-i\tan\delta)$, $\varepsilon_3 = 3.2$

b) $\lambda \sim$ dm, $\varepsilon_2 = 80(1-i\tan\delta)$, $\varepsilon_3 = 3.2$

c) $\lambda \sim$ m, $\varepsilon_2 = 80(1-i\tan\delta)$, $\varepsilon_3 = 3.2$

1) $\tan\delta = 0$; 2) $\tan\delta = 0.2$

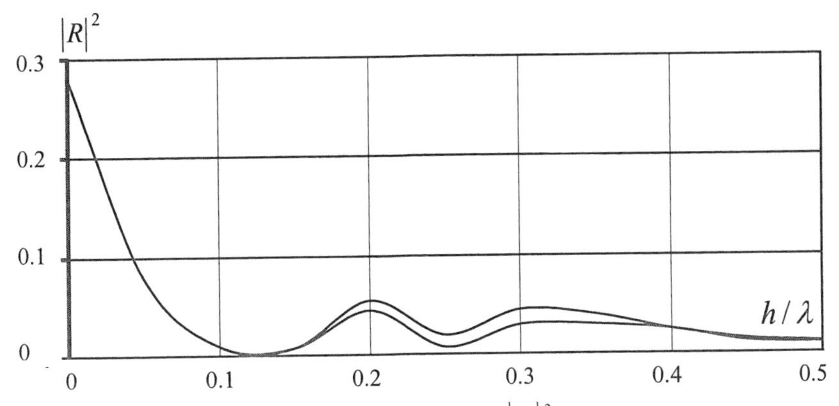

Fig. 8.19 Dependence of $|R|^2$ on h/λ

$\varepsilon_2 = 1.56; \quad \tan\delta_2 = 0 \div 0.01; \quad \varepsilon_3 = 8.$

Upper curve $\tan\delta_3 = 0.2$; lower curve $\tan\delta_3 = 0.001$

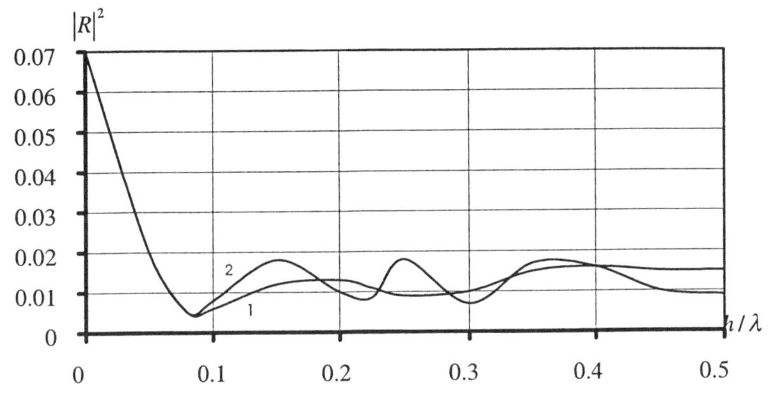

Fig. 8.20 Dependence of $|R|^2$ on h/λ

$\varepsilon_2 = 1.56; \quad \tan\delta_2 = 0 \div 0.001; \quad \varepsilon_3 = 3.2$ 1) $\tan\delta_3 = 0.2;$ 2) $\tan\delta_3 = 0.001$

8.2.3 Single layer with a polynomial permittivity profile

For some types of earth surface covered by vegetation, ice top-soil, frozen top layer and marshes [i.e. tracts of soft wet land with grass or cattails (plants with long flat leaves)] the dielectric permittivity can be well approximated by polynomial functions as $(az+b)^n$. The parameters a and b are determined from matching the permittivity at the boundaries of the layer. The exponent n characterizes the rate of change in the dielectric permittivity. For example, for ice n is dependent upon humidity, salinity and the thickness of the transition layer. For dry and non-salt ice with significant thickness and thin ice layers permeated with water, the parameter n is large. For new ice saturated with water, the parameter n is small.

The polynomial dependence of the dielectric permittivity is shown in Figs 8.21 and 8.22.

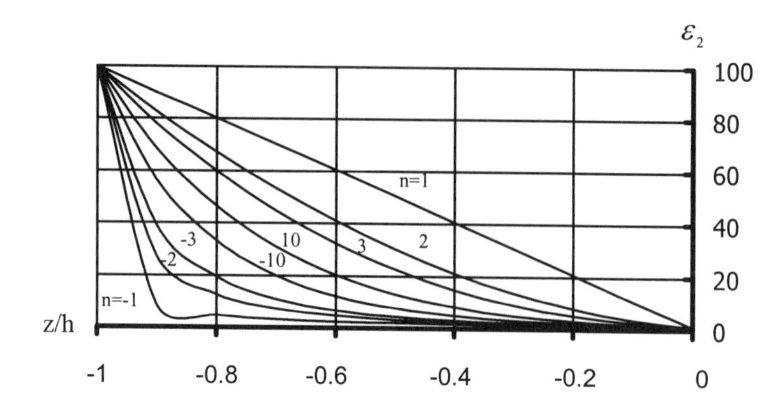

Fig. 8.21 Function $\varepsilon_2 = (az + b)^n$ *vs.* z/h

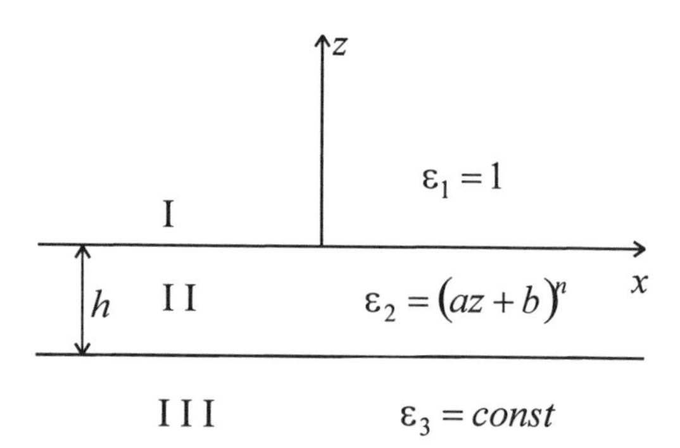

Fig. 8.22 Layer with polynomial permittivity profile

The electric fields in media I and III are written as

$$\left.\begin{aligned} E_1 &= e^{ik_0 z} + R \cdot e^{-ik_0 z} \\ E_2 &= T \cdot e^{ik_0 \sqrt{\varepsilon_3} z} \end{aligned}\right\} \tag{8.41}$$

for vertical incidence $(\beta = 0)$. In medium II the electric field intensity satisfies the Helmholtz equation

$$\frac{d^2 E_2}{dz^2} + k_0^2 (az+b)^n E_2 = 0 \tag{8.42}$$

Introducing the variable $\eta = (az+b)$ results in the following equation:

$$\frac{d^2 E_2}{d\eta^2} + \frac{k_0^2}{a^2}\eta^n E_2 = 0 \tag{8.43}$$

A general solution to this equation is given by

$$E_2 = \sqrt{az+b}\left[C_1 J_{\frac{1}{n+2}}\left(-\frac{2k_0 (az+b)^{\frac{n+2}{2}}}{a(n+2)}\right) + C_2 J_{-\frac{1}{n+2}}\left(-\frac{2k_0 (az+b)^{\frac{n+2}{2}}}{a(n+2)}\right)\right] \tag{8.44}$$

assuming that $a \neq 0, n \neq -2$. The reflection coefficient R is derived from the boundary conditions of tangential continuity in $\vec{E}$ and $\vec{H}$ fields. We find, specifically,

$$R = \frac{(\sqrt{\varepsilon_3}u_4 + u_2) + i(\sqrt{\varepsilon_3}u_1 - u_3)}{(\sqrt{\varepsilon_3}u_4 - u_2) - i(\sqrt{\varepsilon_3}u_1 + u_3)} \tag{8.45}$$

where

$$\left.\begin{aligned} u_1 &= \alpha_{22}\alpha_{33} - \alpha_{32}\alpha_{23} \\ u_2 &= \alpha_{23}\alpha_{42} - \alpha_{22}\alpha_{43} \\ u_3 &= \alpha_{13}\alpha_{42} - \alpha_{12}\alpha_{43} \\ u_4 &= \alpha_{12}\alpha_{33} - \alpha_{13}\alpha_{32} \end{aligned}\right\} \tag{8.46}$$

The significance of the terms in Eqs (8.45) and (8.46) is given in Appendix E. Eq. (8.45) is intractable to further analysis and needs to be evaluated numerically.
The computed results for R (magnitude and phase) for ice without absorption are indicated in Figs 8.23 to 8.28.

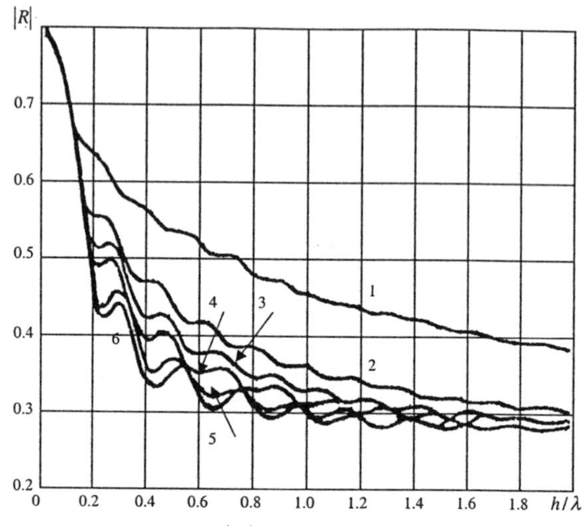

Fig. 8.23 $|R|$ as a function of h/λ

$\varepsilon_2(0)=3$; $\varepsilon_3=80$. 1.$n=1$; 2.$n=2$; 3.$n=3$; 4.$n=4$; 5.$n=10$; 6.$n=20$.

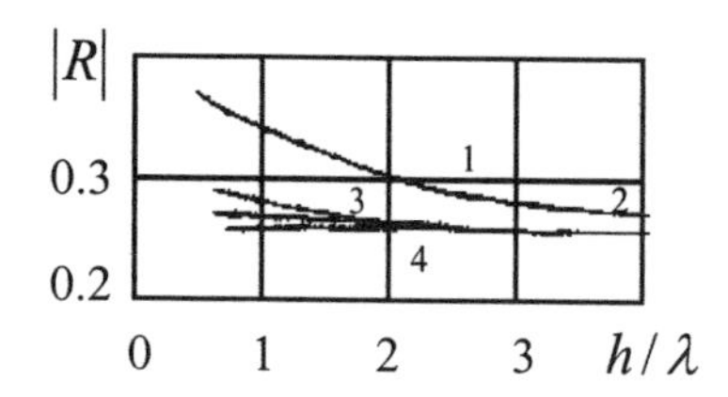

Fig. 8.24 $|R|$ as a function h/λ

$\varepsilon_3=80$; $\varepsilon_2(0)=3$ $1-n=1$; $2-n=2$; $3-n=5$; $4-n=10$

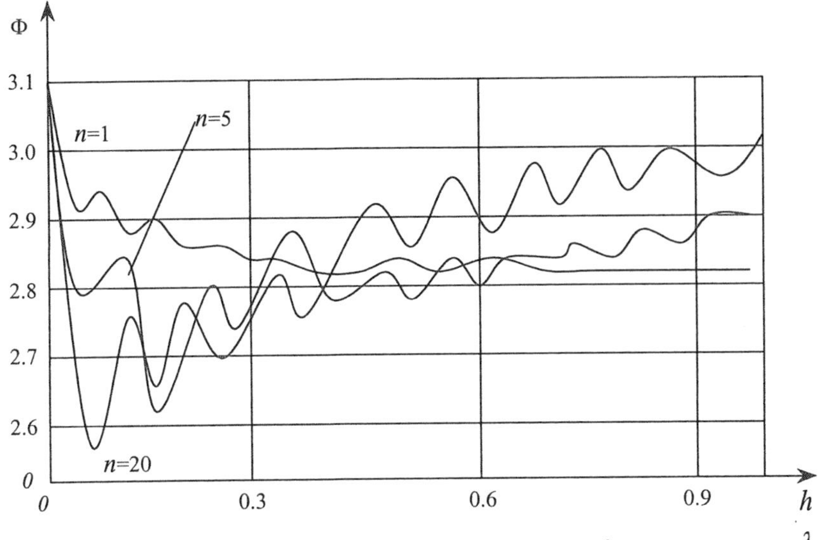

Fig. 8.25 The argument of R as a function of h/λ

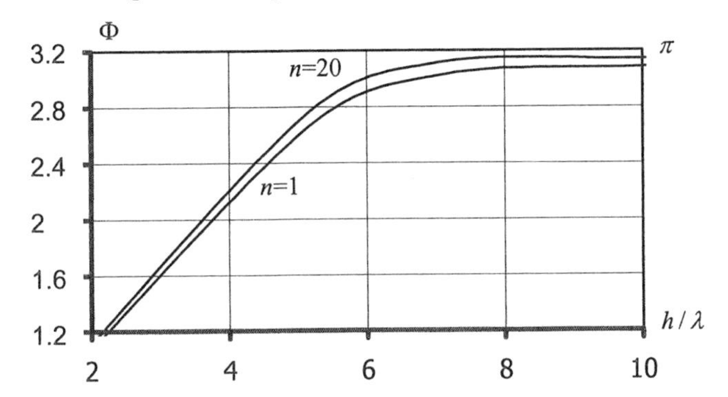

Fig. 8.26 The argument of R as a function of h/λ

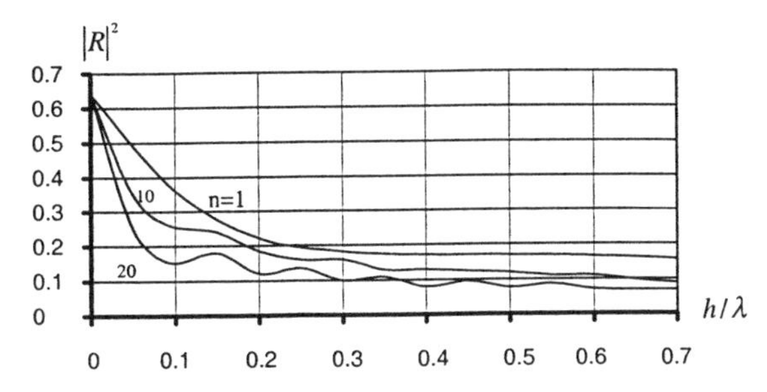

Fig. 8.27 $|R|^2$ as function h/λ. $\varepsilon_3 = 80$; $\varepsilon_2(0) = 3$.

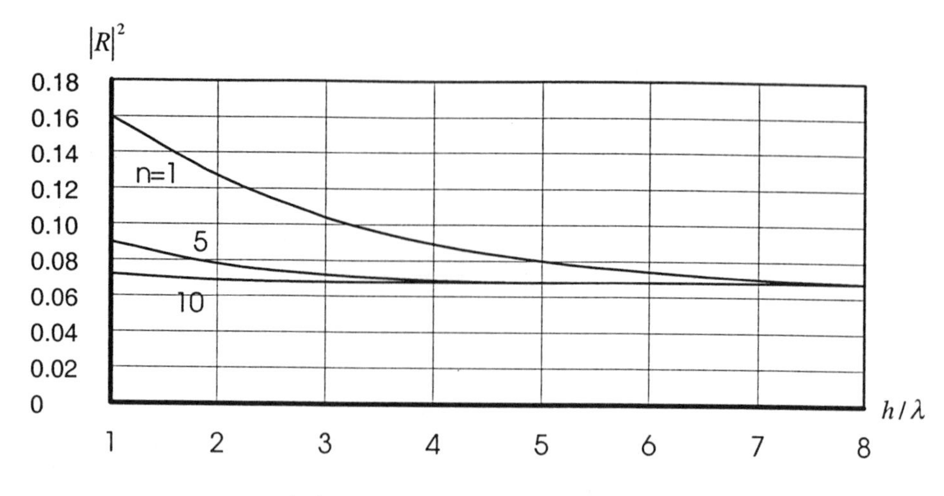

Fig. 8.28 $|R|^2$ as function of h / λ. $\varepsilon_3 = 80$; $\varepsilon_2(0) = 3$.

From Figs 8.23 to 8.28, we can derive appropriate approximations for small and large ice thickness. For example, for $h/\lambda > 2$, $|R|$ practically coincides with the case where $h/\lambda >> 1$.

We note, that except for n = 1 (linear layer), R has only a weak dependence upon n. It enables us to choose a convenient approximation for the unknown n, knowing that we do not make large errors.

8.3 Stochastic case of three layers with flat boundaries

8.3.1 Integral equation approach

Consider the geometry shown in Fig. 8.29. A plane wave propagating in medium I (free space) is incident normally ($\beta = 0$) at the upper flat boundary of medium II consisting of a dielectric layer having variable permittivity $\varepsilon(z)$ and thickness h. This layer is backed by medium III of constant permittivity ε_3.

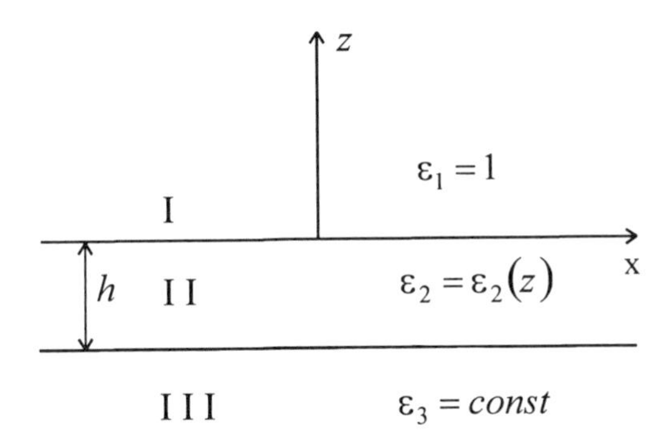

Fig. 8.29 Geometry of three layers with flat boundaries

The permittivity of medium II is characterized by

$$\varepsilon_2 = \varepsilon_{02} + \nu(z) \tag{8.47}$$

where $\nu(z)$ is a zero-mean random function of z, that is

$$\overline{\nu(z)} = 0 \tag{8.48}$$

All three media are assumed to be unbounded in the x- and y-directions. Media I and III extend to infinity from the +z and -z directions, respectively. As all media are assumed to be non-magnetic, the relative magnetic permeability is everywhere equal to one.

For vertical incidence, E_s, the *y*- component of the electrical field in medium *s* (*s* = 1(I), 2(II), 3(III)) is governed by the Helmholtz equation

$$\frac{d^2 E_s}{dz^2} + k_0^2 \varepsilon_s E_s = 0 \tag{8.49}$$

As media I and III are homogeneous, the electric fields in these media can be written in standardized form as

$$\left. \begin{aligned} E_1 &= e^{ik_0 z} + R \cdot e^{-ik_0 z} \\ E_3 &= T \cdot e^{ik_0 \sqrt{\varepsilon_3} z} \end{aligned} \right\} \tag{8.50}$$

In medium II, the electric field obeys the 1-D stochastic Helmholtz equation

$$\frac{d^2E_2}{dz^2}+k_0^2\varepsilon_2E_2=0 \tag{8.51}$$

The latter is solved by assuming that the field consists of two parts:

$$E_2=E_{02}+E_2' \tag{8.52}$$

E_{02} is the electric field in a layer in which ε_2 is constant and equal to ε_{02}. This is the solution of the unperturbed problem and satisfies the equation

$$\frac{d^2E_{02}}{dz^2}+k_0^2\varepsilon_{02}E_{02}=0 \tag{8.53}$$

with the solution

$$E_{02}=C_5e^{ik_0\sqrt{\varepsilon_{02}}\,z}+C_6e^{-ik_0\sqrt{\varepsilon_{02}}\,z} \tag{8.54}$$

The coefficients C_5 and C_6 are calculated from the boundary conditions for the reflection of a plane wave from a homogeneous layer with $\varepsilon=\varepsilon_{02}$:

$$\left.\begin{aligned}C_5&=-\frac{2e^{2ik_0\sqrt{\varepsilon_{02}}h}}{\left(1-\sqrt{\varepsilon_{02}}\right)-\left(1+\sqrt{\varepsilon_{02}}\right)e^{2ik_0\sqrt{\varepsilon_{02}}h}}\\ C_6&=\frac{2}{\left(1-\sqrt{\varepsilon_{02}}\right)-\left(1+\sqrt{\varepsilon_{02}}\right)e^{2ik_0\sqrt{\varepsilon_{02}}h}}\end{aligned}\right\} \tag{8.55}$$

By substituting (8.52) in (8.49) we obtain

$$\frac{d^2E_2'}{dz^2}+k_0^2\varepsilon_{02}E_2'=k_0^2\nu(z)E_{02}+k_0^2\nu(z)E_2' \tag{8.56}$$

This expression is a non-uniform Helmholtz equation, in which the right-hand part contains the random function $\nu(z)$. To solve this equation we use Rytov's method [*Rytov et al, Part 2,* 1978], which is summarized in the following.

Consider ν(z) as a given function of z. Then, Eq. (8.56) for E_s (s =1,2,3) can be solved in principle, taking into account the boundary conditions. In the solutions E_s some functional dependence on ν(z) has to be present. The various statistical averages necessitate working with statistics of products of E_s fields. The outcomes will be described by expressions containing various correlations of functionals of the random function ν(z).

Following this method, the functions $\nu(z)E_{02}, \nu(z)E_2'$ on the right-hand side of Eq. (8.56) are expressed as integrals; specifically,

$$\left.\begin{aligned} \nu(z)E_{02} &= \int_{-\infty}^{\infty} A(p)\cos pz dp \\ \nu(z)E_2' &= \int_{-\infty}^{\infty} D(p)\cos pz dp \\ E_2'(z) &= \int_{-\infty}^{\infty} C(p)\cos pz dp \end{aligned}\right\} \tag{8.57}$$

The coefficients $A(p), D(p)$ and $C(p)$ are determined as follows:

$$\left.\begin{aligned} A(p) &= \frac{1}{\pi}\int_{-h}^{0} \nu(z)E_{02}\cos pz dz \\ C(p) &= \frac{1}{\pi}\int_{-h}^{0} E_2'\cos pz dz \\ D(p) &= \frac{1}{\pi}\int_{-h}^{0} C(p')F(p,p')dp' \end{aligned}\right\} \tag{8.58}$$

$F(p,p')$ is defined explicitly as

$$F(p,p') = \int_{-h}^{0} \nu(z)\cos pz\cos p'z dz \tag{8.59}$$

The substitution of Eqs (8.58) and (8.59) into Eq. (8.56) results in the following integral equation for $C(p)$:

$$C(p)=\frac{k_0^2 A(p)}{p^2-k_0^2\varepsilon_{02}}+\frac{k_0^2}{\pi}\int_{-\infty}^{\infty}\frac{C(p')F(p,p')}{p'^2-k_0^2\varepsilon_{02}}dp' \tag{8.60}$$

This is a Fredholm equation of the second kind. Its kernel, $C(p')F(p,p')/(p'^2-k_0^2\varepsilon_{02})$, is unknown in actual cases. Therefore, Eq. (8.60) can only be solved by approximation-methods. An iterative method is used. In the first iteration, we have

$$C(p)=\frac{k_0^2 A(p)}{p^2-k_0^2\varepsilon_{02}}=C_0(p) \tag{8.61}$$

Substituting this expression into Eq. (8.60), we obtain in the next iteration

$$C(p)=C_0(p)+C_1(p) \tag{8.62}$$

where

$$C_1(p)=\frac{k_0^2}{\pi}\int_{-\infty}^{\infty}\frac{k_0^2 A(p)}{p'^2-k_0^2\varepsilon_{02}}\cdot\frac{F(p,p')}{p^2-k_0^2\varepsilon_{02}}dp' \tag{8.63}$$

A continuation of the iteration procedure yields

$$C(p)=C_0(p)+\sum_{i=1}^{\infty}C_i(p) \tag{8.64}$$

where $C_0(p)$ is given in Eq. (8.61) and $C_i(p)$ can be written as

$$C_i(p)=\left(\frac{k_0^2}{\pi}\right)^{i+1}\frac{1}{p^2-k_0^2\varepsilon_{02}}\int_{-\infty}^{\infty}\frac{dp_1}{p_1^2-k_0^2\varepsilon_{02}}\int_{-\infty}^{\infty}\frac{dp_2}{p_2^2-k_0^2\varepsilon_{02}}\cdots\int_{-\infty}^{\infty}\frac{dp_i}{p_i^2-k_0^2\varepsilon_{02}}\times$$
$$\times\int_{-h}^{0}\int_{-h}^{0}\int_{-h}^{0}\int_{-h}^{0}\cdots\int_{-h}^{0}E_{02}(z)v(z)v(z_1)\ldots v(z_i)\times \tag{8.65}$$
$$\times\cos p_1 z\cos p_1 z_1\cos p_2 z_1\ldots\cos p_i z_{i-1}\cos p_i z_i dz dz_1 dz_2\ldots dz_i$$

The expression for $C(p)$ in Eq. (8.64), together with the last relationship in Eq. (8.57), allows us to determine, in principle, the particular solution of Eq. (8.56). Combining Eqs (8.52), (8.54) and (8.57), we obtain a formal solution to the stochastic Helmholtz Eq. (8.51), viz.,

$$E_2(z) = C_5 e^{ik_0\sqrt{\varepsilon_{02}}\,z} + C_6 e^{-ik_0\sqrt{\varepsilon_{02}}\,z} + \int_{-\infty}^{\infty} C(p)\cos pz\,dp \tag{8.66}$$

for each realization of the random function $\nu(z)$. We shall make use of this solution in the next section in order to determine the average reflection coefficient.

8.3.2 Reflection from layers with constant average permittivity

If, in the analysis carried out in the previous subsection, the deviations in permittivity from its average value are insignificant (as for many types of ice), a lowest-order approximation becomes possible.

We shall demonstrate that a necessary condition for approximating $C(p)$ by the first iteration term $C_0(p)$ [cf. Eq. (8.61)] is given as

$$\left|\frac{\sigma_\nu}{\varepsilon_{02}}\right| << 1 \tag{8.67}$$

where σ_ν is the standard deviation of ν(z). For instance, the condition (8.67) for ice topsoil means that the spread in permittivity should not exceed 10-15%. For the validity of the approximation implied in Eq. (8.61), it is necessary that

$$\left.\begin{aligned} |C_0(p)| &>> |C_1(p)| \\ |C_i(p)| &>> |C_{i+1}(p)| \end{aligned}\right\} \tag{8.68}$$

Using Eqs (8.58), (8.61) and (8.63), these two inequalities can be rewritten as

$$\left|\frac{k_0^2}{p^2-k_0^2\varepsilon_{02}}\cdot\frac{1}{\pi}\int_{-h}^{0}\nu(z)E_{02}(z)\cos pz dz\right|$$
$$>>\left|\frac{k_0^4}{\pi^2\left(p^2-k_0^2\varepsilon_{02}\right)}\int_{-\infty}^{\infty}\frac{dp_1}{p_1^2-k_0^2\varepsilon_{02}}\int_{-h}^{0}\nu(z)E_{02}(z)\cos p_1 z dz\int_{-h}^{0}\nu(z_1)\cos p_1 z_1\cos pz_1 dz_1\right| \tag{8.69}$$

or as

$$\left|\frac{k_0^2}{\pi}\int_{-\infty}^{\infty}\frac{dp_1}{p_1^2-k_0^2\varepsilon_{02}}\int_{-h}^{0}\nu(z)\cos p_1 z_1\cos pz_1 dz_1\right|<<1 \tag{8.70}$$

The latter is rephrased approximately as

$$\left|\frac{k_0^2 n\sigma_\nu}{\pi}\int_{-\infty}^{\infty}\frac{dp_1}{p_1^2-k_0^2\varepsilon_{02}}\int_{-h}^{0}\cos p_1 z_1\cos pz_1 dz_1\right|<<1 \tag{8.71}$$

which differs from Eq. (8.70) in the sense that ν(z) is replaced by an overestimate, namely the standard deviation $n\sigma_\nu$ (n =1,2,3,..).

Carrying out the integrations over z_1 and p_1 in Eq. (8.71), we obtain

$$\left|\frac{k_0^2\sigma_\nu e^{-ik_0h\sqrt{\varepsilon_{02}}}}{2\sqrt{\varepsilon_{02}}}\left(\frac{e^{-iph}}{p+k_0\sqrt{\varepsilon_{02}}}-\frac{e^{iph}}{p-k_0\sqrt{\varepsilon_{02}}}\right)\right|$$
$$\leq\left|\frac{k_0 n e^{-k_0h\alpha_2}}{2}\right|\cdot\left|\frac{\sigma_\nu}{\sqrt{\varepsilon_{02}}}\right|\cdot\left|\frac{2ip\sin ph-2k_0\sqrt{\varepsilon_{02}}\cos ph}{p^2-k_0^2\varepsilon_{02}}\right| \tag{8.72}$$

where $\alpha_2=\mathrm{Im}\sqrt{\varepsilon_{02}}$.

Since Eq. (8.72) has to be valid for any value p (even for p close to zero), it is necessary that

$$ne^{-k_0 h \alpha_2} \left| \frac{\sigma_v}{\varepsilon_{02}} \right| << 1 \tag{8.73}$$

from which it follows that

$$\left| \frac{\sigma_v}{\varepsilon_{02}} \right| << \frac{e^{k_0 h \alpha_2}}{n} \tag{8.74}$$

Later on we shall illustrate the use of the above approximation in calculating the reflections from several types of ice. Experiments show that for marine ice the fluctuations in permittivity, as a rule, do not exceed 14%. It is clear, then, that in actual engineering practice the reduced restriction

$$\left| \frac{\sigma_v}{\varepsilon_{02}} \right| << 1 \tag{8.75}$$

really occurs. However, it should be pointed out that even when this condition is satisfied, the first-order approximation appears to be unrealistic if p equals $\mathrm{Re}\left\{k_0 \sqrt{\varepsilon_{02}}\right\}$. This can happen for a non-attenuating layer. In this case, Eq. (8.60) assumes the form

$$A\left(k_0 \sqrt{\varepsilon_{02}}\right) + \frac{1}{\pi} \int_{-\infty}^{\infty} C(p') F\left(k_0 \sqrt{\varepsilon_{02}}, p'\right) dp' = 0 \tag{8.76}$$

wherein the second part cannot be neglected.

Let us return to the level of the first-order Rytov approximation. In the light of our discussion, we assume that

$$C(p) \approx C_o(p) = \frac{k_0^2 A(p)}{p^2 - k_0^2 \varepsilon_{02}} \tag{8.77}$$

Substituting this expression in Eq. (8.66) and using the continuity of the tangential field components at the boundary, we obtain at the interface with medium III the following system of equations for the determination of the desired reflection coefficient R:

$$\left.\begin{aligned} 1+R &= C_3 + C_4 + k_0^2 I_1 \\ 1-R &= \sqrt{\varepsilon_{02}}\left(C_3 - C_4\right) \\ C_3 e^{-ik_0 h\sqrt{\varepsilon_{02}}} + C_4 e^{ik_0 h\sqrt{\varepsilon_{02}}} &= -k_0^2 I_2 \end{aligned}\right\} \tag{8.78}$$

Here,

$$\left.\begin{aligned} I_1 &= \int_{-\infty}^{\infty} \frac{A(p)}{p^2 - k_0^2 \varepsilon_{02}} dp \\ I_2 &= \int_{-\infty}^{\infty} \frac{A(p)\cos ph}{p^2 - k_0^2 \varepsilon_{02}} dp \end{aligned}\right\} \tag{8.79}$$

In Eq. (8.78), a high conductivity for medium III is allowed, together with the use of the Leontovich boundary conditions. Solving this system of equations we obtain

$$R = R_0 - \frac{k_0^2 \varepsilon_{02}}{D_1}\left(\gamma I_1 - 2I_2\right) \tag{8.80}$$

In this expression, R_0 is the Fresnel reflection coefficient with no change in ε_{02} and vertical incidence; furthermore,

$$\left.\begin{aligned} \gamma &= e^{ik_0 h\sqrt{\varepsilon_{02}}} + e^{-ik_0 h\sqrt{\varepsilon_{02}}} \\ D_1 &= 2e^{-ik_0 h\sqrt{\varepsilon_{02}}} - \gamma\left(1+\sqrt{\varepsilon_{02}}\right) \end{aligned}\right\} \tag{8.81}$$

Taking Eq. (8.48) into account, the averaged reflection coefficient becomes

$$\overline{R} = R_0 \tag{8.82}$$

i.e., within the limits of small deviations in permittivity ε_{02}, layer II does not affect the averaged reflection coefficient. The average reflected power is proportional to $\overline{|R|^2}$, where

$$\overline{|R|^2} = |R_0|^2 + \frac{k_0^4|\varepsilon_{02}|}{|D_1|^2}\left\{|\gamma|^2 \int_{-\infty}^{\infty}\int_{-\infty}^{\infty} \frac{\overline{A(p)A^*(p')}}{(p^2-k_0^2\varepsilon_{02})(p'^2-k_0^2\varepsilon_{02}^*)}dpdp' + \right.$$
$$\left. +4\int_{-\infty}^{\infty}\int_{-\infty}^{\infty} \frac{\overline{A(p)A^*(p')\cos ph\cos p'h}}{(p^2-k_0^2\varepsilon_{02})(p'^2-k_0^2\varepsilon_{02}^*)}dpdp' - 4\,\mathrm{Re}\left[\gamma\cdot\int_{-\infty}^{\infty}\int_{-\infty}^{\infty} \frac{\overline{A(p)A^*(p')\cos ph}}{(p^2-k_0^2\varepsilon_{02})(p'^2-k_0^2\varepsilon_{02}^*)}dpdp'\right]\right\} \tag{8.83}$$

The average $\overline{A(p)A^*(p')}$ entering into the last equation can be expressed in terms of the correlation function $K(z-z')=\overline{\nu(z)\nu(z')}$:

$$\overline{A(p)A^*(p')} = \frac{1}{\pi}\int_{-h}^{0}\int_{-h}^{0} E_{02}(z)E_{02}^*(z')K(z-z')\cos pz\cos pz'dzdz' \tag{8.84}$$

For a well-chosen correlation function $K(z-z')$ it is possible to calculate analytically the average reflected power. From this it becomes clear that (within the limits of the applied approximations) knowledge of the medium can be gained from the reflected power.

It is possible to show that Eq. (8.83) can be transformed into (Appendix F)

$$\overline{|R|^2} = |R_0|^2 + \frac{16k_0}{\left|\sqrt{\varepsilon_{02}}\cos kh\sqrt{\varepsilon_{02}} + i\sin kh\sqrt{\varepsilon_{02}}\right|^4} \times\int_0^h\int_0^h K(z-z')\sin^2\left(k\sqrt{\varepsilon_{02}}z\right)\sin^2\left(k\sqrt{\varepsilon_{02}}z'\right)dzdz' \tag{8.85}$$

As an example, for marine ice lying on water, we assume a correlation function of the form

$$K(z-z') = \frac{\sigma_\nu^2}{2r}e^{-\frac{|z-z'|}{r}} \tag{8.86}$$

r being the correlation radius and σ_ν^2 the variance of $\nu(z)$. It should be noted that $K(z-z')\to\delta(z-z')$ as $r\to 0$, while $K(z-z')\to 0$ as $r\to\infty$.

Eq. (8.83) is now written as:

$$\overline{|R|^2} = |R_0|^2 + \Delta \tag{8.87}$$

The additional term Δ is caused by the randomness of the permittivity. For

$$\left.\begin{array}{c} \varepsilon_{02} = \varepsilon'_{02}\left(1 - i\tan\delta_2\right) \\ \varepsilon'_{02} = 4.0; \quad \tan\delta_2 = 0.1 - 2.5; \\ \dfrac{h}{\lambda} = 0 - 2; \quad \dfrac{r}{\lambda} = 0 - 20; \quad \sigma_v = 0;05\left|\varepsilon_{02}\right| \end{array}\right\} \tag{8.88}$$

numerical results are shown in Figs 8.30 to 8.34.

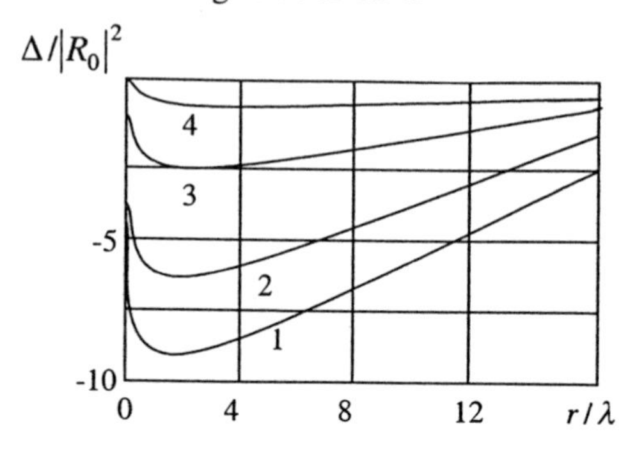

Fig. 8.30 $\Delta/|R_0|^2$ [%] as a function of r/λ. $h/\lambda = 0.1$; $\varepsilon = 40$
1) $\tan\delta = 0.1$ 2) $\tan\delta = 0.2$ 3) $\tan\delta = 0.5$ 4) $\tan\delta = 0.6$

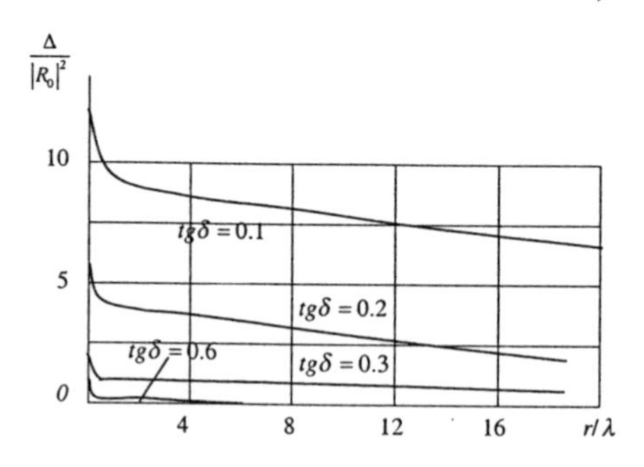

Fig. 8.31 $\Delta/|R_0|^2$ % as a function of r/λ. $h/\lambda = 0.2$; $\varepsilon = 40$

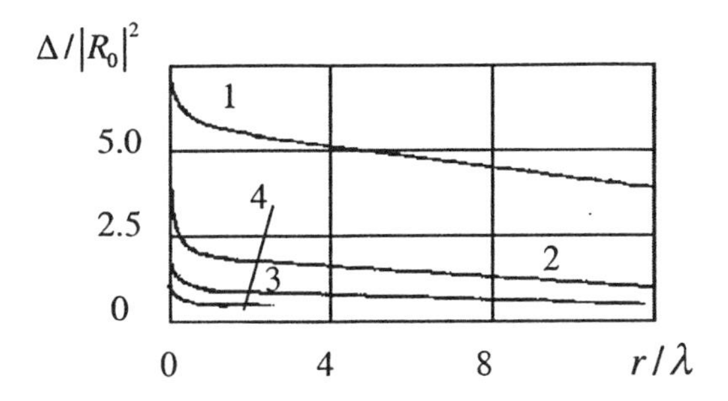

Fig. 8.32 $\Delta/|R_0|^2$[%] as a function of r/λ. $\varepsilon = 40$; $h/\lambda = 0.3$
1) $\tan\delta = 0.1$ 2) $\tan\delta = 0.2$ 3) $\tan\delta = 0.3$ 4) $\tan\delta = 0.4$

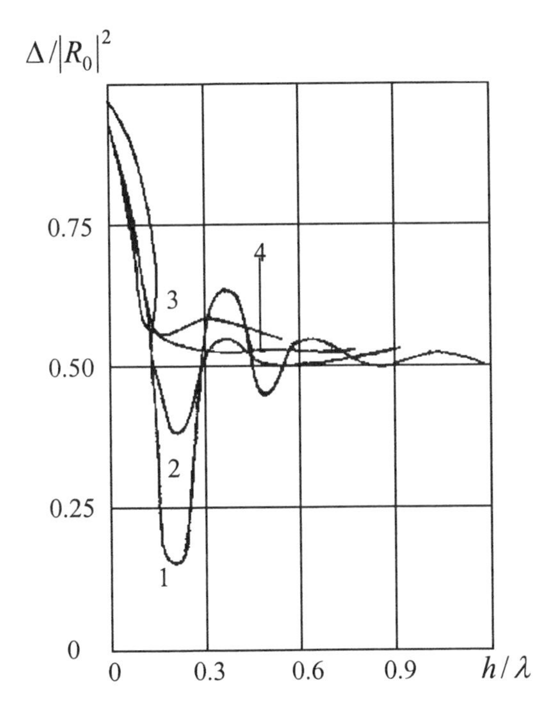

Fig. 8.33 $\Delta/|R_0|^2$ [%] as function of h/λ. $r/\lambda > 2$, $\varepsilon = 40$
1) $\tan\delta = 0.1$ 2) $\tan\delta = 0.2$ 3) $\tan\delta = 0.3$ 4) $\tan\delta = 0.8$

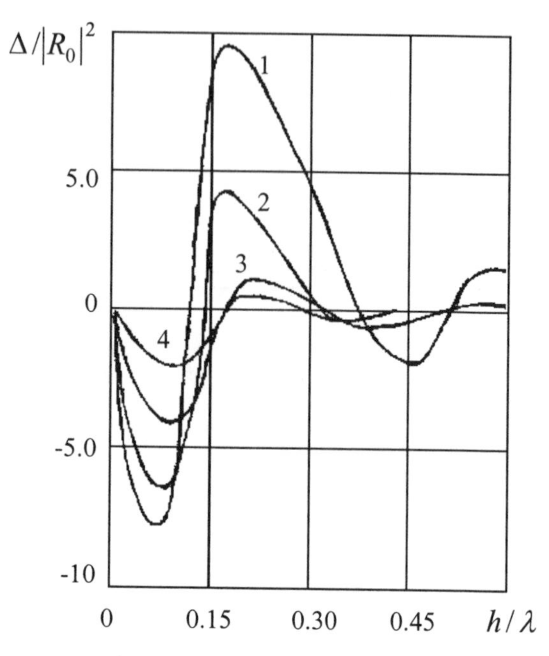

Fig. 8.34 $\Delta/|R_0|^2$ [%] as function of h/λ. $\varepsilon = 40$; $0.4 \le r/\lambda \le 2$
1) $\tan\delta = 0.1$ 2) $\tan\delta = 0.2$ 3) $\tan\delta = 0.3$ 4) $\tan\delta = 0.5$

After analyzing the results, we may draw the following conclusions:

- A small change in permittivity can appreciably affect the reflected power (for example, a 5% root-mean square deviation in ε_2 may result in a change of $\Delta/|R_0|^2$ up to 10%).

- The largest difference takes place when the correlation function approaches the δ function ($r \to 0$).

- At some r/λ, h/λ, the quantity $\Delta/|R_0|^2$ becomes zero. This means, that for given $\frac{r}{\lambda}$ and ice thickness relative to wavelength, changes in ε_2 do not affect the average reflected power.

- An increase of $\frac{r}{\lambda}$ and $\frac{h}{\lambda}$ results into a higher average reflected power.

In Fig. 8.35a, the relation $\left(|R|^2\right)$ is shown, for a homogeneous ice layer as function of its thickness relative to wavelength and various tanδ values. A similar plot is shown in Fig. 8.35b for $\left(|R|^2/|R_0|^2\right)$.

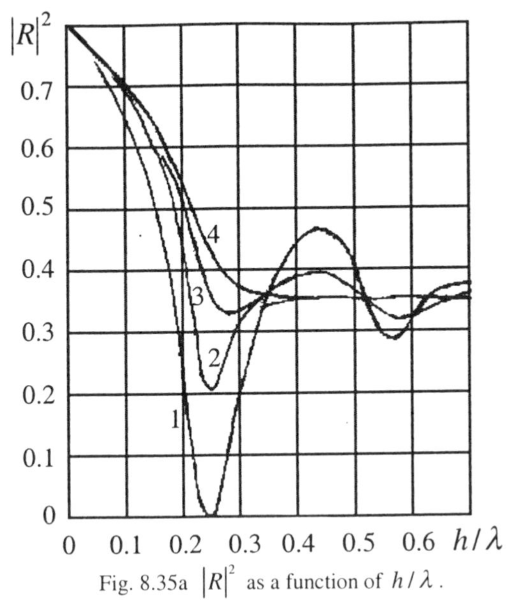

Fig. 8.35a $|R|^2$ as a function of h/λ.

1) $\tan\delta = 0.1$ 2) $\tan\delta = 0.2$ 3) $\tan\delta = 0.3$ 4) $\tan\delta = 0.8$

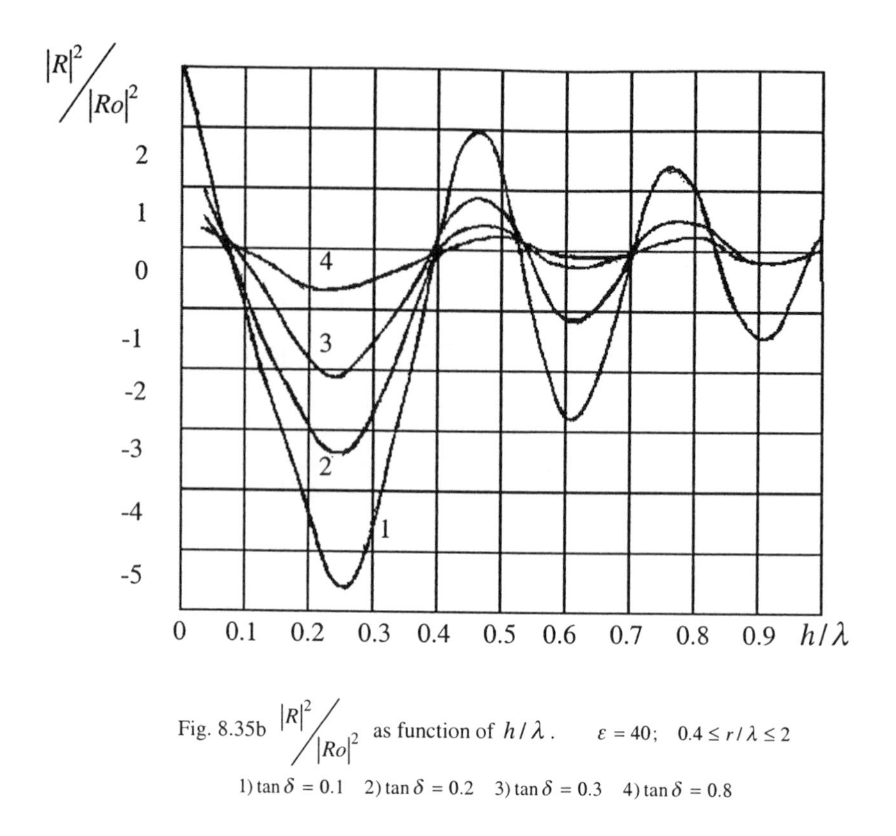

Fig. 8.35b $|R|^2 / |Ro|^2$ as function of h/λ. $\varepsilon = 40$; $0.4 \le r/\lambda \le 2$

1) $\tan\delta = 0.1$ 2) $\tan\delta = 0.2$ 3) $\tan\delta = 0.3$ 4) $\tan\delta = 0.8$

The results do not appear to depend strongly on r/λ for values ranging from 0.4 up to 2.0, while a higher $\tan\delta$ results in reduction of the reflection coefficient.

At a certain ice thickness, the envelope of the reflected wave attains extreme values. (For example, at a thickness of a quarter wave length there is a minimum).

In Figs 8.36 and 8.37, the relation $\left(|R|^2/|R_0|^2\right)$ as function of the ratio between correlation radius and wavelength (r/λ) is indicated for a fixed layer thickness to wavelength ratio (h/λ) and variable parameter $\tan\delta$. Fig. 8.36 shows results for $h/\lambda = 0.2$ and $\tan\delta = 0.1$, 0.2, 0.8 and 1. Fig. 8.37 shows results for $h/\lambda = 0.3$ and $\tan\ \delta\ =\ 0.1$, 0.2 and 0.3. The relative permittivity in both figures is assumed to be constant, $\varepsilon = 40$.

An increase in the correlation radius leads to a reduction in difference compared to a homogeneous layer. Thus, due to an increase in correlation radius r, the changes in the permittivity become smoother. At low r, the correlation grows and has its maximum at $r = 0$.

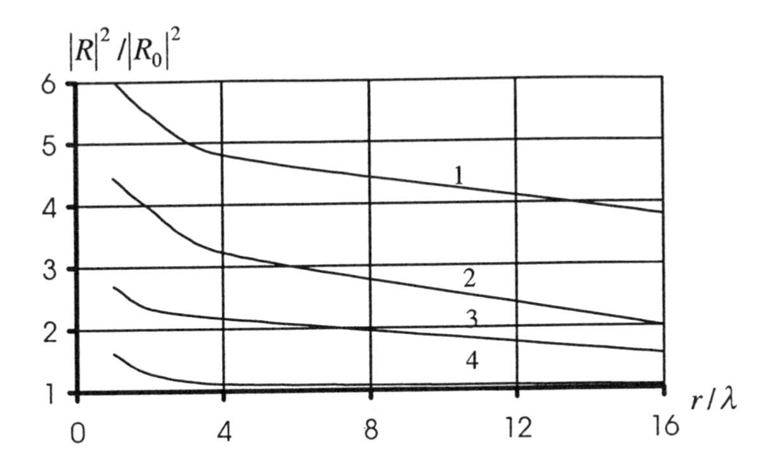

Fig. 8.36 Dependences of $|R|^2/|R_0|^2$ on r/λ.

$\varepsilon = 40$; $h/\lambda = 0.2$. 1) $\tan\delta = 1$ 2) $\tan\delta = 0.8$ 3) $\tan\delta = 0.2$ 4) $\tan\delta = 0.1$

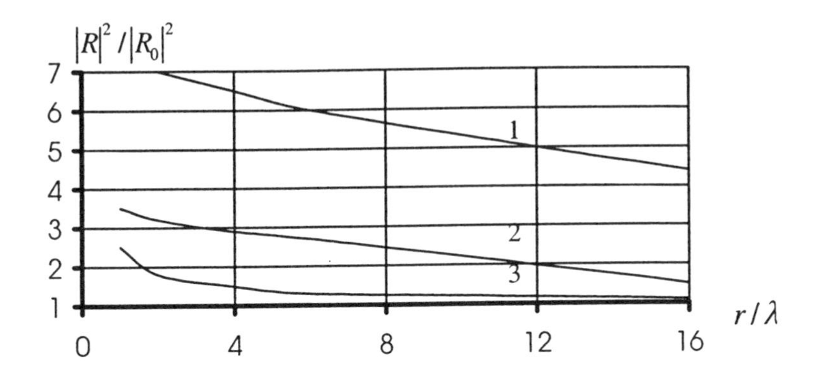

Fig. 8.37 Dependences of $|R|^2/|R_0|^2$ on r/λ.

$\varepsilon = 40$; $h/\lambda = 0.3$ 1) $\tan\delta = 0.3$ 2) $\tan\delta = 0.2$ 3) $\tan\delta = 0.1$

The applicable boundaries of the model can be established through analysis of Eq. (8.85). It can be shown that for rather large h and r values the expression $\left(1-|R|^2/|R_0|^2\right)$ will be determined by :

$$\left(1-|R|^2/|R_0|^2\right)=\sigma_v^2\frac{d_{sl}}{r}\frac{1}{\sqrt{|\varepsilon_{02}|}\left(1+|\varepsilon_{02}|+2\sqrt{|\varepsilon_{02}|}\cos\frac{\delta_2}{2}\right)} \tag{8.89}$$

where d_{sl} is the skin depth. For small correlation radii, $\left(1-|R|^2/|R_0|^2\right)$ converges to the limit:

$$\left(1-\langle|R|^2\rangle/|R_0|^2\right)=\sigma_v^2\frac{kd_{sl}}{\left(1+|\varepsilon_{02}|+2\sqrt{|\varepsilon_{02}|}\cos\frac{\delta_2}{2}\right)} \tag{8.90}$$

The result can be generalized for waves incident on the layer at an arbitrary β angle. In the case of horizontal polarization the electrical field $\vec{E}_s=\vec{j}E_{ys}$ (s = 1, 2, 3) in various media is represented by

$$E_{ys}=e^{ik_0x\sin\beta}U_s(z) \tag{8.91}$$

where $U_s(z)$ satisfies Eq. (8.49), in which it is necessary to replace ε_s by $\left(\varepsilon_s-\sin^2\beta\right)$. Repeating all previous calculations, we obtain an expression similar to Eq. (8.80), i.e.,

$$\overline{R}=R_0-\frac{k_0^2\left|\varepsilon_{02}-\sin^2\beta\right|}{D_2}\left|\gamma I_1-2I_2\right| \tag{8.92}$$

Here, R_0 is the reflection coefficient from a layer of constant permittivity ε_{02} and

$$\left.\begin{aligned} &\gamma = 2\cos\left(k_0 h\sqrt{\varepsilon_{02} - \sin^2\beta}\right) \\ &D_2 = 2e^{-ik_0 h\sqrt{\varepsilon_{02}-\sin^2\beta}} - \gamma\cos\beta - \gamma\sqrt{\varepsilon_{02} - \sin^2\beta} \\ &I_1 = \int_{-\infty}^{\infty} \frac{A(p)}{p^2 - k_0^2\left(\varepsilon_{02} - \sin^2\beta\right)} dp \\ &I_2 = \int_{-\infty}^{\infty} \frac{A(p)\cos ph}{p^2 - k_0^2\left(\varepsilon_{02} - \sin^2\beta\right)} dp \end{aligned}\right\} \tag{8.93}$$

Eq. (8.56) should then be replaced by an appropriate expression for waves incident at an angle β. Using Eqs (8.92) and (8.93) and expressions for C_5 and C_6, together with a particular correlation function, it is possible to derive $\overline{|R|^2}$ for arbitrary β following an approach similar to that for $\beta = 0$.

Results permit us to draw a number of conclusions. The average reflected power from a non-uniform layer depends on electrical, geometrical and statistical parameters. Computations are possible by using the correlation function of the permittivity. The largest difference of average reflected power compared with the case of a homogeneous layer takes place for small correlation radii. The differences are enlarged with increasing layer thickness while an increase in $\tan\delta$ reduces the differences. At certain frequencies a minimum difference occurs. An increase in correlation radius gives an increase in the reflection coefficient magnitude and is proportional to layer thickness and correlation radius.

For a reduced correlation radius, this increase is of limited significance and becomes proportional to the correlation radius and the wavelength.

8.3.3 Reflection from a surface as volume scattering

Under certain conditions, surface scattering can be profitably considered as volume scattering of an incident electromagnetic wave passing through a randomly inhomogeneous medium. The medium is characterized by fluctuations of the dielectric constant with respect to its average [*Andreev*, 1988; *Andreev*, 1989; *Slutsky*, 1989]. The discontinuities in the medium dielectric constant are described by the following relationship:

$$\delta\varepsilon(\vec{r}) = \begin{cases} \varepsilon_2 - \varepsilon_1, & \text{if} \quad 0 \le z \le h(\vec{\rho}) \\ \varepsilon_1 - \varepsilon_2, & \text{if} \quad z \le 0 \quad \text{or} \quad z \ge h(\vec{\rho}) \end{cases} \tag{8.94}$$

Here, h is the surface height profile and $\vec{\rho}$ is the vector difference between two position vectors on a surface described by a Cartesian function $z = f(x, y)$. The relationship given in Eq. (8.94) means that if a real rough surface is elevated in some area over an average surface $z = 0$, then the fluctuation on the average permittivity can be described by the upper equation of relationship (Eq. 8.94), otherwise by the lower equation of Eq. (8.94). Permittivity ε_1 corresponds to the uppermost region, ε_2 to the lowest region.

Such an approach results in an integral equation for an average field. The solution of this equation can be expressed in the form of a series of the parameter: $k\sqrt{<h^2>}$, where h is the surface elevation. We can distinguish three parameter regions:

a) $kh << 1;\quad (\varepsilon_1 - \varepsilon_2)/\varepsilon_1 \approx 1$ (8.95a)

b) $kh \sim 1;\quad (\varepsilon_1 - \varepsilon_2)/\varepsilon_1 << 1$ (8.95b)

c) $kh >> 1;\quad (\varepsilon_1 - \varepsilon_2)/\varepsilon_1 << 1$ (8.95c)

Region (a) (the thin transition layer) is considered below:

The reflection coefficients for linearly polarized incident waves have been calculated. As the solution is based on the application of an iteration procedure, the choice of a zero-th order approximation is significant. [*Slutsky*, 1989] has analyzed the standard problem of reflection from the dielectric discontinuity

$$\varepsilon(z) = \varepsilon_1 + (\varepsilon_1 - \varepsilon_2)\chi(z) \tag{8.96}$$

where $\chi(z)$ is the Heaviside function.

Another proposed model [*Andreev*, 1988, 1989] seems to be more flexible. In this model, the scattering from surface acoustic waves in a crystal was analyzed, with the following relationship chosen as zero-th order approximation:

$$\varepsilon(z) = 1 + \frac{\alpha}{2}\tanh(\beta z) \tag{8.97}$$

There is no rationale given for the choices of the permittivity discontinuities mentioned so far; however, more recent investigations have shown their utility [*Kozlov*, 1992, *Lax*, 1951]. For this reason, we shall discuss these problems in a little more detail. The approach proposed by [*Kuznetsov*, 1991] can be considered as a development of the *Rayleigh* method, whereby a plane wave interacts with a surface and is transformed into radiation with a specified angular spectrum. The amplitude of the angular spectrum of the scattered field is found from the boundary conditions on the surface, by matching the solutions for the scattered field and the incident field under the assumption that the magnitude of roughness (the thickness of a rough layer) is small.

A transition layer of finite thickness in which an angular spectrum is varying smoothly is proposed. The equations describing the transformation of this angular spectrum for incident waves (transmitted through the second medium), as well as for reflected waves within a transition layer, are constructed. The medium is considered to be a volume scatterer. With this approach, an elementary layer – the scattering within this layer can be calculated by means of the Born approximation – with thickness Δz is selected within the transition layer. Knowing the angular spectrum of waves entering this layer, we can calculate the amplitude of the plane waves leaving the elementary layer. Passing to the limit $\Delta z \to 0$, the finite-difference equations become differential equations.

The scattering from a two-scale surface representing a regular cylindrical saw tooth surface characterized by slight perturbations in amplitude was considered by [*Kuznetsov*, 1991; 1993]. The spatial period Λ and the amplitude A of the regular surface have a restricted range of values. The angular spectrum can be calculated using an integral equation approach [*Kuznetsov*, 1993]. Applying an iteration procedure in order to solve the integral equation for the electric field, it is seen that in the zero-th order approximation the incident wave propagates through a planar transition layer with the following effective dielectric constant:

$$\varepsilon_{eff} = \left[\varepsilon_1 - (\varepsilon_2 - \varepsilon_1)\frac{z}{A}\right] \tag{8.98}$$

This relationship, derived as a result of the solution of the scattering problem, is characterized better by the model given in Eq. (8.97) than that in Eq. (8.96). The series convergence for the model given in Eq. (8.97) will be quicker. The result given in Eq. (8.98) can easily be realized by taking into account estimations of the effective dielectric constant as representative of a composite of media with different ε.

8.4 Conclusions and applications

In this chapter, we have discussed the reflection of electromagnetic waves from layered structures characterized by different laws of variation of the permittivity with layer depth. The cases of deterministic and random variation of the permittivity with layer depth have been considered separately.

As a first model of a deterministic variation of the permittivity we assumed an exponential law in one of the layers of the four-layer structure. The other layers were assumed to be homogeneous with constant permittivity. Such a model may be used for earth surfaces covered with vegetation, for sea surface covered with ice, etc. In this model, we considered both vertical and horizontal polarizations for the incident wave. The derived expressions for the reflection coefficients turned out to be quite complicated; for that reason, we undertook a numerical analysis for a four-layer structure in the form of air – sea ice – transition layer – water. The results showed that the largest changes in the reflection coefficient occurred when the ratio of layer depth to wave length was in the range of 0.1 to 1.0. For thick transition layers, specific information on the inner structure of the layer is practically unavailable.

The derived results allowed us to ascertain the dependencies of the reflection coefficients of the electromagnetic waves on the angle of incidence, the thickness of the layer, mode of polarization, etc. In addition, these results were derived for various ranges of wavelengths, specifically, for centimeter, decimeter and meter ranges. Although the results did not differ quantitatively, they exhibited substantial qualitative differences.

In a similar manner, a model was considered whereby the permittivity changed with layer depth according to a polynomial law. In that case, we also derived numerical results for ice covers because the analytic expressions for the wave reflection coefficients turned out to be very complicated. The main conclusion of that analysis was that the order of the polynomial weakly affected the value of the reflection coefficients and that allowed us to choose not very high orders of polynomials confining ourselves to simple relations.

The aforementioned two models, being deterministic, give only a very approximate pattern of scattering of electromagnetic waves from layered surfaces, but in some cases they may turn out to be useful for certain structures and give rough quantitative results.

Stochastic models of three-layer media with flat boundaries represent a more general case. The permittivity is a certain random quantity, which changes stochastically with depth. The problem of determination of the reflection coefficients is to be solved by the Rytov method, as a result of application of which Fredholm equations of second order are derived. The Fredholm equations may be solved by the iteration method. Then, we assume the condition that the average value of scattering of a random quantity describing the permittivity does not exceed 10-15%. This statement is valid for sea ice and allows us to derive quantitative results for such a medium. These results demonstrate quite substantial changes in power of the reflected wave with relatively small mean-square values of the permittivity variance. The derived relations for the reflected wave power are complicated in character when there are changes in the layer depth and the correlation radius of the random quantity relative to the incident wavelength. With some ratios of layer depth to incident wave length, the reflected signal power reaches extreme values. An increase in the correlation radius results in a decrease in differences relative to reflection from a homogeneous layer.

Possible fields of application of the results in this chapter are as follows. First of all, with the use of them, we may solve direct problems of remote sensing for those surfaces that may be described by multiple layer models (e.g., three-layer and four-layer models). The aforementioned relations and numerical calculations allow us to determine the reflection coefficients of electromagnetic waves and the reflected power as functions of the depth of a layer in which a deterministic or stochastic variation of the permittivity with depth takes place. The random variation of permittivity is characterized by its correlation radius, for various ranges of wavelengths. Most realistic surfaces are described exactly by such multiple layer models of electromagnetic wave reflection.

As it has been mentioned in previous chapters, the solution of inverse problems of remote sensing often is based on the results of the solution of direct problems. Therefore, the results presented in this chapter have practical significance for the solution of inverse problems. We have to mention several restrictions of the derived results. These restrictions are connected with the consideration of flat boundaries without taking account of roughness. However, the questions of influence of surface roughness on the reflection of electromagnetic waves will be considered separately later on.

It has been pointed out, the results in this chapter on the reflection of electromagnetic waves from multiple layer media may be used for the investigation of sea ice, ice located on the soil surface, for surfaces covered with vegetation, for agricultural lands, etc. The wide variety of possible objects that may be remotely sensed makes the presented results quite useful in practical applications.

CHAPTER 9

Radiowave Reflection from Structures with Internal Ruptures

9.1 Introduction

In some radar remote sensing problems connected with investigations of the earth's surface, it is necessary to investigate reflections from structures with internal ruptures (e.g., fractures in ice ravines), or with all kinds of hollow spaces. Application of electrodynamic models of ruptures (fractures) in the form of an endless deep pit with vertical walls and with a reflection coefficient equal to zero, turns out to be ill-defined. Experimental data shows that the reflection coefficient may have quite a significant magnitude. A slight deviation of the pit walls parallel orientation can also cause this effect. Analysis based on models therefore must always be extended very carefully.

The reflection of electromagnetic waves from structures characterized by internal ruptures is studied in this chapter under different polarization conditions. Three specific two-dimensional structures are considered in detail: a symmetric and an asymmetric wedge-shaped fracture, a parallel wall's fracture with a finite depth and a pit in the shape of a spherical surface. Characteristic dimensions are assumed to be much larger than the wavelength so that geometrical optics approximations are valid.

In general, a radar senses a structure with a rupture at a specific angle using a given antenna pattern. Relationships between the reflection coefficient with the observation angle θ, the geometrical dimensions, and the electrophysical parameters of the pit must then be determined. Strictly speaking, diffraction theory is needed to solve such problems; however, if we assume that the dimensions of a pit are much larger than the wavelength, geometrical-optics approximations turn out to be very fruitful.

9.2 Reflection from a symmetrical wedge-shaped fracture

We start with a plane wave incident upon the structure under consideration at an angle β. The reflection coefficient of this wave from the walls of a wedge-shaped fracture has to be determined. After repeated reflections, the wave will leave the wedge-shaped fracture at an angle θ with respect to the horizon. This angle can be determined by the geometrical consideration of repeated reflections from the walls of the fracture, as is shown in Fig. 9.1a. The spherical pit model and a model of a rectangular pit with a finite depth are shown in Figs 9.1b and 9.1c.

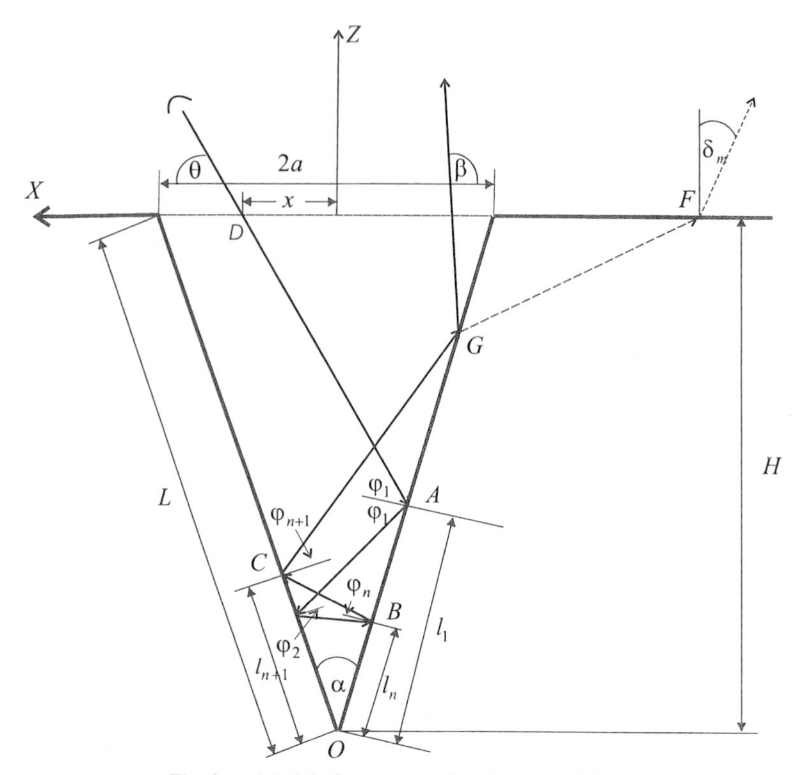

Fig. 9.1a Model of symmetrical wedge-shaped fracture

In Fig. 9.1a, the angle of the wedge is α, the depth of the pit is H, the dimension of a sidewall is L, the angle of the antenna beam with the horizon is θ and the distance between the vertex O of the wedge and point A is l_1. The reflection coefficient of the symmetric wedge can now be defined by

$$R = R(\varphi_1) R(\varphi_2) ... R(\varphi_m) = \prod_{j=1}^{M} R(\varphi_j) \tag{9.1}$$

where φ_j is the angle of an incident beam on the wedge wall after the *j-th* reflection, and φ_1 corresponds to the last emergent beam.

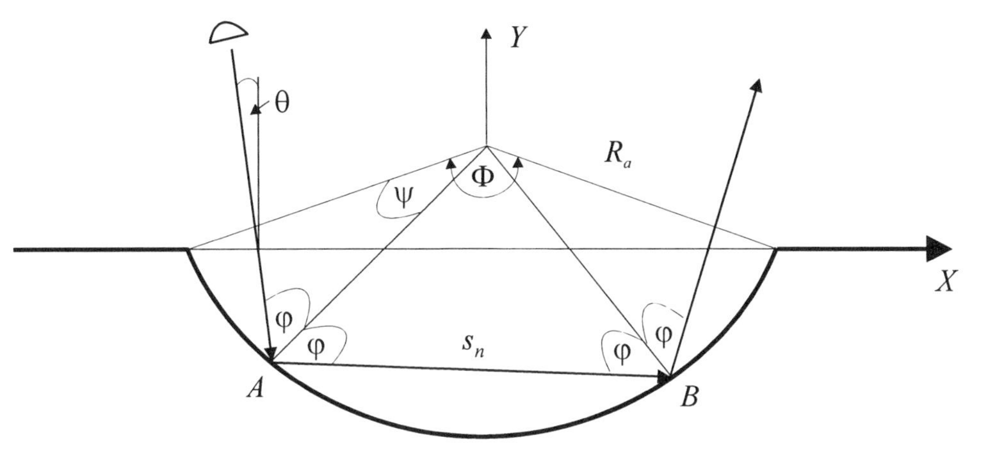

Fig. 9.1b Model of a pit with spherical form

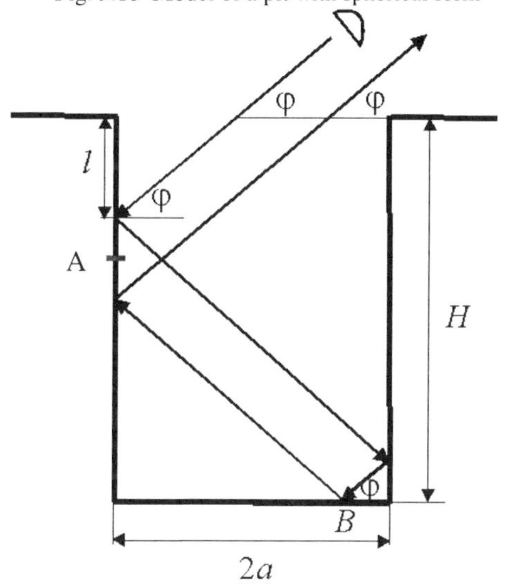

Fig. 9.1c Model of rectangular pit with a finite depth

By $R(\varphi)$ we mean the plane-wave reflection coefficient from that boundary at an incidence angle φ. $R(\varphi)$ depends on the complex permittivity of the wall material.

Thus, the task requires the determination of the reflection angles φ_j and the total number of reflections M.

From triangle BOC in Fig. 9.1a, we can easily find the relationship between angles φ_n and φ_{n+1} :

$$\varphi_{n+1} = \varphi_n - a \tag{9.2}$$

This recurrent expression allows us to derive the angle φ_n for an initial angle φ_1, and after n reflections we have

$$\varphi_n = \varphi_1 - (n-1)\alpha \tag{9.3}$$

In order to define the number of reflections, it is necessary to know l_n, i.e., the distance from the point of *n-th* reflection B to the vertex O.
In BOC we see the recurrent expression

$$l_{n+1} = \frac{l_n \cos\varphi_n}{\cos(\varphi_n - \alpha)} \tag{9.4}$$

Using Eqs (9.3) and (9.4) we get

$$l_{n+1} = \frac{l_1 \cos\varphi_1}{\cos(\varphi_1 - n\alpha)} \tag{9.5}$$

The last reflection will take place at a number $n = M$ at which the following inequalities hold:

$$\left.\begin{aligned} l_{n+1} > L \\ l_n \leq L \end{aligned}\right\} \tag{9.6}$$

These inequalities are transformed into

$$\left.\begin{aligned} n\alpha > \varphi_1 + \arccos(\xi\cos\varphi_1) \\ (n-1)\alpha \leq \varphi_1 + \arccos(\xi\cos\varphi_1) \end{aligned}\right\} \tag{9.7}$$

where $\xi = l_1 / L$.

Now we may determine the number of reflections M :

$$M = \left[\frac{\varphi_1}{\alpha} + \frac{\arccos(\xi \cos\varphi_1)}{\alpha} \right] + 1 \tag{9.8}$$

Here, the bracket [...] indicates the integer part of the enclosed expression.

The parameter ξ can be expressed more naturally using the coordinate of the intersection of the sensor beam with the entrance plane of the pit. In this case, however, the formula becomes more complicated; specifically, we find

$$M = \left[\frac{\theta}{\alpha} - \frac{1}{2} + \frac{\arccos\left(\cos\theta \cos\frac{\alpha}{2} + \eta \sin\theta \right)}{\alpha} \right] + 1 \tag{9.9}$$

where: $-\sin\frac{\alpha}{2} \le \eta = \frac{x}{L} \le \sin\frac{\alpha}{2}$.

The (composite) reflection coefficient is then given by

$$|R|^2 = \prod_{j=1}^{M} \left| R[\varphi_1 - (j-1)\alpha] \right|^2 \tag{9.10}$$

with the following relationship between the angles φ_1 and θ:

$$\varphi_1 = \theta - \frac{\alpha}{2} \tag{9.11}$$

From this we can easily derive that only one reflection takes place for:

$$\varphi_1 \le 0, \text{ i.e., } \theta \le \frac{\alpha}{2} \tag{9.12}$$

The maximum number of reflections is found from Eq. (9.8), for $\varphi_1 = 90^\circ$:

$$M_{\max} = \left[\frac{180^o}{\alpha}\right] + 1 \tag{9.13}$$

This happens when

$$\theta = 90^o - \frac{\alpha}{2} \tag{9.14}$$

In Fig. 9.2 we show $|R(\varphi)|^2$ as a function of the incidence angle and for characteristic wall materials.

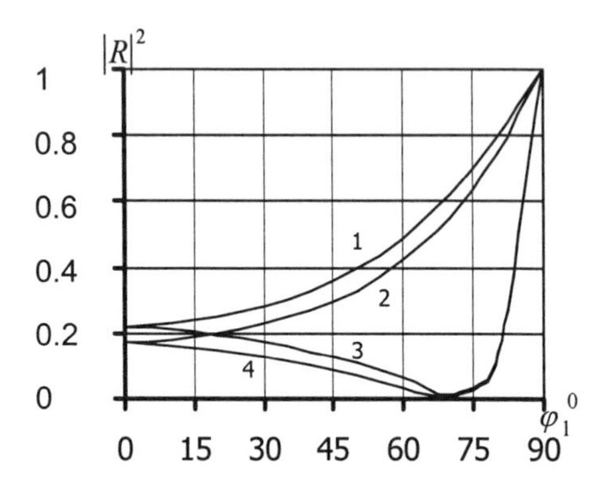

Fig. 9.2 Dependence $|R|^2$ and φ_1^0 HP – 1,2; VP – 3,4
1,3 – ground $\varepsilon = 8.0 \cdot (1 - i \cdot 0.1)$ 2,4 – concrete $\varepsilon = 6.25 \cdot (1 - i \cdot 0.01)$

The relationships are shown for horizontal (HP) and vertical (VP) polarizations. Values of the complex permittivities are taken:

$$\begin{aligned} \varepsilon_g &= 8.0(1 - i0.1) \quad \text{(earth)} \\ \varepsilon_c &= 6.25(1 - i0.01) \quad \text{(concrete)} \end{aligned} \tag{9.15}$$

The abscissa in the figure marks the angle $\varphi_1 = \left|\theta - \frac{\alpha}{2}\right|$. Intervals equal to α also mark the number of reflection points (M-1), etc. The vertical axis gives the product in Eq. (9.10).

9.2.1 Vertical probing

For vertical probing ($\theta = 90^o, \varphi = 90^o - \frac{\alpha}{2}$), the number of reflections defined by (9.9) becomes

$$M = \left[\frac{90^o}{\alpha} - \frac{1}{2} + \frac{\arccos\eta}{\alpha} \right] \tag{9.16}$$

Analysis of this expression, when geometrical optics is valid, yields:

$$M = \left[\frac{180^o}{\alpha} \right] \tag{9.17}$$

This means, that for vertical probing the reflection coefficient of a wedge-shaped fracture is determined by the wedge angle α only and does not depend on the point of probing. However, for angles $\alpha \approx 95^\circ - 110^\circ$, this conclusion is not true, because at these angles the number of reflections appears to be dependent on the coordinate x. The relationship between the magnitude $\left(\left|R(\alpha)\right|^2\right)$ and the wedge angle α for concrete fractures is shown in Fig. 9.3 for vertical and horizontal polarizations.

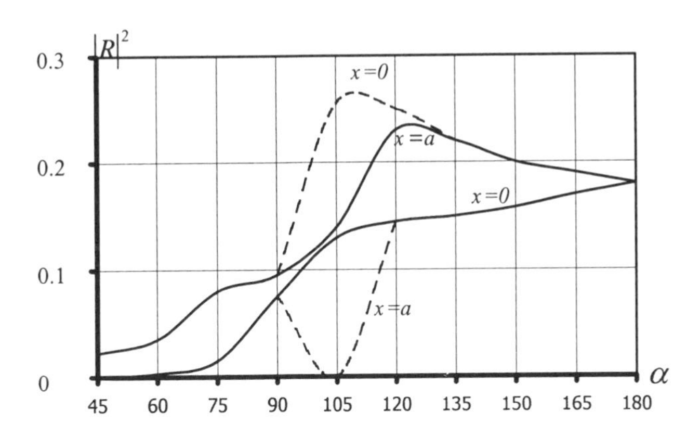

Fig. 9.3 Dependence of $|R|^2$ on α
Lower curves – VP, Upper curves – HP

The rise in the curves for HP and in the interval $\alpha < 105^{\circ}$ is due to the decrease of the number of reflections and the drop for $\alpha < 105^{\circ}$ is due to the decrease of the reflection coefficient associated with the decrease of the angle φ_1 for the same number of reflections. The angle $\alpha < 108^{\circ}$ corresponds to a smooth surface.

The cross-hatched area in the plots reflects the relationship between the magnitude ($|R(\varphi)|^2$) and the coordinate x. The lower bound of this area corresponds to $x = 0$ and the upper bound to $x = a$. There is no such relationship for other intervals of the curves.

Fig. 9.3 shows that for vertical probing a fracture can be considered as a blackbody for angles up to $\alpha \approx 40^{\circ} - 45^{\circ}$ for horizontal polarization and up to $\alpha \approx 75^{\circ}$ for vertical polarization.

9.2.2 Probing at low grazing angles

Now we discuss observations at angles θ, where the beams pass in the vicinity of point A (see Fig. 9.4a). From geometric considerations, we know that $\varphi_1 = \theta - \frac{\alpha}{2}$ and the number of reflections becomes

$$M = \left[\frac{2\theta}{\alpha}\right] + 1 \tag{9.18}$$

The relationship between the magnitude $\left(1 - |R|^2\right)$ and $\frac{2\theta}{\alpha}$ for different angles α and vertical and horizontal polarizations is illustrated in Fig. 9.4b. Curves are smooth for $\alpha > 90^{\circ}$ because there is only one reflection. The sharp rise of the curves is due to the appearance of the second reflection.

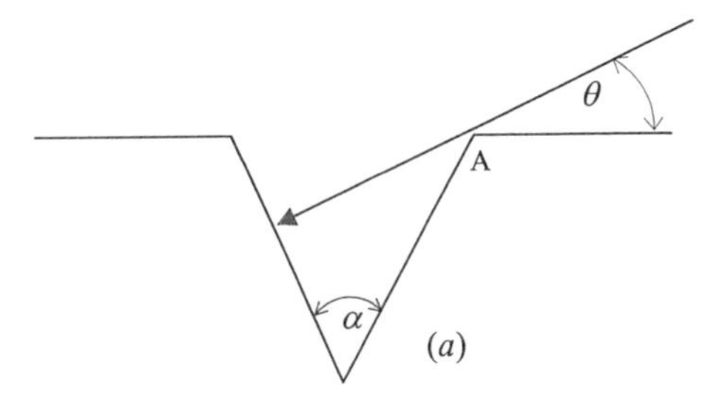

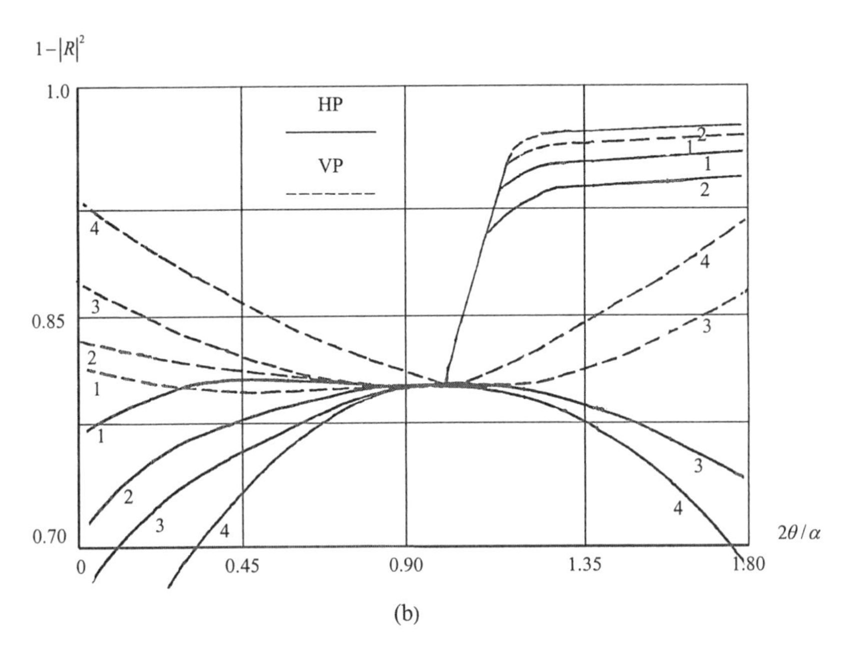

(b)

Fig. 9.4 a: Geometry of beam incidence
b: Dependence $\left(1-|R|^2\right)$ and $2\theta/\alpha$ (for concrete)
$1.\alpha = 30^0$; $2.\alpha = 60^0$; $3.\alpha = 90^0$; $4.\alpha = 120^0$

9.2.3 Restrictions

As was mentioned above, the approximations of geometrical optics remain valid as long as the dimensions of the pit and the distance between adjacent reflections S_n (for example, BC in Fig. 9.1a) are large compared to the wavelength, i.e.,

$$S_n >> \lambda, l_{n+1} >> \lambda \tag{9.19}$$

From triangle *BCO* in Fig. 9.1a, it is easy to find that

$$S_n = \frac{l_{n+1} \sin \alpha}{\cos \varphi_n} \tag{9.20}$$

Taking into account the recurrent relationships (9.2) and (9.4), condition (9.19) can be rewritten as

$$\left.\begin{aligned} S_n &= \frac{l_1 \sin \alpha \cos \varphi_1}{\cos(\varphi_1 - n\alpha)\cos[\varphi_1 - (n-1)\alpha]} >> \lambda \\ l_{n+1} &= \frac{l_1 \cos \varphi_1}{\cos(\varphi_1 - n\alpha)} >> \lambda \end{aligned}\right\} \tag{9.21}$$

so that

$$\left.\begin{aligned} \frac{l_1}{\lambda} &>> \frac{ctg\dfrac{\alpha}{2}}{2\cos\varphi_1} \\ \frac{l_1}{\lambda} &>> \frac{1}{\cos\varphi_1} \end{aligned}\right\} \tag{9.22}$$

For $ctg\dfrac{\alpha}{2} > 2$, i.e., when $\alpha < 53^{\circ}$, the restrictions in the method are determined by the first inequality in Eq. (9.22); otherwise, the second condition gives the determining restriction.

9.3 Reflection from an asymmetric wedge-shaped fracture

Results derived in Sec. 9.2 can easily be extended to non-symmetrical wedge-shaped fractures. All formulae remain practically unchanged. The only difference is that in this case the number of reflections is defined by the smaller of the numbers *M* or *N*,

$$\left.\begin{aligned} M &= \left[\frac{\varphi_1}{\alpha} + \frac{\arccos(\xi_1 \cos\varphi_1)}{\alpha}\right] + 1 \\ N &= \left[\frac{\varphi_1}{\alpha} + \frac{\arccos(\xi_2 \cos\varphi_2)}{\alpha}\right] + 1 \end{aligned}\right\} \tag{9.23}$$

where $\xi_1 = \frac{l_1}{L_1}$; $\xi_2 = \frac{l_2}{L_2}$; L_1 and L_2 are the dimensions of the pit side walls.

9.4 Reflection from a pit with spherical form

With reference to Fig. 9.1b, the radius of the spherical pit is R_a, the opening angle is Φ, the receiving antenna "looks" at point A and the "enhancement" field is directed at point B.

Taking into account the equality of all incidence and reflection angles, it is clear that

$$R_\Sigma = \left|R(\varphi)\right|^{2N} \tag{9.24}$$

The number of reflections N is derived from the following condition:

$$\Psi + \left(180^\circ - 2\varphi\right)N > \Phi \tag{9.25}$$

i.e.,

$$N = \left[\frac{\Phi - \Psi}{180^\circ - 2\varphi}\right] + 1 \tag{9.26}$$

By means of simple geometric considerations the following relationships for θ, x, φ, R_a can be found:

$$\left.\begin{aligned} \Psi &= \frac{\Phi}{2} + \theta - \varphi \\ \varphi &= \arcsin\left(\xi \cos\varphi\right) \\ \xi &= \frac{x}{R} \end{aligned}\right\} \tag{9.27}$$

For vertical probing ($\theta = 0^\circ$) of a "half-sphere" pit ($\Phi = 180^\circ$), we find that

$$N = \left[\frac{90^\circ + \arcsin\xi}{2\left(90^\circ - \arcsin\xi\right)}\right] + 1 \tag{9.28}$$

When the pit is filled with a medium having a complex permittivity ε it is also necessary to take the attenuation into account.

The relationships between the magnitude of the reflection coefficient ($|R|^2$) and parameter ξ for earth and concrete in the case of vertical probing ($\theta = 0^\circ$) with different angles Φ are shown in Figs 9.5 to 9.8. The complicated character of the curves can be explained by the change in the number of reflections. This change is caused by a variation of the point of sighting.

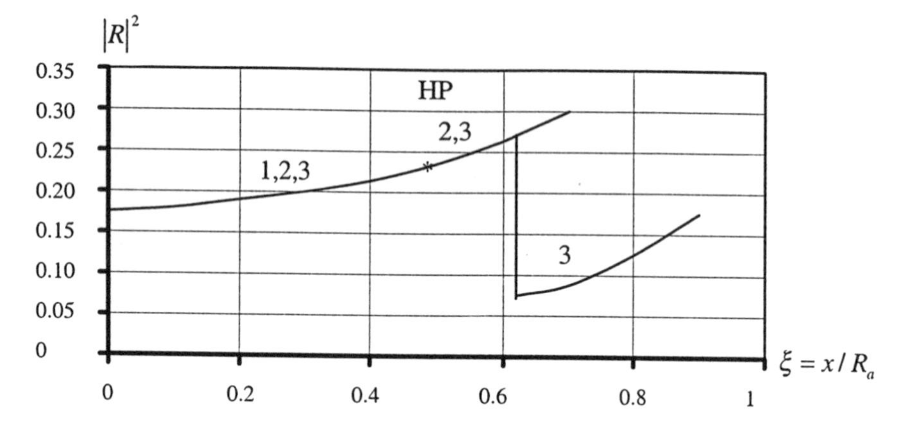

Fig. 9.5a Dependence of $|R|^2$ on $\xi = x / R_a$
Concrete $\varepsilon = 6.25 \cdot (1 - i \cdot 0.01)$
1.$\Phi = \pi / 3$, 2.$\Phi = \pi/2$, 3.$\Phi = 2\pi/3$

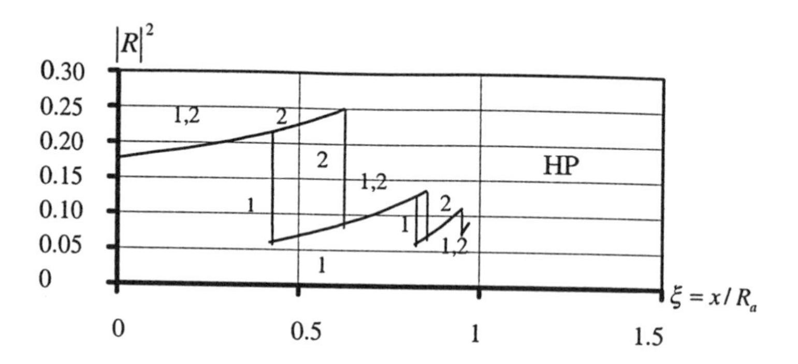

Fig. 9.5b Dependence of $|R|^2$ on $\xi = x / R_a$
Concrete $\varepsilon = 6.25 \cdot (1 - i \cdot 0.01)$
1.$\Phi = \pi$, 2.$\Phi = 3\pi/4$

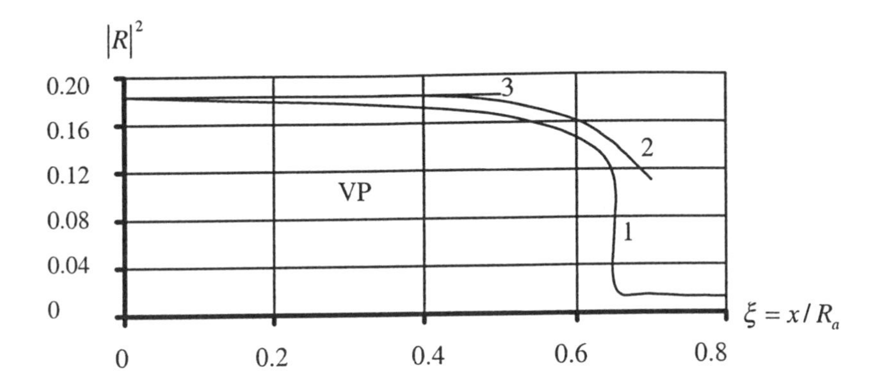

Fig. 9.6a Dependence of $|R|^2$ on $\xi = x / R_a$
Concrete $\varepsilon = 6.25 \cdot (1 - i \cdot 0.01)$
1.$\Phi = 2\pi / 3$; 2.$\Phi = \pi/2$; 3.$\Phi = \pi/3$

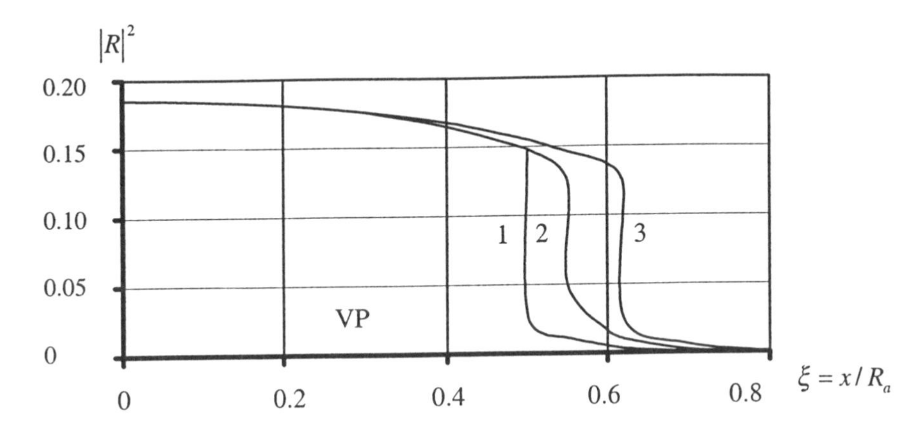

Fig. 9.6b Dependence of $|R|^2$ on $\xi = x / R_a$
Concrete $\varepsilon = 6.25 \cdot (1 - i \cdot 0.01)$
1.$\Phi = \pi$; 2.$\Phi = 5\pi/6$; 3.$\Phi = 3\pi/4$

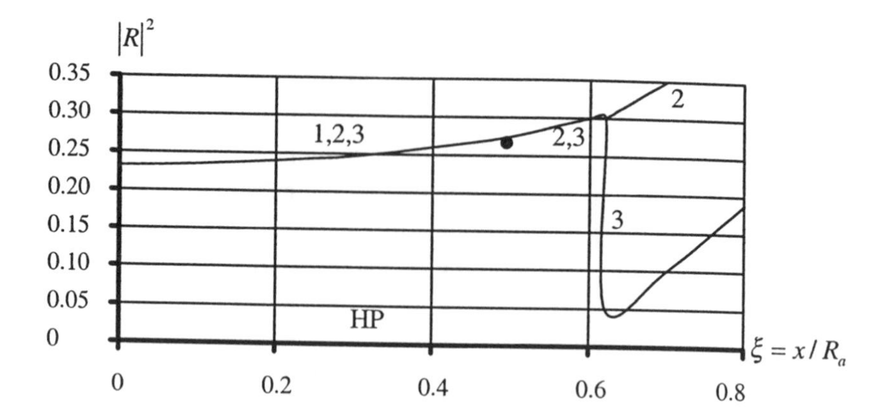

Fig. 9.7a Dependence of $|R|^2$ on $\xi = x / R_a$
Ground $\varepsilon = 8 \cdot (1 - i \cdot 0.1)$
1.$\Phi = \pi / 3$; 2.$\Phi = \pi/2$; 3.$\Phi = 2\pi/3$

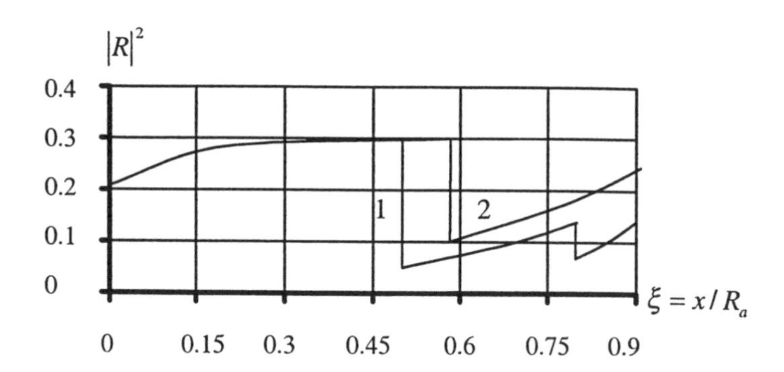

Fig. 9.7b Dependence of $|R|^2$ on $\xi = x / R_a$
Ground $\varepsilon = 8 \cdot (1 - i \cdot 0.1)$
1.$\Phi = \pi$; 2.$\Phi = 3\pi/4$

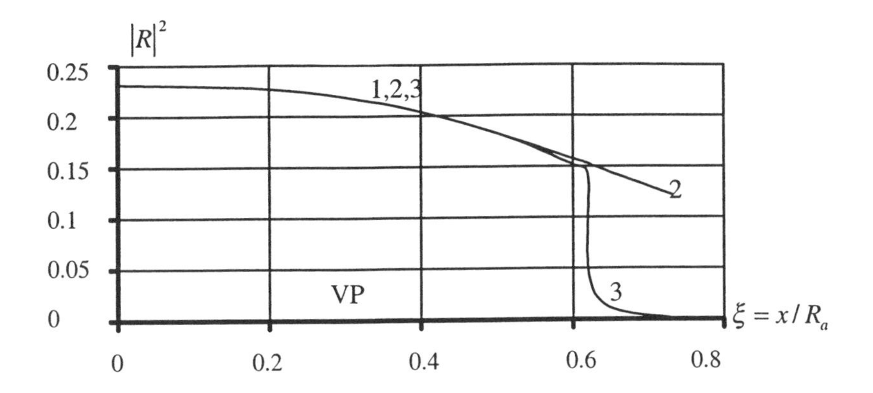

Fig. 9.8a Dependence of $|R|^2$ on $\xi = x / R_a$
Ground $\varepsilon = 8 \cdot (1 - i \cdot 0.1)$
1.$\Phi = \pi / 3$; 2.$\Phi = \pi/2$; 3.$\Phi = 2\pi/3$

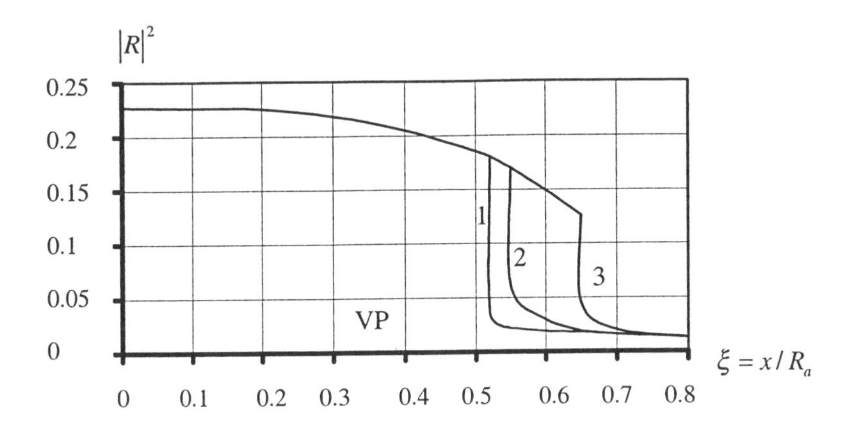

Fig. 9.8b Dependence of $|R|^2$ on $\xi = x / R_a$
Ground $\varepsilon = 8 \cdot (1 - i \cdot 0.1)$
1.$\Phi = \pi$; 2.$\Phi = 5\pi/6$; 3.$\Phi = 3\pi/4$

9.5 Reflection from a rectangular pit with finite depth

When the reflected wave is received at angle φ from a surface element situated near point A (see Fig. 9.1c) and located on a side wall of a rectangular pit having width equal to $2a$ and depth equal to H, the direction of "enhancement of wave propagation" is defined by means of geometrical optics. In order to calculate the number of reflections it is advisable to extend the pit towards the bottom direction up to a depth of $2H$.

The conditions for beam emergence from the pit after n reflections from the walls become

$$\left.\begin{array}{l} n \cdot 2a \cdot tg\varphi > 2H - l \\ (n-1) \cdot 2a \cdot tg\varphi < 2H - l \end{array}\right\} \tag{9.29}$$

From (9.29), it follows that the number of reflections M from the pit walls equals

$$M = \left[\frac{2H - l}{2a} ctg\varphi \right] + 1 \tag{9.30}$$

In addition, there is the reflection from point B at an angle $(90^0 - \varphi)$. The total reflection coefficient is determined by

$$|R_\Sigma|^2 = |R(\varphi)|^{2M} \cdot |R(90^0 - \varphi)|^2 \tag{9.31}$$

The relationship (9.30) shows that for an infinitely deep pit, i.e., for $H \to \infty$, one has $M \to \infty$ and, therefore, $|R|^2 \to 0$.

Curves expressing $(|R|^2)$ as function of the probing angle φ for a filled with concrete pit are shown in Fig. 9.9.

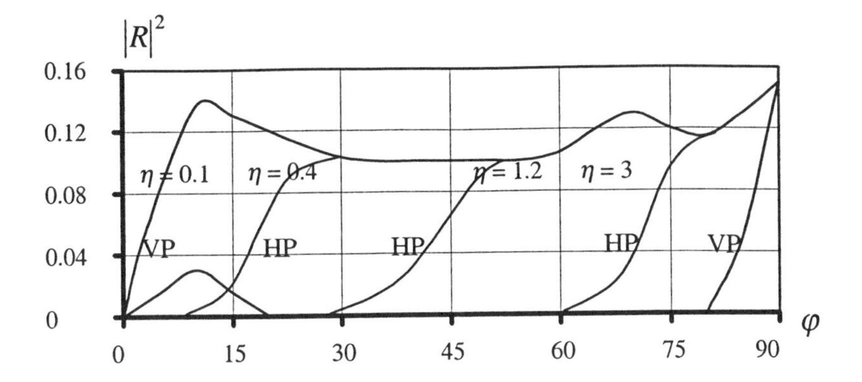

Fig. 9.9 Dependence of $|R|^2$ on φ ; $\eta = (2H - l)/2a$

The parameter η in this figure is defines as follows:

$$\eta = \frac{2H - l}{2a} \tag{9.32}$$

It is seen from the Fig. 9.9 that for horizontal polarization shallow and wide pits have larger backscattering and that the magnitude $|R|^2$ significantly differs from zero even at small angles φ. For narrow and deep pits and for angles φ up to 70° to 80°, the pit can be considered as a black body.

For vertical polarization, the finite depth does give information on the Brewster's angle effects. That is why for vertical polarization the reflection from a pit with rectangular walls can nearly always be considered as absent.

9.6 Antenna pattern and fracture filling effects

The receiving-antenna pattern gives an angular weighting over the pit walls which are illuminated. If the fractures have a dielectric filling, the latter are taken into consideration by additional reflections at the boundary with air and also by multiplication of $|R_\Sigma|^2$ with $\exp\{-ky \cdot \mathrm{Im}\sqrt{\varepsilon}\}$ due to the losses (y is the length of the ray path in the dielectric).

9.7 Combined model

The next step in modeling is the simultaneous introduction of electrical and geometrical inhomogeneities. One of the simplest models is shown in Fig. 9.10.

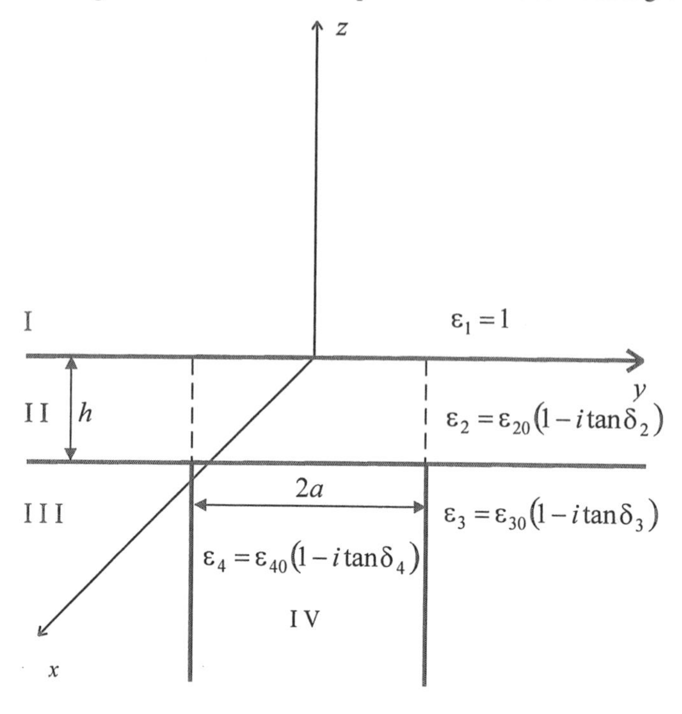

Fig. 9.10 Combined model

According to this model there are four media:

Medium I $(z > 0)$ is air

Medium II a layer of homogenous dielectric with thickness h (ice, earth, concrete)

Medium III $(z < -h; y > a; y < -a)$ is a semi-infinite space of another homogenous dielectric (ice, earth, concrete)

Medium IV $(z < -h; |y| < a)$ is a rectangular pit with width $2a$ and filled with a homogenous dielectric (air, concrete)

In the general case, losses can be present in media II, III and IV; the media are characterized by the complex relative permittivities $\varepsilon_2, \varepsilon_3$ and ε_4, respectively. For medium I, $\varepsilon_1 = 1$. All media are assumed to be nonmagnetic.
For this model, we want to find the variation of the reflection coefficient with respect to the point of observation γ, the incidence angle β, the mode of polarization and the parameters characterizing the considered model, i.e., the thickness h of layer II, the pit width $2a$ and the electrophysical properties of media II, III and IV. In the sequel, it will be assumed that the pit width $2a$ is much larger than the wavelength λ. Internal ruptures in ice or in earth and ravines can be considered within the framework of this model.

9.7.1 Computation of the reflection coefficient

A plane wave is incident at angle β on the planar boundary between media I and II.

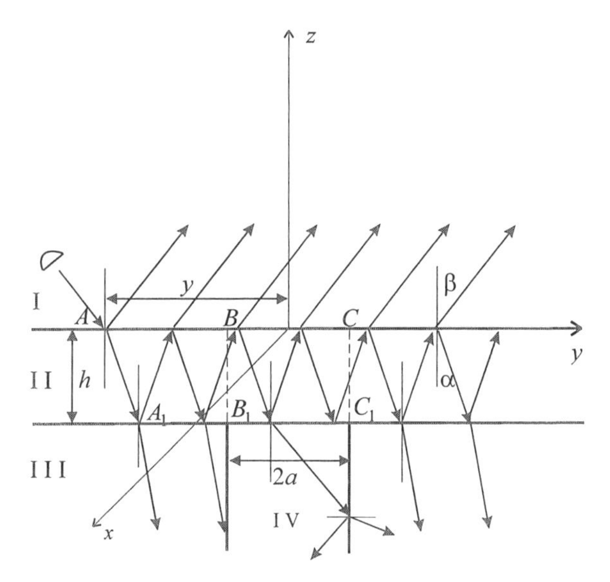

Fig. 9.11 Model of radiowave reflection. Observation left from pit.

In order to define the reflection coefficient of the "illuminating wave," we apply geometrical optics. This is valid, as pointed out earlier, as long as $a >> \lambda$. Let the ray of the receiving antenna "look" at point A, which is located at a distance y from the middle of the pit (cf. Fig. 9.11). The field in this direction is defined by means of the algebraic addition of three types of rays:

- Type I: Rays which undergo reflection at the boundary between II–III and deflection at the boundary I-II.
- Type II: Rays which undergo (multiple) reflections from boundary II–III.
- Type III: Rays which undergo (multiple) reflections from boundary II–IV.

Naturally, the directly reflected ray at point A must be added to the others. Fig. 9.11 shows that, in the framework of ray theory, there are no other rays reaching point A. The contribution of rays of type I (which is affected by j reflections at the boundary II-III) is given by

$$J_{1j} = \left(1 - R_{12}^2\right) R_{32}^j R_{12}^j \gamma^{2j} \tag{9.33}$$

where R_{12}, R_{23} are the Fresnel reflection coefficient from boundaries II-I and III-II and

$$\gamma^2 = \exp\left\{2ikh\sqrt{\varepsilon_2 - \sin^2\beta}\right\} \tag{9.34}$$

Let there be S reflections within the interval A_1B_1. Then, the total contribution from rays of type I is given as follows:

$$I_1 = \sum_{j=1}^{S} J_{1j} = \left(1 - R_{12}^2\right) R_{32}\gamma^2 \sum_{j=1}^{S} \left(R_{32}R_{12}\gamma^2\right)^{j-1} = \left(1 - R_{12}^2\right) R_{32}\gamma^2 \frac{1 - \left(R_{32}R_{12}\gamma^2\right)^S}{1 - \left(R_{32}R_{12}\gamma^2\right)} \tag{9.35}$$

The number of reflections, S, is found from geometrical considerations:

$$S = \left[\frac{(y-a)ctg\alpha}{2h}\right] \tag{9.36}$$

Here, the bracket [...] indicates the integer value of the enclosed expression, and the angle α is connected to the angle β according to the relationship:

$$\sin\beta = \sqrt{(\mathrm{Re}\sqrt{\varepsilon_2 - \sin^2\beta})^2 + \sin^2\beta \sin\alpha} \tag{9.37}$$

Rays incident on the surface B_1C_1 have S reflections from A_1B_1 and M reflections from B_1C_1. The contribution of these rays assumes the form

$$J_2 = \left(1 - R_{12}^2\right)\left(R_{32}R_{12}\gamma^2\right)^S \sum_{j=1}^{M} R_{42}^j R_{12}^{j-1}\gamma^{2j} = \left(1 - R_{12}^2\right)\left(R_{32}R_{12}\gamma^2\right)^S R_{42}\gamma^2 \frac{1-\left(R_{42}R_{12}\gamma^2\right)^M}{1 - R_{42}R_{12}\gamma^2} \tag{9.38}$$

where R_{42} is the Fresnel reflection coefficient from boundary II-IV and

$$M = \left[\frac{a \cdot ctg\alpha}{h}\right] \tag{9.39}$$

Rays incident on an interval further to the right (right of C_1) have S reflections on A_1B_1, M reflections on B_1C_1 and 1, 2, 3 ... reflections on the interval $\left[C_{1,}...\infty\right]$.The contribution of these rays is

$$J_3 = \left(1 - R_{12}^2\right)\left(R_{32}R_{12}\gamma^2\right)^S \left(R_{42}R_{12}\gamma^2\right)^M R_{32}\gamma^2 \sum_{j=1}^{\infty}\left(R_{32}R_{12}\gamma^2\right)^{j-1} =$$
$$= \left(1 - R_{12}^2\right)\left(R_{32}R_{12}\gamma^2\right)^S \left(R_{42}R_{12}\gamma^2\right)^M \frac{R_{32}\gamma^2}{1 - R_{32}R_{12}\gamma^2} \tag{9.40}$$

The total ray contribution becomes

$$I_1 = -R_{12} + J_1 + J_2 + J_3 \tag{9.41}$$

Using the Eqs (9.35), (9.38), (9.40) and some algebraic transformations, Eq. (9.41) can be written as

$$I_1 = Z_{123} + \left(Z_{123} - Z_{124}\right)\left(R_{32}R_{12}\gamma^2\right)\left[1 - \left(R_{12}R_{42}\gamma^2\right)^M\right] \tag{9.42}$$

where Z_{123} and Z_{124} are the Fresnel reflection coefficients from the homogeneous layer-like structure. Z_{123} for example, is defined as

$$Z_{123} = \frac{-R_{12} + R_{32}\gamma^2}{1 - R_{12}R_{32}\gamma^2} \tag{9.43}$$

If $S \to \infty$ (far from the pit) and when $M \to 0$ (an infinitely narrow pit), Eq. (9.42) shows that $I_1 \to Z_{123}$, as expected. In the case that $M \to \infty$ and $S \to 0$ (a very wide pit), $I_1 \to Z_{124}$.

Now, let us consider the case whereby the receiving antenna "looks" at the pit projection, i.e., $S = 0$. (cf. Fig. 9.12).

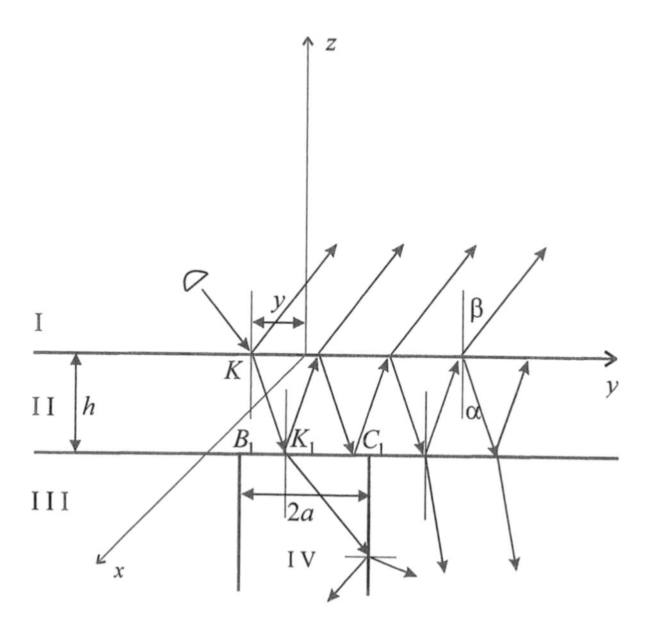

Fig. 9.12 Model of radiowave reflection. Observation at the pit

By analogy to the previous case, corresponding relationships can be derived. Incident rays on the interval K_1C_1 have N reflections. That is why the field defined by these rays is given by

$$J_4 = \left(1 - R_{12}^2\right) R_{42} \gamma^2 \sum_{j=1}^{N} \left(R_{42} R_{12} \gamma^2\right)^{j-1} = \left(1 - R_{12}^2\right) \frac{1 - \left(R_{12} R_{42} \gamma^2\right)^N}{1 - R_{12} R_{42} \gamma^2} \tag{9.44}$$

where N is defined by an expression similar to that in Eq. (9.36), viz.,

$$N = \frac{(a - y)ctg\alpha}{2h} \tag{9.45}$$

Incident rays on an interval further to the right of point C_1 have N reflections on K_1C_1 and 1, 2, 3 .... reflections on the interval $[C_1, \infty]$. The resulting field defined by these rays is

$$J_5 = (1 - R_{12}^2)(R_{12}R_{42}\gamma^2)^N R_{32}\gamma^2 \sum_{j=1}^{\infty}(R_{12}R_{32}\gamma^2)^{j-1} = (1 - R_{12}^2)(R_{12}R_{42}\gamma^2)^N \frac{R_{32}\gamma^2}{1 - R_{12}R_{32}\gamma^2} \tag{9.46}$$

The total field yields

$$I_2 = -R_{12} + J_4 + J_5 \tag{9.47}$$

which can be transformed into

$$I_2 = Z_{124} + (Z_{123} - Z_{124})(R_{12}R_{42}\gamma^2)^N \tag{9.48}$$

As follows from Eq. (9.48) when $N \to \infty$ (an infinitely wide pit), $I_2 \to Z_{124}$ and when $N \to 0$ (an infinitesimal narrow pit), $I_2 \to Z_{123}$.

Let us finally consider the case where the antenna "looks" further to the right of the pit (see Fig. 9.13).

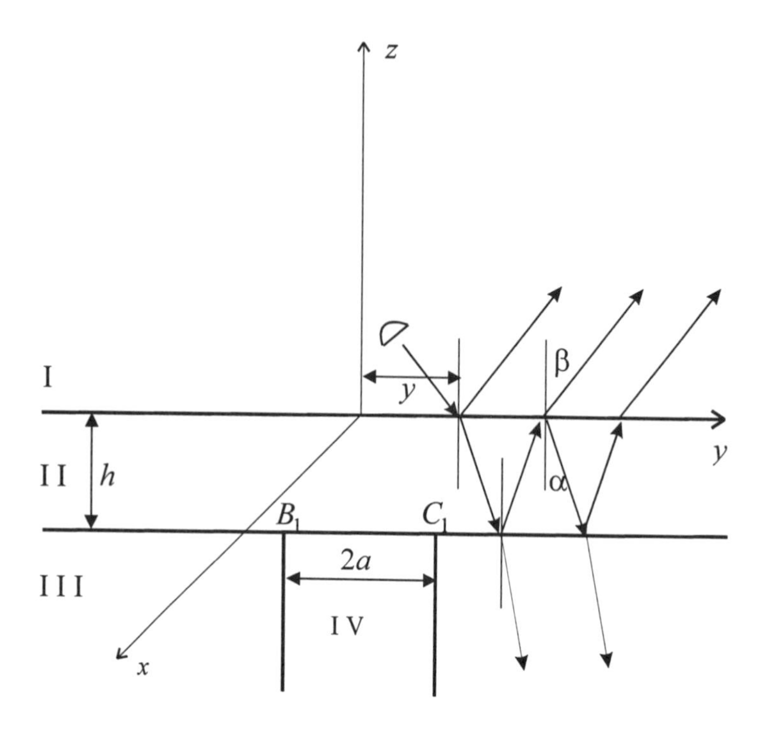

Fig. 9.13 Model of radiowave reflection. Observation right from pit

The expression for the total field can be written as

$$
\begin{aligned}
I_3 &= -R_{12} + \left(1 - R_{12}^2\right) R_{32}\gamma^2 \sum_{j=1}^{\infty} \left(R_{12}R_{32}\gamma^2\right)^{j-1} = \\
&= -R_{12} + \left(1 - R_{12}^2\right) \frac{R_{32}\gamma^2}{1 - R_{12}R_{32}\gamma^2} = \frac{-R_{12} + R_{32}\gamma^2}{1 - R_{12}R_{32}\gamma^2} = Z_{123}
\end{aligned}
\tag{9.49}
$$

As expected, we obtain an expression which coincides with the known reflection coefficient from a homogeneous layer.

Relations (9.42), (9.48) and (9.49) allow us to compute the reflection coefficients for structures in the proposed model within the framework of ray optics. The reflection depends on angle and line of sight of the receiving antenna, on the electrophysical properties of the various media and on the mode of polarization.

9.8 Conclusions and applications

This chapter deals with an investigation of electromagnetic wave reflection from surfaces that have cracks, fractures and other irregularities. The determination of the reflection coefficients from such types of surfaces is effected by means of the methods of the diffraction theory. However, if the assumption is made that the dimensions of the surface irregularities are much larger than the wavelength, then there is a real opportunity to solve the above-mentioned problems using geometrical optics methods.

As a first model, we considered a symmetrical wedge-like crack, which was illuminated (at a random angle) by a plane electromagnetic wave. When solving that problem, we took into consideration the effect of multiple reflection of the wave from the crack walls. We also included the fact that local coefficients of reflection depend on the complex permittivity of the reflecting walls and the angle of incidence of the wave on them. The solution of the problem showed that the total (integral) coefficients of reflection for vertical and horizontal wave polarizations substantially differed from each other. With vertical sensing, the crack may be considered in terms of reflections as a black body for wedge angles of 40-45 degrees in the case of horizontal polarization and for wedge angles of 75 degrees in the case of vertical polarization. When considering a nonsymmetrical wedge-like crack, the structure of the derived formulas is not affected, but the number of local reflections from the crack walls is subject to change.

As a second model, we considered spherical-form hollows. In that case, also, the final reflection coefficients of radio waves with different polarizations substantially differ from each other. The reflection coefficient depends on both the complex permittivity of the pit walls and the ratio of the pit depth to its radius.

As a third model, we considered a rectangular-form crack with finite depth. For a horizontal polarization of radio waves, the coefficient of reflection substantially differed from zero even with small angles of incidence for non-deep and wide hollows. Narrow and deep hollows (with the angle of incidence up to $70^0 - 80^0$) may be considered as a blackbody. For a vertically polarized radio wave, the model of the hollow with the form of a rectangular pit with finite depth resulted in no reflections.

We analyzed a combined model that took into account the presence of four media. As such models, we considered air, an electrically uniform layer with finite depth, an electrically uniform semi-infinite space with different permittivity and a rectangular-form hollow within the semi-infinite space having its own permittivity. It was also assumed that the width of the hollow was much larger than the wavelength. When solving that problem we took into account reflections of three types: reflection from the boundary of the second and third media, reflection from the inner surface of the hollow and reflection from the outer surface of the hollow.

For the combined model we derived design relations which allowed us to determine the reflection coefficients of electromagnetic waves from the analyzed structure for different modes of wave polarization, different permittivities of the media, and also the angle, at which the medium was analyzed. The final reflection coefficients were expressed in terms of local Fresnel reflection coefficients for homogeneous media.

Thus, we derived the reflection coefficients of radio waves with different modes of polarization. The derived results included reflections from a wide class of natural earth covers (both of artificial and natural origin), with various geometrical irregularities. The variation of the reflection coefficients strongly depends, in general, on the angle of illumination, the illuminated area, the mode of radio wave polarization and other factors; so it is potentially possible, when carrying out earth surface sensing on the basis of the results of measurements of the reflection coefficients, to identify the types of earth surface at least for wedge-like cracks and spherical and rectangular hollows. Our results may have practical applications, for example, when carrying remote sensing of sea ice, especially in Arctic and Antarctic regions. It is important for scientific expeditions is such regions to be informed of the presence of fractures and their structure on the surface of both continental and sea ice. It is, also, necessary in terms of working safety, provisions and the possibility of finding places of landing of airplanes and other aircraft.

The derived results may be also useful when investigating the state of drainage systems for agricultural needs, various irrigation systems, canals and other water development facilities. Terrain topographical map-making for the purpose of finding irregularities in surface smoothness is not a less significant field of application.

It should be noted that the derived relations to a certain extent may help to solve problems of determination of possible depth and width of fractures and pits located on the earth surface by means of remote sensing methods. It is connected with the fact that electromagnetic reflection coefficients differ substantially for narrow, deep and small, and wide rectangular hollows. This fact may be used in mountain regions, in canyons and other types of earth surface with complicated terrain. It is evident that many of the enumerated results may be used when carrying out various rescue operations, particularly in the aforementioned mountain regions.

CHAPTER 10

Scattering of Waves by a Layer with a Rough Boundary

10.1 Introduction

In this chapter, the analysis of electromagnetic scattering from a surface layer with rough borders is carried out. It is well known that the degree of roughness essentially influences the process of interaction of the radio wave with the surface. It is assumed that the surface roughness is described by some random function of the space coordinates. Appropriate equations for the scattered fields and their solutions are discussed. These are illustrated by examples using the first and the second-order approximations discussed in Chapter 5. Algorithms for higher-order approximations are constructed carefully. Strategies for construction of scattering diagrams are considered and some particular results are indicated.

A method for evaluation of the parameters characterizing the vegetation by an ensemble of coaxial cylinders is carried out under the assumption of small-scale roughness. An important role is played by wave polarization aspects as described in a special section.

The coherent scattering of horizontally polarized waves by a finite layer of vegetation covering the ground is studied. The layer itself, as well as the ground, are considered to be homogeneous, isotropic, lossy, dielectric media. However, the vegetation-air interface is modeled as a randomly rough surface. Separately, the randomness of the vegetation layer itself is considered by means of a model involving an ensemble of co-directional cylinders.

10.2 Initial equations and solutions

Here, the problem is considered of determining the field reflected by a rough surface as function of electrophysical, statistical and geometrical properties of the structure.

The following electrodynamic model is considered. There is a homogeneous layer with thickness h (medium II, Figure 10.1) on an underlying half-space (medium III). On the other side, the dielectric has its boundary with medium I (air). This boundary is described by a random function:

$$z = \xi(x, y) \tag{10.1}$$

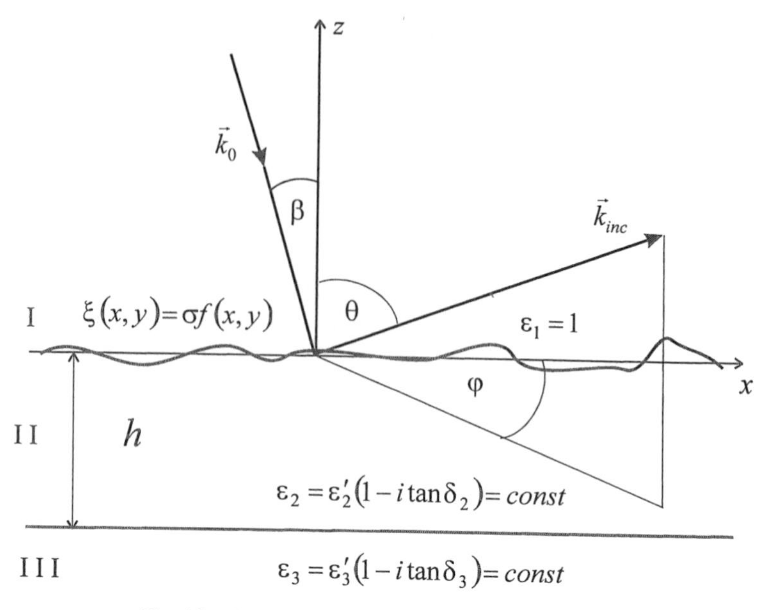

Fig. 10.1 Layer model with rough upper boundaries

All media are assumed infinite in the x and y direction and their relative permeabilities are assumed to be equal to 1. A plane electromagnetic wave is incident from medium I at an arbitrary angle β with respect to the vertical axis z.

The reflected field in medium I has to be determined. It is known from experimental data that small-scale roughness (relative to wavelength) does not influence the reflection appreciably for vertical polarization. That is why the example of a horizontally polarized wave is considered here, with

$$\vec{E} = \vec{j} e^{iK_0 (x \sin \beta - z \cos \beta)} \tag{10.2}$$

As the relative permittivities of all media are assumed to be constant $(\varepsilon_s = const,\ s = 1, 2, 3)$, the fields to be determined satisfy the following equation:

$$\Delta \vec{E}_s + K_s^2 \vec{E}_s = 0 \tag{10.3}$$

For solving (10.3) the small-perturbation method will be used, i.e., the influence of roughness on media I-II boundary will be considered as a perturbation superimposed on the field $\vec{E}_0$ obtained under the assumption that there is no roughness, i.e.,

$$\vec{E}_s = \vec{E}_{0s} + \sigma\vec{E}_{1s} + \sigma^2\vec{E}_{2s} \tag{10.4}$$

where σ is a small dimensionless parameter (σ << 1).

We construct the following equations and inequalities:

$$\begin{cases} \xi = \sigma f(x,y) << 1 \\ \dfrac{\partial\xi}{\partial x} = \sigma\dfrac{\partial f(x,y)}{\partial x} << 1 \\ \dfrac{\partial\xi}{\partial y} = \sigma\dfrac{\partial f(x,y)}{\partial y} << 1 \end{cases} \tag{10.5}$$

The conditions at the interface between media I and II will take the form

$$\begin{aligned} &\left[\left(\vec{E}_1 - \vec{E}_2\right)\cdot x\vec{n}\right] = \vec{0} \\ &\left[\left(\vec{H}_1 - \vec{H}_2\right)\cdot x\vec{n}\right] = \vec{0}, \ \ at \ \ z = \xi(x,y) \end{aligned} \tag{10.6}$$

where $\vec{n}$ is the outwardly directed normal unit vector to the surface $z = \xi(x,y)$. The conditions (10.6) can be written explicitly as

$$\left.\begin{aligned} &\left(E_{1y} - E_{2y}\right)n_z - \left(E_{1z} - E_{2z}\right)n_y = 0 \\ &\left(E_{1z} - E_{2z}\right)n_x - \left(E_{1x} - E_{2x}\right)n_z = 0 \\ &\left(H_{1y} - H_{2y}\right)n_z - \left(H_{1z} - H_{2z}\right)n_y = 0 \\ &\left(H_{1z} - H_{2z}\right)n_x - \left(H_{1x} - H_{2x}\right)n_z = 0 \end{aligned}\right\} \tag{10.7}$$

Note that two more equations result from (10.6); however, they are not needed for the following discussion. The components of the normal vector $\vec{n}$ to surface $\xi(x,y)$ are equal to

$$\left.\begin{aligned} n_x &= \frac{\dfrac{\partial \xi}{\partial x}}{\sqrt{1+\left(\dfrac{\partial \xi}{\partial x}\right)^2+\left(\dfrac{\partial \xi}{\partial y}\right)^2}} \\ n_y &= \frac{\dfrac{\partial \xi}{\partial y}}{\sqrt{1+\left(\dfrac{\partial \xi}{\partial x}\right)^2+\left(\dfrac{\partial \xi}{\partial y}\right)^2}} \\ n_z &= \frac{1}{\sqrt{1+\left(\dfrac{\partial \xi}{\partial x}\right)^2+\left(\dfrac{\partial \xi}{\partial y}\right)^2}} \end{aligned}\right\} \tag{10.8}$$

With these equations, (10.7) becomes:

$$\left.\begin{aligned} &\left(E_{1y}-E_{2y}\right)-\left(E_{1z}-E_{2z}\right)\sigma\frac{\partial f}{\partial y}=0 \\ &\left(E_{1z}-E_{2z}\right)\sigma\frac{\partial f}{\partial x}-\left(E_{1x}-E_{2x}\right)=0 \\ &\left(H_{1y}-H_{2y}\right)-\left(H_{1z}-H_{2z}\right)\sigma\frac{\partial f}{\partial y}=0 \\ &\left(H_{1z}-H_{2z}\right)\sigma\frac{\partial f}{\partial x}-\left(H_{1x}-H_{2x}\right)=0 \end{aligned}\right\} \tag{10.9}$$

We now present the total fields $\vec{E}_s$ ($\vec{H}_s$ respectively) in the form

$$\vec{E}_s(x,y,z)=\vec{E}_s(x,y,0)+\xi\left.\frac{\partial \vec{E}_s(x,y,z)}{\partial z}\right|_{z=0}+\frac{1}{2}\xi^2\left.\frac{\partial^2 \vec{E}_s(x,y,z)}{\partial z^2}\right|_{z=0}+\ldots \tag{10.10}$$

If one substitutes (10.10) into (10.9), the discontinuity of the fields $\vec{E}_s$ (and $\vec{H}_s$) are determined on the surface which coincides with the plane $z=0$.

The roughness of the upper boundary of medium II is taken into account via the boundary conditions (10.9) which contain the fields as well as the derivatives of ξ. In addition to the boundary conditions (10.9), the fields must satisfy the boundary conditions at $z = -h$; specifically, the continuity in the tangential components of $\vec{E}$ and $\vec{H}$, viz.,

$$\left.\begin{aligned} E_{2x} = E_{3x}, E_{2y} = E_{3y}, \\ H_{2x} = H_{3x}, H_{2y} = H_{3y}. \end{aligned}\right\} \qquad (10.11)$$

Furthermore, Maxwell's equations require that

$$div\vec{E}_s = div\vec{H}_s = 0 \qquad (10.12)$$

The problem, thus, is reduced to solving equations (10.3) and (10.12) under conditions (10.9) and (10.11), with an incident wave described by (10.2).

A sequential approximation method is used with the non-perturbed field:

$$\left.\begin{aligned} \vec{E}_{01} &= \vec{j}e^{iK_o x\sin\beta}\left(e^{iK_o z\cos\beta} + Re^{-iK_o z\cos\beta}\right) \\ \vec{E}_{02} &= \vec{j}e^{iK_o x\sin\beta}\left(D_1 e^{iK_o z\sqrt{\varepsilon_2 - \sin^2\beta}} + D_2 e^{-iK_o z\sqrt{\varepsilon_2 - \sin^2\beta}}\right) \\ \vec{E}_{03} &= \vec{j}e^{iK_o x\sin\beta}Te^{iK_o z\sqrt{\varepsilon_3 - \sin^2\beta}} \end{aligned}\right\} \qquad (10.13)$$

The reflection coefficient R, the transmission coefficient T and the constants D_1 and D_2 are connected by

$$\left.\begin{aligned} 1 + R = D_1 + D_2 \\ (1 - R)\cos\beta = (D_1 - D_2)\sqrt{\varepsilon_2 - \sin^2\beta} \end{aligned}\right\} \qquad (10.14)$$

relations resulting from the boundary conditions.

10.2.1 First-order approximation

The first-order approximation

$$\vec{E}_s \cong \vec{E}_{0s} + \sigma\vec{E}_{1s} \qquad (10.15)$$

is substituted into (10.3). We find, then,

$$\Delta^2\vec{E}_{0s} + K_0^2\varepsilon_s\vec{E}_{0s} = 0 \qquad (10.16)$$

$$\Delta^2 \vec{E}_{1s} + K_0^2 \varepsilon_s \vec{E}_{1s} = 0 \tag{10.17}$$

For solving (10.17), the following transverse spectral representations of $\vec{E}_{1s}$ are needed:

$$\left.\begin{aligned} \vec{E}_{11} &= \int\!\!\int_{-\infty}^{+\infty} \vec{C}_{11}(z) e^{i(p_1 x + p_2 y)} dp_1 dp_2 \\ \vec{E}_{12} &= \int\!\!\int_{-\infty}^{+\infty} \vec{A}_{12}(z) e^{i(p_1 x + p_2 y)} dp_1 dp_2 \\ \vec{E}_{13} &= \int\!\!\int_{-\infty}^{+\infty} \vec{L}_{13}(z) e^{i(p_1 x + p_2 y)} dp_1 dp_2 \end{aligned}\right\} \tag{10.18}$$

Substituting these equations into equation (10.17) results in

$$\frac{d^2 \vec{M}(z)}{dz^2} + \left[K_0^2 \varepsilon_s - \left(p_1^2 + p_2^2 \right) \right] \vec{M}(z) = 0 \tag{10.19}$$

where $\vec{M}(z)$ represents $\vec{C}_{11}, \vec{A}_{12}$, or $\vec{L}_{13}$. The solution of (10.19) can be written as

$$\vec{M}(z) = \vec{M}_1 e^{i\mu_1 z} + \vec{M}_2 e^{i\mu_2 z} \tag{10.20}$$

where $\vec{M}_1, \vec{M}_2$ are vectors dependent only on p_1 and p_2 and

$$\mu_1 = -\mu_2 = \sqrt{K_0^2 \varepsilon_s - \left(p_1^2 + p_2^2 \right)} \tag{10.21}$$

The radiation conditions in regions I and III and the interference of forward $(+z)$ and backward $(-z)$ components in region II allows us to rewrite Eq. (10.18) as follows:

$$\left.\begin{aligned}
\vec{E}_{11} &= \int_{-\infty}^{+\infty}\int \vec{C}_{11} e^{i\vec{p}\cdot\vec{r}} dp_1 dp_2 \\
\vec{E}_{12} &= \int_{-\infty}^{+\infty}\int \left(\vec{A}_{12} e^{i\vec{t}\cdot\vec{r}} + \vec{B}_{12} e^{i\vec{t}'\cdot\vec{r}}\right) dp_1 dp_2 \\
\vec{E}_{13} &= \int_{-\infty}^{+\infty}\int \vec{L}_{13} e^{i\vec{s}\cdot\vec{r}} dp_1 dp_2
\end{aligned}\right\} \tag{10.22}$$

where

$$\left.\begin{aligned}
\vec{p} &= \left(p_1, p_2, -\sqrt{K_0^2 - \left(p_1^2 + p_2^2\right)}\right) \\
\vec{t} &= \left(p_1, p_2, \sqrt{K_0^2 \varepsilon_2 - \left(p_1^2 + p_2^2\right)}\right) \\
\vec{t}' &= \left(p_1, p_2, -\sqrt{K_0^2 \varepsilon_2 - \left(p_1^2 + p_2^2\right)}\right) \\
\vec{s} &= \left(p_1, p_2, \sqrt{K_0^2 \varepsilon_3 - \left(p_1^2 + p_2^2\right)}\right) \\
\vec{r} &= x\vec{i} + y\vec{j} + z\vec{k}
\end{aligned}\right\} \tag{10.23}$$

and $\vec{C}_{11}, \vec{A}_{12}, \vec{B}_{12}, \vec{L}_{13}$ are vectors dependent only on p_1 and p_2. Since ε_3 is complex, in general, care must be taken so that $\mathrm{Im}\{s_z\} > 0$ in region III.

Since

$$\vec{H} = -\frac{1}{i\mu_0\omega} \nabla x \vec{E} \tag{10.24}$$

we obtain $\vec{H}_{1s}$

$$\left.\begin{aligned}
H_{11j} &= -\frac{1}{\omega\mu_0}\int_{-\infty}^{+\infty}\int \left(P_{j+1}C_{11\,j-1} - P_{j-1}C_{11\,j+1}\right) e^{i\vec{p}\cdot\vec{r}} dp_1 dp_2 \\
H_{12j} &= -\frac{1}{\omega\mu_0}\int_{-\infty}^{+\infty}\int \left(T_{j+1}A_{12\,j-1} - T_{j-1}A_{12\,j+1}\right) e^{i\vec{t}\cdot\vec{r}} + (T'_{j+1}B_{12\,j} - T'_{j-1}B_{12\,j+1}) e^{i\vec{t}\cdot\vec{r}} dp_1 dp_2 \\
H_{13j} &= -\frac{1}{\omega\mu_0}\int_{-\infty}^{+\infty}\int \left(S_{j+1}L_{11\,j-1} - S_{j-1}L_{11\,j+1}\right) e^{i\vec{s}\cdot\vec{r}} dp_1 dp_2
\end{aligned}\right\} \tag{10.25}$$

where the index j means x, y, z.

From (10.22) (b), we write

$$\vec{E}_{12} = \vec{E}'_{12} + \vec{E}''_{12} \tag{10.26}$$

where

$$\begin{cases} \vec{E}'_{12} = \iint\limits_{-\infty}^{+\infty} \vec{A}_{12} e^{i\vec{t}\,\vec{r}}\, dp_1 dp_2 \\ \vec{E}''_{12} = \iint\limits_{-\infty}^{+\infty} \vec{B}_{12} e^{i\vec{t}'\vec{r}}\, dp_1 dp_2 \end{cases} \tag{10.27}$$

are derived from Eq.(10.22).

Consider, next, the divergence equation $\nabla \cdot \vec{E}_s = \left(\vec{E}_{0s} + \vec{E}_{1s}\right) = 0$. Assuming that

$\nabla \cdot \vec{E}_{0s} = 0$, we obtain

$$\nabla.\left(\vec{E}'_{12} + \vec{E}''_{12}\right) = 0 \tag{10.28}$$

There is no restriction in requiring that

$$\nabla.\vec{E}'_{12} = \nabla \cdot \vec{E}''_{12} = 0 \tag{10.29}$$

Substituting (10.22) into (10.11) and taking into account (10.13), (10.14) and (10.23), we obtain

$$\left(\vec{p} \cdot \vec{C}_{11}\right) = \left(\vec{t} \cdot \vec{A}_{12}\right) = \left(\vec{t}' \cdot \vec{B}_{12}\right) = \left(\vec{s} \cdot \vec{L}_{13}\right) = 0 \tag{10.30}$$

and

$$\left.\begin{aligned} A_{12x}\alpha + \frac{1}{\alpha'} B_{12x} &= \gamma L_{13x} \\ A_{12y}\alpha + \frac{1}{\alpha'} B_{12y} &= \gamma L_{13y} \end{aligned}\right\} \tag{10.31}$$

where

$$\left.\begin{aligned} &\alpha = e^{-it_3 h} \\ &\gamma = e^{-is_3 h} \\ &\alpha' = e^{-it_3' h} \end{aligned}\right\} \quad (10.32)$$

The unknown constants can be determined by using the boundary conditions at $z = 0$, *viz.*,

$$\left.\begin{aligned} &E_{11y} - E_{12y} = 0 \\ &E_{11x} - E_{12x} = 0 \\ &H_{11y} - H_{12y} = 0 \end{aligned}\right\} \quad (10.33)$$

Substitution of (10.22), (10.25) and (10.26) into (10.33) yields

$$\left.\begin{aligned} &C_{11y} - A_{12y} - B_{12y} = 0 \\ &C_{11x} - A_{12x} - B_{12x} = 0 \\ &(t_3 A_{12x} - p_1 A_{12z}) - (t_3 B_{12x} + p_1 B_{12z}) + (p_3 C_{11x} + p_1 C_{11z}) = 0 \\ &(p_2 C_{11z} + p_3 C_{11y}) - (p_2 A_{12z} - t_3 A_{12y}) - (p_2 B_{12z} + t_3 B_{12y}) = \alpha_1 \end{aligned}\right\} \quad (10.34)$$

where $\alpha_1 = iK_0^2(1+R)(1-\varepsilon_2)N(p_1, p_2)$ and $N(p_1, p_2)$ is defined through the Fourier transformation

$$f(x, y)e^{iK_0 x \sin\beta} = \iint_{-\infty}^{+\infty} N(p_1, p_2)e^{i(p_1 x + p_2 y)} dp_1 dp_2 \quad (10.35)$$

There are 12 equations in (10.30), (10.31) and (10.34) for determining the 12 unknown values $\vec{A}_{12}, \vec{B}_{12}, \vec{C}_{11}, \vec{L}_{13}$.

10.2.2 Second-order approximation

The field in a second-order approximation is described by

$$\vec{E}_s = \vec{E}_{0s} + \sigma\vec{E}_{1s} + \sigma^2 \vec{E}_{23} \quad (10.36)$$

where for second order terms it is necessary to replace in (10.22), (10.25), (10.30) and (10.31) $\vec{E}_{1s}(\vec{H}_{1s})$ by $\vec{E}_{2s}(\vec{H}_{2s})$. Eq. (10.9) will then take the form

$$\left.\begin{aligned}
E_{21y}-E_{22y} &= (E_{11z}-E_{12z})\frac{\partial f}{\partial y}-f\left(\frac{\partial E_{11y}}{\partial z}-\frac{\partial E_{12y}}{\partial z}\right)+\frac{1}{2}f^2K_0^2(1+R)(1-\varepsilon_2)e^{iK_0x\sin\beta}\\
E_{21x}-E_{22x} &= -(E_{11z}+E_{12z})\frac{\partial f}{\partial x}-f\left(\frac{\partial E_{11x}}{\partial z}-\frac{\partial E_{12x}}{\partial z}\right)\\
H_{21x}-H_{22x} &= (H_{11z}-H_{12z})\frac{\partial f}{\partial x}-f\left(\frac{\partial H_{11x}}{\partial z}-\frac{\partial H_{12x}}{\partial z}\right)+\frac{1}{2}f^2K_0^2(1+R)(1-\varepsilon_2)e^{iK_0x\sin\beta}\\
H_{21y}-H_{22y} &= -(H_{11z}+H_{12z})\frac{\partial f}{\partial y}-f\left(\frac{\partial H_{11y}}{\partial z}-\frac{\partial H_{12y}}{\partial z}\right)
\end{aligned}\right|_{z=0}\quad (10.37)$$

If $\vec{E}_{1s}$ and $\vec{H}_{1s}$ have been determined first, then by using (10.30), (10.31) and (10.34), the second order terms will follow.

10.2.3 Scattering diagram

In the case of scattering from a plane surface, the angle of reflection equals the angle of incidence. Reflected fields are observed also in other directions due to the presence of surface roughness between media I and II.

A plot of the angle dependence of the reflected power is called the surface scattering diagram. This scattering diagram defines D as the statistically averaged power $\Delta\overline{P}$ reflected in an infinitesimal space angle $\Delta\Omega$, i.e.,

$$D=\frac{\Delta\overline{P}}{\Delta\Omega}\quad (10.38)$$

For an incident wave with horizontal polarization, the reflected field vector component co-linear with the incident field vector is given by

$$E_{1r}=E_{1x}\sin\varphi+E_{1y}\cos\varphi\quad (10.39)$$

where the angle φ is the scattering angle. The statistically averaged power $\overline{P}$ in a first-order approximation is determined from the reflected field as follows:

$$\overline{P}=\overline{|E_{1r}|^2}=\sigma^2\overline{|E_{1x}\sin\varphi+E_{1y}\cos\varphi|^2}\quad (10.40)$$

With reference to (10.22), $\overline{P}$ may be expressed through the correlation function of $C_{11}(\vec{p})$. However, (10.30), (10.31) and (10.34) lead to rather cumbersome equations for C_{11}. Taking into account (10.22) and (10.35), the averaged power can also be written in the form

$$\overline{P} = \sigma^2 \int_{-\infty}^{+\infty} \iiint \Phi(p_1, p_2, \varphi)\Phi^*(p_1', p_2', \varphi)\overline{N(p_1, p_2)N^*(p_1', p_2')} \times \\ \times \exp\left(i\left[(p_1 - p_1')x + (p_2 - p_2')y - (p_3 - p_3')z\right]\right) dp_1 dp_2 dp_1' dp_2' \tag{10.41}$$

where $\Phi(p_1, p_2, \varphi)$ is defined later. From (10.35), we know that

$$\overline{N(p_1, p_2)N^*(p_1', p_2')} = \frac{1}{(2\pi)^2} S(p_1, p_2)\delta\left(p_2 - p_1', p_2 - p_2'\right) \tag{10.42}$$

where $\delta(p_1', p_2')$ is the Dirac delta function and

$$S(p_1, p_2) = \int_0^\infty \int_0^\infty R(x, y)\cos\left[K_0(x\sin\beta - p_1 x - p_2 y)\right]dxdy \tag{10.43}$$

with

$$R(x, y) \equiv R(|x_1 - x_2|, |y_1 - y_2|) = \overline{f(x_1, y_1)f^*(x_2, y_2)} \tag{10.44}$$

the correlation function of the surface roughness (assumed a stationary random process).

Substitution of (10.42) in (10.41) gives

$$\overline{P} = \frac{\sigma^2}{(2\pi)^2} \int_{-\infty}^{+\infty}\int S(p_1, p_2)\left|\Phi(p_1, p_2, \varphi)\right|^2 dp_1 dp_2 \tag{10.45}$$

with

$$\left.\begin{aligned} p_1 &= K_0 \sin\theta\cos\varphi \\ p_2 &= K_0 \sin\theta\sin\varphi \end{aligned}\right\} \tag{10.46}$$

where the angle θ is the spatial angle. The Jacobian of the variable transformation yields

$$dp_1 dp_2 = K_0^2 \sin\theta \cos\theta d\theta d\varphi = K_0^2 \cos\theta d\Omega \tag{10.47}$$

Comparing (10.38) and (10.47), the scattering diagram D becomes

$$D = \frac{\sigma^2 K_0^2}{(2\pi)^2} s(\theta,\varphi) |\Phi(\theta,\varphi)|^2 \cos\theta \tag{10.48}$$

In the plane $\varphi = 0$, we write

$$|\Phi(\theta,\varphi)|^2 = |C'_{11}(y)|^2 \tag{10.49}$$

where $C'_{11}(y)$ coincides with $C_{11}(y)$ under the condition that the value α_1 in (10.34) is replaced by:

$$\alpha'_1 = iK_0^2 (1+R)(1-\varepsilon_2) \tag{10.50}$$

With (10.32), (10.41), (10.46), (10.47) and (10.49), we find

$$|\Phi|^2 = \frac{K_0^2 |1+R|^2 |1-\varepsilon_2|^2}{\left| \cos\theta + \sqrt{\varepsilon_2 - \sin^2\theta} \dfrac{1-\alpha^2(\theta) R_{h23}(\theta)}{1+\alpha^2(\theta) R_{h23}(\theta)} \right|^2} \tag{10.51}$$

where

$$\alpha(\theta) = e^{-iK_0 h \sqrt{\varepsilon_2 - \sin^2\theta}} \tag{10.52}$$

and

$$R_{h23}(\theta) = \frac{\sqrt{\varepsilon_2 - \sin^2\theta} - \sqrt{\varepsilon_3 - \sin^2\theta}}{\sqrt{\varepsilon_2 - \sin^2\theta} + \sqrt{\varepsilon_3 - \sin^2\theta}} \tag{10.53}$$

One can show that:

$$|1+R|^2 = \frac{4\cos^2\beta}{\left|\cos\beta + \sqrt{\varepsilon_2 - \sin^2\beta}\,\frac{1-\alpha^2(\beta)R_{h23}(\beta)}{1+\alpha^2(\beta)R_{h23}(\beta)}\right|^2} \quad (10.54)$$

Therefore,

$$D(\beta,\theta) = \frac{4K_0^2\sigma^2}{(2\pi)^2}|1-\varepsilon_2|^2\cos^2\beta\cos^2\theta \times$$

$$\times \frac{1}{\left|\cos\theta + \sqrt{\varepsilon_2 - \sin^2\theta}\,\frac{1-\alpha^2(\theta)R_{h23}(\theta)}{1+\alpha^2(\theta)R_{h23}(\theta)}\right|^2} \times \quad (10.55)$$

$$\times \frac{\int_0^\infty\int_0^\infty R(x,y)\cos[K_0 x(\sin\theta - \sin\beta)]dxdy}{\left|\cos\beta + \sqrt{\varepsilon_2 - \sin^2\beta}\,\frac{1-\alpha^2(\beta)R_{h23}(\beta)}{1+\alpha^2(\beta)R_{h23}(\beta)}\right|^2}$$

For a particular correlation function, one can derive the explicit dependence of the scattering diagram D on the statistical properties of the surface. The expression for the 'inverse' scattering diagram can be obtained if $\theta = -\beta$ is substituted in (10.55). As an example we assume no loss and a surface-roughness correlation function

$$R(x,y) = \frac{1}{4\pi r^2} e^{-\frac{x^2+y^2}{r^2}} \quad (10.56)$$

where r is the correlation radius. For simplicity, the calculation is only done for a structure with a rough upper boundary and the result becomes

$$D(\beta,\theta) = \frac{K_0^2\sigma^2}{(2\pi)^2}|1-\varepsilon_2|^2\cos^2\beta\cos^2\theta\exp\left(-\left(\frac{\pi r}{\lambda}\right)^2(\sin\theta - \sin\beta)^2\right) \times$$

$$\times \frac{1}{\left|\cos\theta + \sqrt{\varepsilon_2 - \sin^2\theta}\,\frac{1-\alpha^2(\theta)R_{h23}(\theta)}{1+\alpha^2(\theta)R_{h23}(\theta)}\right|^2 \left|\cos\beta + \sqrt{\varepsilon_2 - \sin^2\beta}\,\frac{1-\alpha^2(\beta)R_{h23}(\beta)}{1+\alpha^2(\beta)R_{h23}(\beta)}\right|^2}$$

(10.57)

Inverse-scattering diagrams are shown in Figs 10.2a and 10.2b. As can be seen from these figures, the inverse-scattering diagram is strongly dependent upon the correlation radius r relative to wavelength λ (see Fig. 10.2a) and upon the layer thickness h relative to wavelength λ (see Fig. 10.2b). The layer thickness effects can only be disregarded at small values of r/λ (see for example: the layer model as shown in Fig. 10.3). As the layer thickness is increased, the inverse-scattering diagram becomes gradually wider. The physical interpretation is that by increasing the layer thickness, the scattering volume is increased. This causes a widening in the scattering diagram.

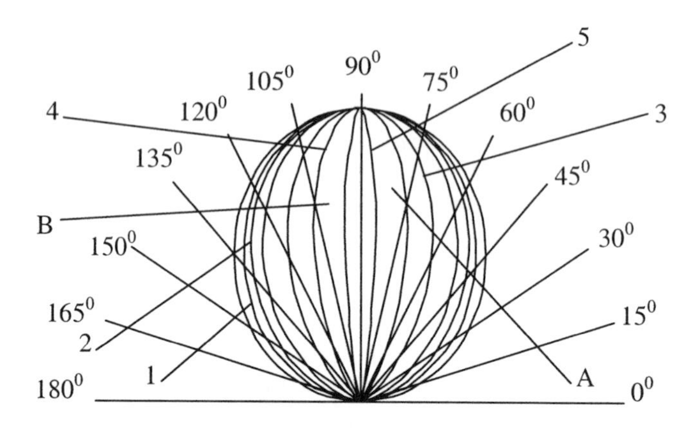

Fig. 10.2a Inverse-scattering diagram A. $\cos\beta$ B. $\cos^2\beta$ $h/\lambda = 2$
1) $r/\lambda=0$ 2) $r/\lambda=0.5$ 3) $r/\lambda=4$ 4) $r/\lambda=20$ 5) $r/\lambda=50$

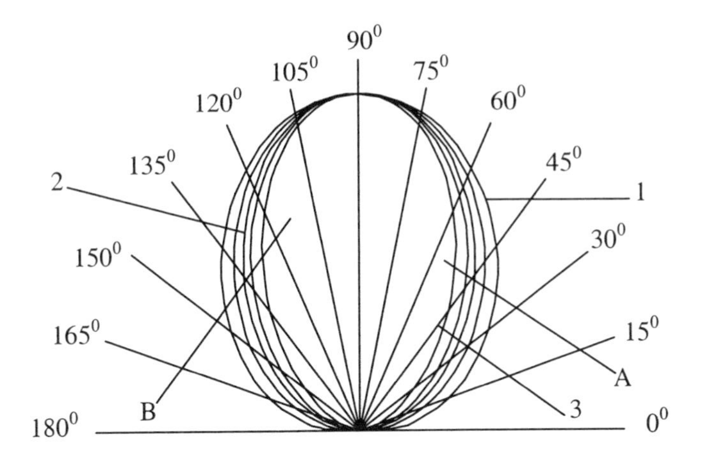

Fig. 10.2b Inverse-scattering diagram $r/\lambda = 0$
A. $\cos\beta$ B. $\cos^2\beta$ 1) $h/\lambda = 500$ 2) $h/\lambda = 2$ 3) $h/\lambda = 0$

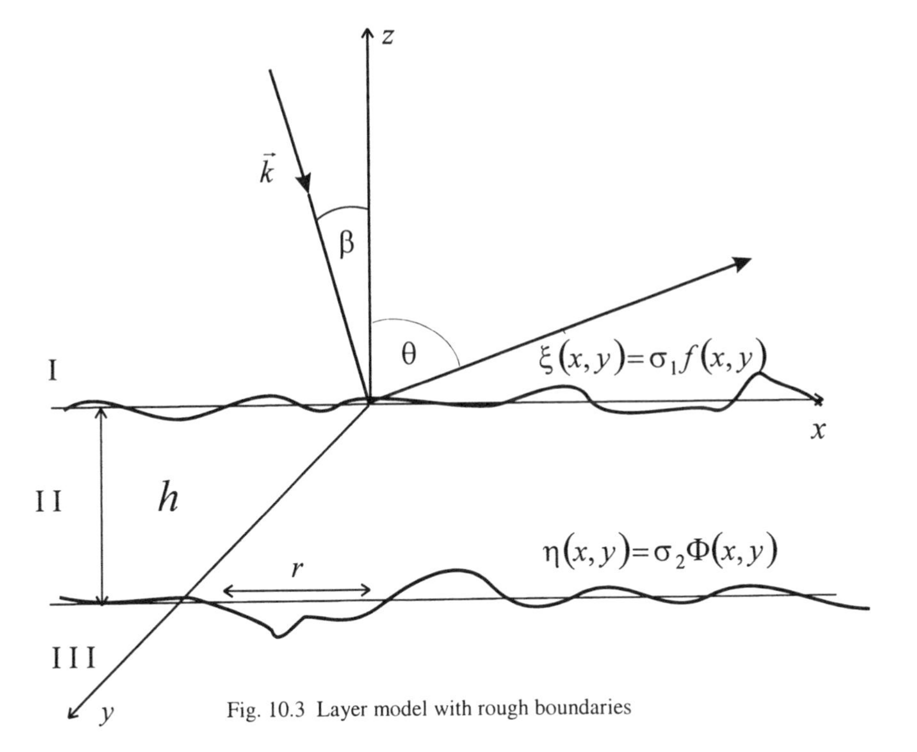

Fig. 10.3 Layer model with rough boundaries

10.3 Model parameters of an ensemble of co-directional cylinders

10.3.1 Radar backscattering matrix of a vegetation-earth two-layer system

The analysis of coherent scattering from a vegetation layer means that a scattering matrix $\hat{S}$ has to be constructed for the (vegetation-earth) two-layer system. This analysis of backscatter is based on the physical layer model described in section 10.2 of this chapter. The monostatic radar backscattering matrix can be written in the form

$$\hat{S} = \hat{T} \cdot \hat{R} \cdot \hat{T} \tag{10.58}$$

Here, $\hat{T}$ describes the polarization state of the coherent field of the incident wave. In the absence of anisotropic attenuation, the expression for matrix $\hat{T}$ in linear polarization can be written in the form

$$\hat{T} = \begin{pmatrix} e^{i\frac{\Delta\varphi}{2}} & 0 \\ 0 & e^{-i\frac{\Delta\varphi}{2}} \end{pmatrix} \cdot e^{-\frac{iH}{L\cos\theta}} \tag{10.59}$$

H is the thickness of the vegetation layer, θ is the incident wave angle, L is the extinction length of the electromagnetic wave in the medium and $\Delta\varphi$ is the additional phase difference between waves with orthogonal polarizations after a single transmission through the vegetation layer.

Matrix $\hat{R}$ describes the wave reflection from the interface with the earth. An explicit form of this matrix can be derived for a rough boundary vegetation earth surface with small-scale irregularities when the scattered field can be evaluated by means of the method of small perturbations. The scattered field has the following form

$$\left(\vec{p} \cdot \vec{E}\right) = \frac{k^2 e^{ikr}}{\pi r} \left[A\left(\vec{p} \cdot \vec{p}_0\right) + B\left(\vec{n} \cdot \vec{p}\right)\left(\vec{n} \cdot \vec{p}_0\right) \right] \cdot \int_{S_0} \nu\left(\vec{\rho}\right) e^{-2i\vec{p} \cdot \vec{\rho}} d\vec{\rho} \tag{10.60}$$

where r denotes the distance between radar and target. In this expression, $\vec{p}_0$ is the vector which determines the field orientation of the radar wave with linear polarization; $\vec{p}$ is the vector corresponding to the component of the scattered field with a polarization which is analyzed; $\vec{n}$ is the normal vector to the averaged surface; $\nu(\vec{\rho})$ characterizes the profile with small-scale roughness ($\langle \nu(\vec{\rho}) = 0 \rangle$). Furthermore,

$$\left.\begin{aligned} A &= (\varepsilon - 1)\left(\frac{a}{a+b}\right)^2 \\ B &= 2(\varepsilon - 1)^2 \frac{a^2 b}{(b + a\varepsilon)^2 (a+b)} \end{aligned}\right\} \tag{10.61}$$

where $a = \cos\theta$; $b = \sqrt{\varepsilon - \sin^2\theta}$, ε being the soil permittivity. S_0 is the projection of the illuminated surface on the averaged interface of the two media.

Omitting common factors, we derive

$$\hat{R} \sim \begin{pmatrix} A+B & B\cos\theta \\ B\cos\theta & A + B\cos\theta \end{pmatrix} \sim \begin{pmatrix} 1 & \nu \\ \nu & \alpha \end{pmatrix} \tag{10.62}$$

This formula for the matrix $\hat{R}$ is symmetrical ($R_{12} = R_{21}$), with unequal diagonal elements. Substituting (10.60) and (10.62) into (10.58) yields

$$\hat{S} \sim \begin{pmatrix} e^{i\Delta\varphi} & \nu \\ \nu & \alpha \cdot e^{-i\Delta\varphi} \end{pmatrix} \tag{10.63}$$

As in (10.62), the common factor is omitted, the non-diagonal elements for arbitrary roughness can be complex.

Information of interest is found in the arguments of the different diagonal elements of (10.63) describing the additional phase of the waves with polarization. The possibility of experimental determination of this value by means of remote sensing methods will allow us to calculate values of a dimensionless parameter characterizing the effective density the biomass layer thickness and the biomass electrodynamic properties (see, also, Chapters 6-7 of this monograph).

10.3.2 Radar polarization effects

For analyzing the phase shift of backscattered waves with orthogonal polarization in a vegetation layer, we make a transition to another polarization basis. The main requirements for the new basis can be formulated in the following way:

- Phase shift $\Delta\varphi$ should contribute (in the new basis) not only in the argument but also in the magnitudes of the matrix elements.

- A new basis should not be exotic; it should be easy (in a technical sense) to realize these polarization states.
- A circular polarization basis meets the above requirements, as shown below. The transition to the new basis can be carried out with the use of an unitary matrix Q:

$$Q = \frac{1}{\sqrt{2}} \begin{pmatrix} 1 & i \\ 1 & -i \end{pmatrix} \tag{10.64}$$

so that

$$S_C = Q S_L Q^+ \tag{10.65}$$

where C stands for circular and L for linear polarizations. Q^+ is the Hermitean conjugate of the matrix Q.

The expression for the scattering matrix in a circular polarization basis becomes

$$S_C = \frac{1}{2} \begin{pmatrix} e^{i\Delta\varphi} + \alpha \cdot e^{-i\Delta\varphi} & e^{i\Delta\varphi} - \alpha \cdot e^{-i\Delta\varphi} + 2iv \\ e^{i\Delta\varphi} - \alpha \cdot e^{-i\Delta\varphi} - 2iv & e^{i\Delta\varphi} + \alpha \cdot e^{-i\Delta\varphi} \end{pmatrix} \tag{10.66}$$

This matrix may be represented by the sum of a diagonal matrix S_C^D and an additional matrix $\Delta_1 S_C$:

$$\left. \begin{aligned} S_C^D &= \frac{1}{2} \left(e^{i\Delta\varphi} \quad e^{-i\Delta\varphi} \right) \begin{pmatrix} 1 & 0 \\ 0 & 1 \end{pmatrix} \\ \Delta_1 S_C &= \frac{1}{2} \left(e^{i\Delta\varphi} \quad -e^{-i\Delta\varphi} \right) \begin{pmatrix} 0 & 1 \\ 1 & 0 \end{pmatrix} + 2iv \begin{pmatrix} 0 & 1 \\ -1 & 0 \end{pmatrix} \end{aligned} \right\} \tag{10.67}$$

S_C^D maintains the polarization of the incident field and $\Delta_1 S_C$ changes the field into the orthogonal polarization.

Receiving and transmitting with the same specified circular polarization means that the scattered field component with the non-diagonal part in the scattering matrix is filtered out. In this case, it is sufficient to change the radar equation so that we multiply the gain of the antenna with the complex factor Λ:

$$\Lambda = \frac{1}{2}\left(e^{i\Delta\varphi} + \alpha \cdot e^{-i\Delta\varphi}\right) \tag{10.68}$$

The squared modulus of Λ is proportional to the received power of an antenna with circular polarization. It can be shown that when applying linear and circular polarizations, the received powers I_L and I_C have the following ratio:

$$f = \frac{I_C}{I_L} = |\Lambda|^2 = \frac{1}{2}\left(1+\alpha^2\right) + \alpha\cos 2\Delta\varphi \tag{10.69}$$

Here, $\alpha = \dfrac{A + B\cos\theta}{A+B}$, A and B are given by (10.61) and I_C and I_L are measured experimentally. This approach allows us to calculate the value of the phase shift $\Delta\varphi$ determined by the vegetation cover. The plot of $f = f_\theta(\Delta\varphi)$ is shown in Fig. 10.4.

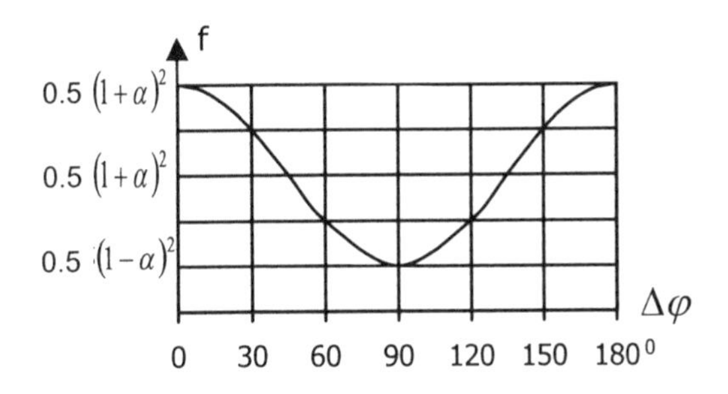

Fig. 10.4 Dependence of f on Δφ

10.4 Conclusions and applications

The investigation of radio wave scattering from a surface layer with the presence of roughness has been the main subject of this chapter. In general, surface roughness, assumed to be described by a random function of the space coordinates, affects to a great degree the process of interaction between radio waves and the surface.

As a whole, the problem was formulated as an investigation of the process of radio wave reflection from a rough surface as a function of the electrical, physical, statistical and geometrical characteristics of the surface. We used a three-layer medium as a surface model. The first layer is air, the second is an electrically uniform layer with rough boundaries and the third one is assumed to be an electrically uniform space. All three layers are assumed to be infinite in the X and Y directions. A plane electromagnetic wave illuminates the surface at a certain random angle relative to the Z axis. Since the influence of small-scale roughness (with respect to the wave length) on a reflected wave with vertical polarization is weak, the investigation was confined only to the study of this influence to a wave with horizontal polarization.

In order to solve the problem, we used the method of small perturbations supplemented by the introduction of a convenient small space parameter. In solving the respective equations, the roughness was taken into account by introducing it into boundary conditions. Then, we used the method of successive approximations in order to derive the solution needed for the calculation of the reflected field from the underlying structure. At first, we used a first-order approximation. As a result, we derived 12 equations for the determination of 12 unknown quantities characterizing the process of reflection. Similarly, we derived and solved the respective equations in a second-order approximation. By carrying out the first-order approximation, we found the scattering diagrams determined as the angular distribution of the averaged power of the reflected wave. In addition, we constructed the diagrams of inverse scattering under the assumption that the correlation function of surface roughness was exponentional in character. The scattering diagrams showed a strong dependence on the ratio of the correlation radius to wavelength. An increase of the layer thickness resulted in a wider diagram of inverse scattering.

We also considered the processes taking place in the case of electromagnetic wave scattering from a two-layer structure. As a specific example of such a model, we analyzed a vegetation layer located on the earth surface.

It is important to determine the phase shift between orthogonally polarized components of the field when passing through a vegetation layer. This is due to the fact that knowledge of such a phase shift gives us a potential opportunity to determine the density and the thickness of biomass of a vegetation layer and its electromagnetic characteristics. In this chapter, we have illustrated that when it is possible to measure the received scattered radio waves corresponding to linear and circular polarizations, the ratio of these quantities is connected with the phase shift by means of an equation, which includes the soil permittivity and the angle of sensing as parameters.

The results derived in this chapter may have a number of practical applications. Most of the results related to problems of remote sensing do not take into account the influence of surface roughness due to analytical complexities. Therefore, in spite of the fact that the proposed approach is complicated and tedious, it allows us to derive a strict solution of the problem which takes the influence of surface roughness on the scattering of radio waves into account. Substituting in the respective equations appropriate random functions describing the surface roughness, we can construct physically useful diagrams of radio wave scattering.

Knowledge of inverse scattering diagrams, determined by the character of surface roughness, is very important for distinguishing the influence of the main factors in the process of radio wave scattering. These factors include the influence of the surface permittivity and the influence of the roughness at the boundary between different media. Since the permittivity is connected with the electrical and physical properties of the surface and its knowledge is the key factor for solving the inverse problems of remote sensing, it is important to be able to distinguish the influence of the variations of the permittivity and the influence of the surface roughness on the scattered field characteristics.

The results presented in this chapter show that, to some extent, we can distinguish the above-mentioned factors of influence when sufficiently mild conditions are met.

The formulated conditions permitting the determination of the phase shift occurring between the orthogonal polarized components of the field when passing through a vegetation layer may have practical applications. As pointed out above, knowledge of this phase shift allows one to determine the density and the thickness of biomass of a vegetation layer, two quantities that can aid one to predict the yield of agricultural crops and determine the degree of their ripening. Furthermore, knowledge of the density and the thickness of biomass of a vegetation layer allow us to determine other significant biometric characteristics.

For solving the above-mentioned problem, it is necessary that the radar of the remote sensing system be able to radiate a linearly polarized wave and a wave with circular polarization. Presently, radars used in remote sensing systems are available to radiate such waves.

CHAPTER 11

Polarimetric Methods for Measuring Permittivity Characteristics of the Earth's Surface

11.1 Introduction

A principally new method for determining the complex dielectric permittivity of layered media arising in remote sensing problems is considered. It is shown that a relative comparison of the voltages and the phases of the signals in the orthogonal channels of receiving devices allows us to determine the desired complex dielectric permittivity for a wide class of layered media. This method permits us to construct a special sphere (referred to as KLL-sphere), each point of which displays a certain type of an earth surface. The distinction of the different types of earth surfaces depends on the complex dielectric permittivities that are involved. The KLL-sphere properties are investigated in detail. In particular, a rule is established for changing the earth surface "images" on the KLL-sphere as the real earth surface physical and chemical characteristics vary.

Environmental studies and ecological monitoring are among the main tasks of natural sciences. Remote sensing is a modern method for solving such problems. The determination of physical, mechanical, chemical, and other properties of the environment and, in particular, of layered media may be carried out by means of remote sensing. In order to solve these problems, we have to analyze the characteristics of radiowaves scattered from sensed objects. The aforementioned characteristics are determined by geometrical parameters and the complex dielectric permittivities of the analyzed objects. At the same time, the complex dielectric permittivities are determined by physical, mechanical, chemical, and other properties of the analyzed objects. Therefore, the knowledge of the complex dielectric permittivity allows us to determine the main characteristics of these objects. Although the determination of these characteristics is the fundamental inverse problem (i.e., such problems belong to the class of ill-defined problems), the only way to solve remote sensing problems consists in the complex dielectric permittivity determination. So, we have to find methods which would allow us to determine the complex dielectric permittivity with maximum accuracy and reliability.

In this chapter, a new approach for determining the complex dielectric permittivity is offered. It is based on the feasibility, established by the authors, of determining these characteristics from the results of relative measurements of signals in orthogonal channels of the receiving device (ratio of voltage and difference in phases). As it is shown below, in order to realize such a feasibility it is necessary to irradiate an investigated surface by an electromagnetic wave with a special kind of a polarization. Thus, the measurement may be confined to only one (!) pulse.

11.2 Determination of the complex permittivity

The available methods for the determination of the complex dielectric permittivity are characterized by a number of principal difficulties. These difficulties are connected with the necessity to conduct absolute measurements, and to know true distances from analyzed objects. As a result, we can only determine the reflection coefficient with a substantial error. Therefore, the permittivity itself can be determined with a substantial error only.

In the present work, we propose a principally new method for the determination of the dielectric permittivity of layered media arising in the problems associated with remote sensing. In order to explain the physics of the proposed method, we first consider a smooth surface. By means of this example, we shall show how it is possible to determine the relative dielectric permittivity (the real and imaginary parts of the complex relative dielectric permittivity) of the analyzed surface using only relative measurements.

The Fresnel formula for the reflection coefficient assumes the form

$$R_{HP} = \frac{\cos\theta - \sqrt{\varepsilon - \sin^2\theta}}{\cos\theta + \sqrt{\varepsilon - \sin^2\theta}} \tag{11.1}$$

$$R_{VP} = -\frac{\varepsilon\cos\theta - \sqrt{\varepsilon - \sin^2\theta}}{\varepsilon\cos\theta + \sqrt{\varepsilon - \sin^2\theta}} \tag{11.2}$$

for horizontal and vertical polarization, respectively. In formula (11.2), a minus sign is placed before the fraction in order that, for vertical surveillance $(\theta = 0)$, when horizontal and vertical polarizations do not differ, formulas (11.1) and (11.2) coincide.

For further analysis it is reasonable to derive the relationship between the Fresnel reflection coefficients for the two polarizations. If in formulas (11.1) and (11.2) we eliminate ε, the desired relation is given by

$$R_{VP} = R_{HP}\frac{\cos 2\theta - R_{HP}}{1 - R_{HP}\cos 2\theta} \tag{11.3}$$

A distinctive feature of this formula consists in the fact that it directly connects the Fresnel reflection coefficients and the surface surveillance angle. We introduce, next, the following notation:

$$f = \frac{R_{VP}}{R_{HP}} \tag{11.4}$$

The parameter f is called the polarization ratio. An important property is that f is equal to the ratio of the voltages of the signals in the orthogonal channels of a receiving device. The polarization ratio can be easily and quite accurately measured. We have to take into account that the polarization ratio is a complex number, i.e., it depends not only on the ratio of the powers of the signals, but it is determined also by the phase difference of the signals in the orthogonal channels of a receiving device. If these channels are identical (in attenuation and phase incursion) and the directional properties of the antenna are the same for the orthogonal components of the wave, the polarization ratio does not depend on the distance from the analyzed surface and the antenna gain. This property allows us to determine the complex dielectric permittivity not knowing the distance and the antenna characteristics. Thus, the reflection coefficients can be uniquely determined using the polarization ratio derived by relative measurements.

Using Eqs (11.3) and (11.4), the reflections coefficients can be expressed in terms of the polarization ratio as follows:

$$\begin{cases} R_{HP} = \dfrac{\cos 2\theta - f}{1 - f\cos 2\theta} \\ R_{VP} = f\dfrac{\cos 2\theta - f}{1 - f\cos 2\theta} \end{cases} \tag{11.5}$$

Knowing the polarization ratio, we can determine the complex dielectric permittivity using formulas (11.1) and (11.2); specifically, we obtain

$$\varepsilon = \left[1 + \frac{4f}{(1-f)^2}\sin^2\theta\right]\tan^2\theta \tag{11.6}$$

Thus, relative measurements allow us to determine both the reflection coefficients and the complex dielectric permittivity (real and imaginary parts of the complex dielectric permittivity) of a layered medium.

Next, we express the polarization ratio in polar form, viz.,

$$f = |f| e^{i\psi} \tag{11.7}$$

We derive the following relations using Eq. (11.6):

$$\begin{cases} \operatorname{Re}\varepsilon = \left[1 + 4|f| \dfrac{\left(1+|f|^2\right)\cos\psi - 2|f|}{\left(1 - 2|f|\cos\psi + |f|^2\right)^2} \sin^2\theta \right] \tan^2\theta \\ \operatorname{Im}\varepsilon = 4|f| \dfrac{\left(1-|f|^2\right)\sin\psi}{\left(1 - 2|f|\cos\psi + |f|^2\right)^2} \sin^2\theta \tan^2\theta \end{cases} \tag{11.8}$$

These formulas allow us to evaluate the errors in determining the desired parameters. The results and the above drawn conclusions can be used for other types of layered media. The following examples will make these statements clear.

In the case of a one-position radar, the reception and transmission are carried out at the same antenna. If a layered medium is a smooth infinite plane, the reflections in the direction of the antenna are absent. The presence of surface roughness results in the appearance of waves propagating in the direction of the antenna. For many types of layered media modeled by rough structures, the reflection coefficient in the direction of antenna is the product of two multipliers [*Bogorodsky,* 1985; *Kozlov,* 1997; *Zhukovsky,* 1979]. The first multiplier, as a rule, is a certain function depending only on geometrical characteristics of the analyzed surface (the statistical parameters of medium-height roughnesses and the tilting angle, their dispersions and correlation distances, the illuminated area boundary, etc.) In most cases, the dependence of this multiplier upon the type of polarization is weak. The second multiplier is, as a rule, the Fresnel reflection coefficient [cf. Eq. (11.5)] corresponding to the "mirror" angle. The polarization ratio is determined by Eq. (11.4). This allows us, also in this case, to apply formulas (11.6) and (11.8) for the determination of the dielectric permittivity.

More complicated models give more complicated relations for the determination of the desired reflection coefficient in the direction of the antenna. However, practically in all cases, this reflection coefficient is connected with the complex dielectric permittivity of a layered medium only by Fresnel's coefficients for the "mirror" angle [*Bogorodsky*, 1985; *Kozlov,* 1997; *Zhukovsky,* 1979]. This allows us (at least in principle) to experimentally determine the complex dielectric permittivity ε using relative measurements of the voltages and the phases of the signals in the orthogonal channels of a receiving device, i.e., using the polarization ratio.

Thus, the results derived in this chapter offer ample scope for the remote determination of the complex dielectric permittivity of layered media. The principal distinction of the proposed method consists in the fact that it is based on relative measurements of the voltages and the phases (amplitudes-phase difference ratio) in the orthogonal channels of a receiving device. Such measurements can provide small-error results. Therefore, the dielectric permittivity calculated using the results of these measurements can be also determined with high accuracy.

In [*Bogorodsky*, 1985], several models of rough surfaces were proposed. Let us show that for some of them it is possible to determine the complex permittivity by means of carrying out the corresponding relative measurements. We begin the consideration with Model 1 in [*Bogorodsky*, 1985]. Model 1 covers the following classes of underlying surfaces: The surfaces $\xi(x,y)$ are large-scale ($\rho_0/\lambda >> 1$), smooth ($\rho_{cor}/\lambda >> 1$), flat-lying ($(\nabla\xi_1)^2 << 1$), with a random roughness σ/λ on the average plane $\overline{\xi}_1 = 0$, where ρ_0, ρ_{cor} are the radii of curvature and correlation of the surface, and σ is the mean square height of roughness. This model describes unplowed fields, hilly terrain without substantial vegetation, water surfaces with high waves, etc.

After some transformation of the formulas indicated in [*Bogorodsky*, 1985] for Model 1, it is possible to write the expressions for the scattering matrix eigenvalues, viz.,

$$\lambda_1 = -\frac{a_2 n_z^2}{1 - a_2 n_z^2}\sqrt{a_1 + \eta^2}\, R_{HP} \tag{11.9a}$$

$$\lambda_2 = -\frac{a_2 n_z^2}{1 - a_2 n_z^2}\sqrt{a_1 + \eta^2}\, R_{VP} \tag{11.9b}$$

The parameters a_1, a_2, η, n_z are determined by the statistical characteristics of the rough surfaces [*Bogorodsky,* 1985]; specifically, we have

$$a_1 = \sin\theta - \gamma\cos\theta\,,\; a_2 = \cos\theta + \gamma\sin\theta\,,\; n_z = \left(1 + \left(\frac{\partial\xi}{\partial x}\right)^2 + \left(\frac{\partial\xi}{\partial y}\right)^2\right)^{-0.5} \tag{11.10a}$$

$$\eta = \frac{\partial\xi}{\partial x}\sin\varphi - \frac{\partial\xi}{\partial x}\cos\varphi\,,\; \gamma = \frac{\partial\xi}{\partial x}\cos\varphi + \frac{\partial\xi}{\partial x}\sin\varphi \tag{11.10b}$$

Therefore, in an eigenpolarization basis, the ratio of voltages in the orthogonal channels of a receiving device determines the parameter f. This allows us to determine the complex permittivity of the underlying surfaces that can be described by Model 1.

The derived relationships determine the principal scheme for the determination of the complex permittivity. The first step consists in the selection of a type of polarization of the radiated wave, such that the cross component of the reflected wave disappears (actually, it becomes very small). This selection may be carried out by means of using the type of total polarization scanning. The second step consists in the selection of a type of polarization of the radio wave, such that the orthogonal components in the eigenpolarization basis of the matrix S are in phase and have equal powers. The third step consists of the measurement of the ratio of the differences of phases and amplitudes of the voltage in the orthogonal channels of a receiving device. The latter allows us to determine the desired ratio f and then to determine the complex permittivity. The knowledge of the polarization ratio also allows us to determine the ratio of statistical parameters η/γ that is given by

$$\left(\frac{\eta}{\gamma}\right)^2 = \tan^2\varsigma = \left(\frac{U_{11}}{U_{22}} + f\right)\left(1 + f\frac{U_{11}}{U_{22}}\right)^{-1} \tag{11.11}$$

where U_{11}, U_{22} are the voltages in the orthogonal channels, measured in the polarization basis (HP-VP).

Let us now consider Model 2. This model covers the following classes of underlying surfaces [*Bogorodsky,* 1985]: The surfaces are small-scale ($\rho_0/\lambda \le 1$), flat-lying $(\nabla\xi_2)^2 << 1$), and slightly rough ($\sigma/\lambda << 1$) on the average plane $\overline{\xi}_2 = 0$. This model describes concrete, asphalt, sand and gravel layers, calm water surfaces with small ripples, smooth steppes with low vegetation in the centimeter and longer wavebands, etc.

In [*Kozlov*, 1998], the following formula for the polarization ratio is derived:

$$f = -\sin^2\theta \pm \sqrt{\sin^4\theta + 2\frac{U_{11}}{U_{22}}\cos^2\theta - 1} \tag{11.12}$$

Thus, for surfaces described by the Model 2, the complex permittivity determination method based on relative measurements in orthogonal channels of a receiving device remains valid. The principal aspects of the application of this method are the same as in the case of analyzing smooth surfaces. As a sensing wave, it is possible to use a radio wave with $LP-45$ or a wave with circular polarization with the corresponding correction of the phase ratios.

Let us now consider Model 3. This model covers the following classes of underlying surfaces [*Bogorodsky,* 1985]: The surfaces are complex (continuous or discontinuous) characterized by small-scale irregularities on top of large-scale irregularities ($\xi_3(x,y) = \xi_1(x,y) + \xi_2(x,y)$). This model describes rough water surfaces, deserts with large sand-hills, plowed hilly surfaces, etc. Unfortunately, it is impossible to derive an explicit expression for the polarization ratio in this case. Nevertheless, if we assume *a priori* information about the statistical characteristics of the underlying surfaces, it is possible to determine the complex permittivity on the basis of relative measurements.

Let us consider, next, the underlying surfaces described by Model 4 (surfaces with different complex geometrical structures). This model describes residential areas, mountains, large forest areas, etc. In the centimeter and decimeter waveband, it is possible to represent these surfaces in the form of random compositions of incoherent independent scatterers. It is impossible to speak about a certain specific value of the complex permittivity in this case. Nevertheless, if we formally conduct the calculation of the complex permittivity by one of the formulas (11.9) or (11.12), it is possible to assign to these surfaces a certain complex permittivity (pseudo-permittivity).

The dielectric permittivity module for some types of surfaces (long-term ice, $\theta = 45^o$; one year ice, $\theta = 45^o$; water surface at wind speed 10-15 m/s, $\theta = 70^o$; field without grass, $\theta = 70^o$) was designed for an illustration (see Table 11.1).

Surfaces	$\langle S_{11} \rangle$	$\langle S_{22} \rangle$	ψ^o	calculated $\lvert\varepsilon\rvert$	$\lvert\varepsilon\rvert$ experimental
long-term ice	0.0356	0.0689	7	2.7	3.0
one-year ice	0.0156	0.0277	5	3.5	3.4
water surface	0.0084	0.0097	30	35	40-70
field	0.283	0.235	10	14	12

Table 11.1 Dielectric permittivity module

The wavelength used was $\lambda = 21\,\text{cm}$. The values of the scattering matrix elements are taken from [*Boerner*, 1997]. The calculated dielectric permittivity modules were compared with other sources [*Kozlov*, 1993; *Finkelshtein*, 1984].

11.3 The KLL-sphere

In order to use formula (11.6), it is necessary to provide the consecutive radiation of the equal-power radio waves with horizontal and vertical polarization. It should be taken into account that providing the consecutive radiation we have to measure (every time) the voltage in the corresponding receiving channel and after that to calculate the parameter f and to use formula (11.6). However, it is possible to proceed in a different way. If a linearly polarized (with the polarization angle 45°) radio wave LP-45 is radiated, we shall have a similar situation. In this case, the in-phase radio waves are simultaneously radiated by both orthogonal channels. Since for the class of surfaces under consideration the cross components do not appear during the reflection, the signals in the orthogonal receiving channels may be used for the determination of the polarization ratio f, which, in this case, describes fully the reflected radio wave polarization ellipse. Thus, the parameter f characterizes the reflected wave polarization under the condition that the surface is illuminated by LP-45. Therefore, a certain polarization, which (as is well known) is one-to-one mapped by the corresponding point of the Poincare sphere, corresponds to every value of the parameter f. On the other hand, the parameter f determines uniquely the complex permittivity, i.e., the type of surface. This means that the points on the Poincare sphere one-to-one correspond to a specific type of underlying surface. Thus, it is possible to construct a sphere, each point of which determines another type of surface. This sphere, referred to as the KLL-sphere is shown in Fig. 11.1.

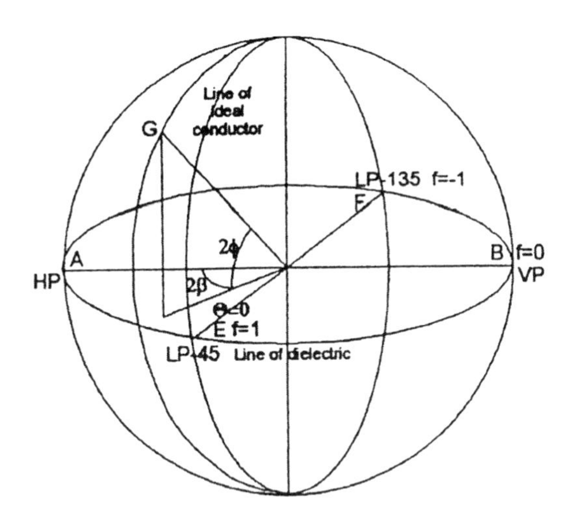

Fig. 11.1 The KLL sphere

The KLL-sphere gives new possibilities for underlying surface classification. The reflected wave polarization type is the main classification characteristic under the condition that the analyzed surface is illuminated by a radio wave with polarization $LP-45$. In order to construct the KLL-sphere, it is not necessary to use $LP-45$ waves. It is also possible to use waves with any other known polarization. In such case, we have to correct the polarization ratio taking into consideration the existing differences in amplitudes and phases of the orthogonal components of the radiated radio waves. In connection with this, the circular polarization is of most interest. The application of circular polarization allows us to avoid inaccuracy in the antenna installations with respect to the earth surface, which is difficult for antennas that use LP-45.

For deriving quantitative relationships, we present the formulas connecting the parameter f and the coordinates of the points on the Poincare sphere that maps the corresponding polarization. We use geographical coordinates. Let 2β denote the longitude of some point D on the KLL-sphere and 2φ its latitude. We then have

$$2\beta = \arctan\frac{2|f|\cos\psi}{1-|f|^2} \tag{11.13a}$$

$$\varphi = \arctan\sqrt{\frac{\tan^2\beta - 2|f|\tan\alpha\cos\psi + |f|^2}{1+2|f|\tan\beta\cos\varphi + |f|^2\tan^2\beta}} \tag{11.13b}$$

Recall that ψ is the argument of the polarization ratio.

In order to use formulas (11.13a, 11.13b) for circular polarization, it is necessary to change the phase difference of signals in the orthogonal channels of a receiving device by 90°. Formula (11.9) shows that with such an approach the angle θ plays a very significant role. Let us consider it briefly. If the surveillance angle $\theta = 0$, the horizontal and vertical polarizations for any type of surface do not differ. Therefore, the parameter $f = 1$. The $LP-45$ corresponds to this value of f. In this case, all types of surfaces are mapped into a single point (point E in Fig. 11.1). The surfaces in this case are not distinguishable. In the second extreme case, when $\theta = \pi/2$, for any type of surface we obtain $f = -1$. In this case, all surfaces are also mapped into a single point (the point F in Fig.11.1). The point F is diametrally opposite to the point E. This point corresponds to the linear polarization $LP-135$. A change of the surveillance angle θ from 0 to $\theta = \pi/2$ causes the points corresponding to different types of surfaces to circumscribe eigenpaths, each starting at E and ending at F.

Let us consider another particular case. Let the analyzed surface represent an ideal dielectric (ε is a real number, $\psi = 0$). In this case, f is a real number. That means that a change in the surveillance angle θ causes the point that maps the type of radio wave polarization to move along the equator of the KLL-sphere. In this case, when the surveillance angle is equal to the Brewster's angle, the moving point reaches the point B (Fig. 11.1), i.e., the point corresponding to a vertical polarization. Thus, dielectrics are mapped on the equator of the KLL-sphere. For this reason, we refer to the equator as the "dielectric" line.

The dependencies of the geographical coordinates ($2\varphi, 2\beta$) of points on the KLL-sphere that map different types of underlying surfaces are shown in Fig. 11.2.

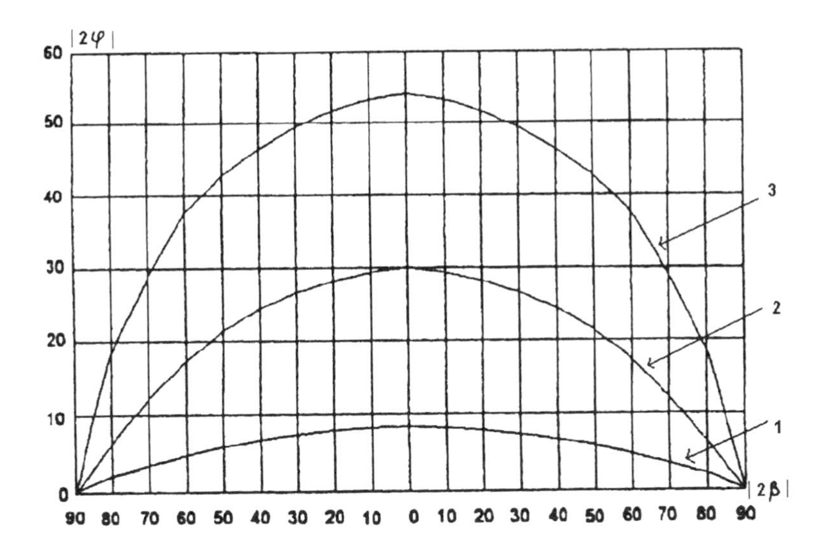

Fig. 11.2 Different types of surface as a function of geographical coordinates of KLL-sphere: 1. wet sand; 2. sea water; 3. corn

This figure shows that as the surveillance angle changes, for each type of underlying surface, the curve starting from the point with the coordinates $2\varphi = 0, 2\beta = 90^o$ and ending at the point with $2\varphi = 0, 2\beta = -90^o$ is plotted on the KLL-sphere. Negative angles 2β with a continuous variation of β correspond to angles from 360° to 270°. The surfaces are chosen so that the ratio of the real and imaginary parts of the complex permittivity is a) much lower than 1; b) much higher than 1; c) about 1. In the representation of such surfaces were chosen:

- agricultural crop (corn), with $\varepsilon = 3 - i15$ (curve 3 in Fig. 11.2);
- sea water (NaCl concentration is equal to 2g/moll liter), with $\varepsilon = 40 - i40$ (curve 2 in Fig. 11.2);
- wet sand (relative weight humidity is equal to 12%), with $\varepsilon = 8 - i1$ (curve 1 in Fig. 11.2).

The points corresponding to the same surveillance angle θ are of interest. Fig. 11.2 clearly shows that it is possible to provide the maximum difference in coordinates of the points on the KLL-sphere, which maps different type of surfaces, by means of selecting the surveillance angle θ. This opens new possibilities for increasing the efficiency of solving the problem of distinguishing underlying surfaces. In addition, the possibility of determining the complex permittivity on the basis of relative measurements, together with the improved quality of distinguishing underlying surfaces, also opens new possibilities for a substantial improvement in solving the inverse problem.

11.4 Conclusions and applications

A new method for the determination of the complex permittivity of layered media in problems of remote sensing has been described in this chapter. It has been shown that a relative comparison of the voltages and the phases of the signals in the orthogonal polarization channels of receiving devices opens the possibility to determine the complex permittivity for a wide class of layered media, with the use of the KLL-sphere.

The application of the KLL-sphere allows us to carry out a rather unique modeling of underlying surfaces. The core of this modeling lies in the fact that there exists a given type of radio wave polarization that may correspond to each type of smooth underlying surface observed at the angle θ. This type of polarization coincides with the reflected wave polarization appearing as a result of the $LP-45$ radio wave reflection from an underlying surface.

This means that for each type of underlying surface observed at the angle θ we can find the corresponding type of the reflected radio wave polarization. This type of polarization coincides with the polarization of the reflected radio wave that appears as a result of reflection of the radio wave with polarization $LP-45$ from the underlying surface. This statement, as it has been said already, means that we can find a point on the unit radius sphere (KLL-sphere), which one-to-one corresponds to a specific type of underlying surface. Any change in the physical or chemical characteristics of the surface causes this point to move on the KLL-sphere, indicating the change in the aforementioned characteristics.

CHAPTER 12

Implementing Solutions to Inverse Scattering Problems: Signal Processing & Applications

12.1 Introduction

We have seen that polarization measurements provide information on characteristics of a medium, e.g. surface roughness, morphology, permittivity, etc. Measurements of time of arrival and spectral contents of the scattered-echo signal allow the determination of the position and speed of a reflecting object. The solution to the inverse problem is to find from the measurements of polarization responses these characteristics of a medium or a scatterer. To this purpose, various types of polarimetric radar are used. Scatterometers are used to measure the surface reflectivity as a function of frequency, polarization and illumination direction. They are used to characterize quantitatively surface roughness. Altimeters are used for topographic mapping applications. Synthetic aperture radars (SAR) are used to produce high resolution images. Applications are for earth remote sensing, e.g., monitoring of vegetation, weather-atmospheric conditions, ocean profiles, terrain roughness, etc. The remote sensing data can also be used to improve adaptive radar techniques (e.g. adaptive clutter suppression).

12.2 Radar imaging

To design an imaging radar, a good knowledge of the expected range of the backscatter cross section "σ" is important. The scatter cross section is the ratio of the scattered power per unit solid angle and the power per unit area incident upon the scatterer. The scattered intensity (per unit solid angle) calculated in the examples on scattering in the previous chapter is proportional to the scattering cross section of the medium.

12.2.1 Processing

The design of the radar requires accurate mathematical modelling of "σ", as well as extensive measurements for:

- scattering from terrain
- scattering from ocean surface
- scattering from the atmosphere
- solid surface sensing (geological structures, soil moisture, forestry inventory, etc.)

- geological mapping
- images (geometry, shape of objects-scatterers)
- scattering from sea and river ice.

Image reconstruction starts from the estimation of the backscatter cross section of the scatterer for a given transmitted signal in a given polarization configuration and a set of recorded returns [*Zebker*, 1991]. The imaging process correlates the set of recorded returns with test functions and gives an "image" function of the scatterer; specifically,

$$b(x,y)=\int_{-\infty}^{+\infty}\int_{-\infty}^{+\infty}B(x_0,y_0)\cdot h(x-x_0,y-y_0)\,dx_0dy_0 \tag{12.1}$$

where $B(x_0,y_0)$ is the wavefield function representing the polarization response of the scatterer (scattering cross section) and $h(x,y)$ is a test function representing the reference data response.

The scattering cross section of the scatterer is given by [*Ulaby*, 1990]

$$\sigma_{pq}(\psi_s,\chi_s;\psi_t,\chi_t)=\lim_{r\to\infty}4\pi r^2\left(\frac{P_p^s}{P_q^t}\right) \tag{12.2}$$

In this expression, P_p^s denotes the scattered power received using polarization "p" at the receiver, P_q^t is the transmitted power (field q-polarized), ψ_s, χ_s are the orientation and ellipticity angles of the receiving antenna polarization, respectively, and ψ_t, χ_t are the respective orientation and ellipticity angles of the transmitting antenna polarization.

For imaging radars, the individual power measurements for each radar resolution element (pixel) are statistically related. Thus, in some radar realizations several (N) power measurements are added to reduce the standard deviation by the ratio $1/\sqrt{N}$, at the expense of loss of spatial resolution. In SAR radars, a large number of returned echoes are used to generate the intensity (brightness) in one image pixel. The average scattering cross section per unit area of a set of N measurements is given by [*Ulaby*, 1990]

$$\sigma_{pq}^0(\psi_s,\chi_s;\psi_t,\chi_t)=\frac{1}{A}\langle\sigma_{pq}^{(n)}(\psi_s,\chi_s;\psi_t,\chi_t)\rangle \tag{12.3}$$

where σ_{pq}^{n} is the scattering cross section of the n-th individual measurement and A is the illuminated area.

The polarization response (or signature) of a remotely sensed scatterer can be represented geographically using the scattering cross section as a function of the ellipticity and the orientation angles of the transmitted electric field. This is done for both co-polar and cross-polar responses. The solution to the inverse problem is to find the characteristics of the scatterer(s) from the known (measured) polarization response. To this purpose, the polarization response "U" is tested against a group of reference data responses (test function "h") before it can be identified. This testing process can be optimized with the use of a polarization filter [*Poelman*, 1981; *Watts*, 1996] which maximizes the contrast between the remotely sensed object and the background medium (clutter) and minimizes the "distance" between the measured (polarization) response and the reference data (maximum correlation)

The reference data needed to reconstruct the image can be a set of polarization responses of targets (scatterers) of known geometry and features. Alternatively, they can be obtained from the prediction - modelling of polarization responses of points (or distributed) targets [*Tatarinov, V. et al.*, 1998]. This processing of the scatterer features is done for each resolution element, or "pixel", of the "image". Features can be the type of scattering mechanism, a target embedded in a medium, or different geographical areas that need to be mapped. A test can be realized by a best fitting processing routine between the results of a mathematical predictive model (propagation modelling, or target modelling) and the observations.

The procedure involved in this routine is to adjust the model parameters (or inputs) to approximate the observed polarization response with minimum error. The model parameters resulting from the best-fit routine between model outputs and observations can give information about the scattering mechanism, e.g., the dominant type of scattering and the types of scatterers. For example, we may introduce in the model different scattering mechanisms. In a vegetation model, for example, we introduce backscattering from branches, trunks, double-bounce scattering from branches and ground, etc. By comparing the relative amount of back-scattered energy for each different mechanism with the experimental results, we can find the type of dominant scattering and deduce the type or distribution of scatterers that have been sensed.

Using the best-fit processing routine between modelling and polarization response from experiments, we can also derive values of surface roughness. The model input parameters, rms height, correlation length and permittivity, are varied until the best fit between the observed and predicted polarization responses is found. In the case of scattering from the sea, for example, varying only the input wind speed in the model allows one to estimate the "in-situ" roughness of the sea surface under investigation from a best-fit analysis with measured polarization response.

A functional block diagram describing the processing for imaging by a polarimetric radar is given in Fig. 12.1

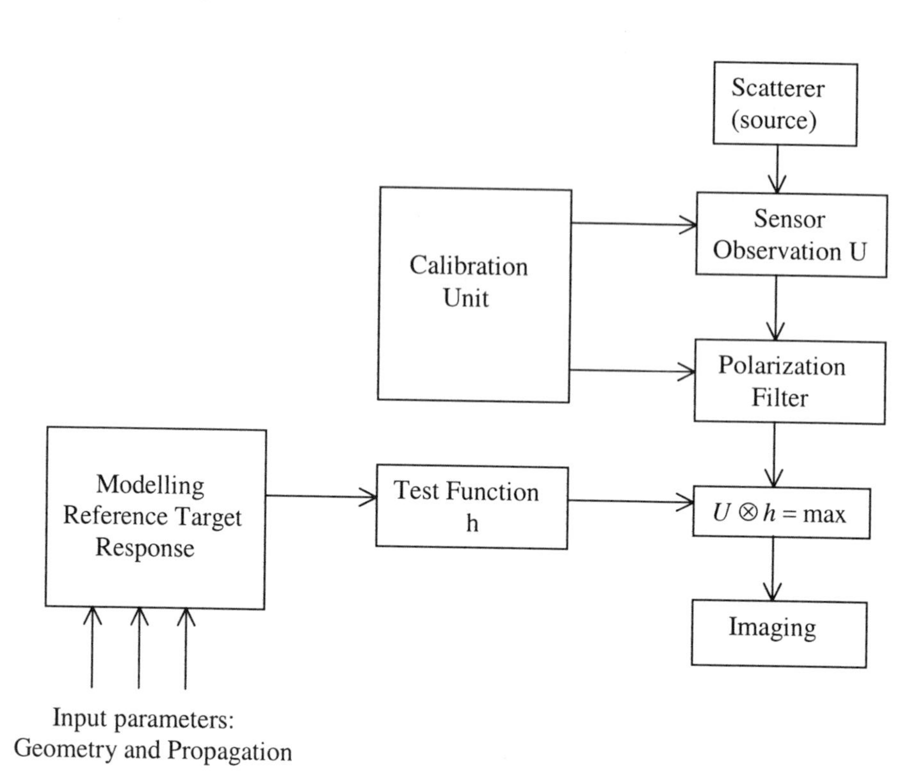

Fig. 12.1 A flow-chart for image processing with polarimetric radar

12.2.2 Examples of classification

The image pixel brightness "b" is a function of the reflected power, or backscatter cross section σ, of the "sensed" surface area, e.g., [*Elachi*, 1987]

$$b = f(\sigma) \tag{12.4}$$

Surface features (e.g., roughness, slope, type or density of vegetation cover, etc.) can be identified as a change in the image brightness in a radar image pattern that reproduces the surface features.

In order to rectify and calibrate the brightness in the image, the function f(σ) needs to be known or measured for each image pixel. In Fig. 12.2, the main elements affecting the image brightness are shown. The transmitted power is measured by a power meter at the input of the antenna. The antenna pattern is usually measured on a test range. The receiver and data-handling system transfer function can be measured by injecting a calibrated signal at the output of the antenna, which is the input of the receiver. A series of signals of increasing amplitude is used in order to "map" the transfer function for different input signals. This procedure of calibration is done for the amplitude and the phase for each polarization channel of the polarimetric radar. This allows one to synthesize known polarization states and scattering matrix for each pixel.

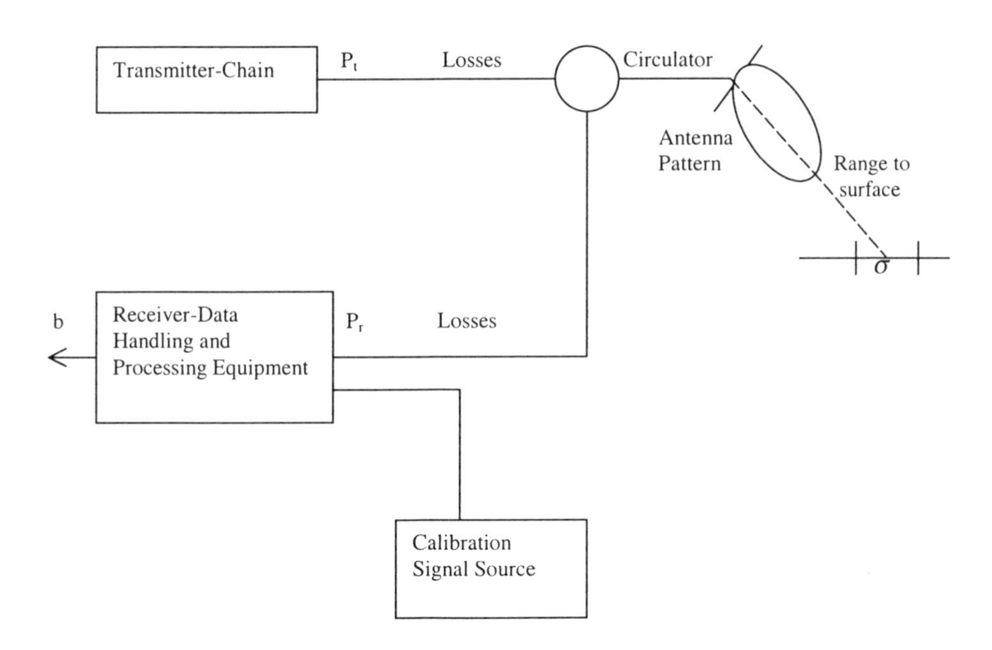

Fig. 12.2 Elements affecting radar image brightness
©IEEE (reproduced with permission of IEEE, C. Elachi, "Spaceborne Radar Remote Sensing: Applications and Techniques", Fig. 5.25, Ch. 5, 1987)

Imaging radars can provide a two-dimensional image of the spatial distribution of the scatterers. They provide, also, a 3-D image by generating two images of the same area from two different incident angles. In oceanographic applications, for example, an imaging radar provides an image of the spatial distribution of small gravity waves and capillary waves which are the main source of backscattered energy. Any surface phenomena affecting the amplitude or spectral distribution of these waves is visible in the radar image. These phenomena include surface swells (which also affect surface slope), internal waves, current, wind eddies and ship wakes.

The small gravity waves and capillary waves are modulated by the ocean swells with a spatial periodicity that "reproduces" the swells pattern. The local incidence angle equals the illumination angle plus the slope of the sea waves in the plane of incidence. These two factors lead to variations in the radar back-scatter return, which spatially "reproduces" the swell pattern, thus allowing to image it with high resolution polarimetric radars. The backscattering cross section of the ocean is a function also of the wind speed. The wind causes an increase in wave heights and introduces a Doppler shift in the back-scattered field. This Doppler effect is a major factor for SAR (Synthetic Aperture Radar) which uses echo-Doppler to generate high resolution images. With SAR radar we can generate a "large antenna aperture" by coherent processing of the received scattered signals. A detailed description of SAR processing is given in the next section. The periodically varying Doppler shift added by the wave motion may, in some situations, enhance the ability to image waves, while in other cases leads to errors that need to be compensated for.

The effect of surface slope plays a major role in imaging undulated terrain (dune fields). The points with zero-slope (maximum points) may produce a pattern of "bright scatterers points" (the high back-scatter creates a bright image tone on the radar image). The high sensitivity of the radar return to surface slope and the partial penetration capability allow the radar to image geologic structural features covered by vegetation canopies. A vegetation canopy is usually of constant height; therefore, the air-vegetation interface "reproduces" the surface morphology.

In applications of sub-surface penetration radars, the nature of the soil, or types of materials embedded, can be investigated based on the "transparency" to certain frequencies (spectral analysis) or polarization states (polarization analysis). This may enhance detection and classification of objects of particular shape buried into the ground. For the imaging of the sub-surface layer, a time-migration processing is shown in Fig. 12.3.

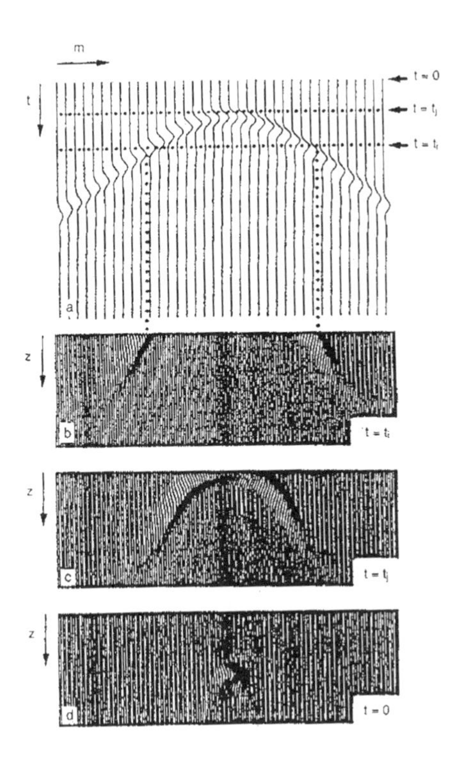

Fig. 12.3 Time-migration process for ground penetrating radars. (Reproduced with permission of IEE, D.J. Daniels, "Surface Penetrating Radar", Fig. 6.29, Ch. 6, 1996)

Another application to inverse radar problems is the classification of scatterers by "imaging in the frequency domain", that is, by measuring the response of the radar at resonance with the natural frequencies of the scatterer under investigation. With this method, we may recognize the target from its own natural frequencies. This may be achieved by comparing the measured resonance frequencies with a "library" of known natural frequencies of scatterers to be sensed. This method is called SEM (singularity Expansion Method) developed in electromagnetic pulse scattering [*Boerner*, 1981; *Baum*, 1991, 1997]. The purpose of this method is to express the electromagnetic signature of the scatterer in terms of singularities of the transfer function (in the complex frequency plane). A broadband pulse excites the corresponding poles, which are referred to as the natural frequencies of the object. It has been shown [*Marin*, 1973, 1974] that whereas the locations of the target's set of natural frequencies in the complex frequency plane is independent of aspect angle, polarization and excitation, the coefficients of the transfer function are not. The transfer function can be expanded in a series of terms for each pole. The coefficients of these terms (residues) are polarization dependent [*Marin*, 1973, 1974].

Multi-frequency and multi-polarization radar imaging is used to acquire detailed information about the surface (roughness) and to allow classification of surface units on a pixel-by-pixel basis by using variations in their spectral or polarimetric signatures. An example of spectral signature is given in Fig. 12.4 and one of polarization signature in Fig. 12.5.

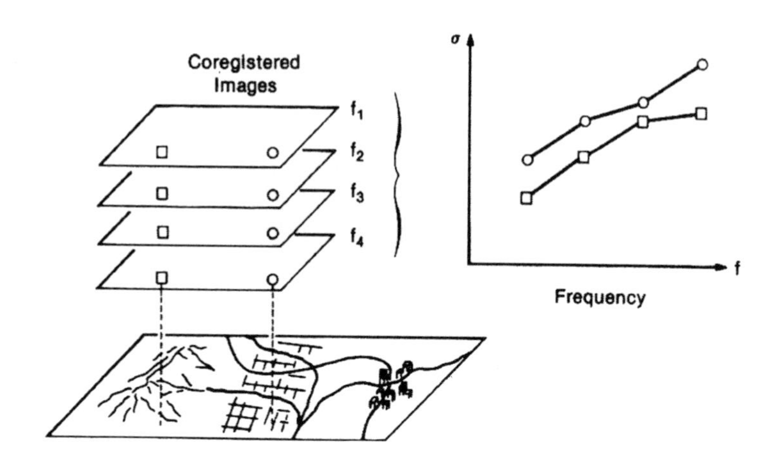

Fig. 12.4 Generation of spectral signature
©IEEE (Reproduced with permission of IEEE, C. Elachi, "Spaceborne Radar Remote Sensing", Fig. 5.28, Ch. 5, 1987)

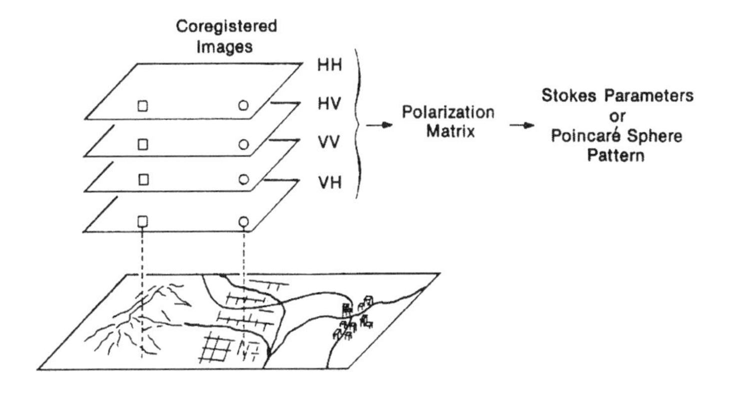

Fig. 12.5 Generation of polarization signature
©IEEE (Reproduced with permission of IEEE, C. Elachi,
"Spaceborne Radar Remote Sensing", Fig. 5.29, Ch. 5, 1987)

A third type of multi-channel radar is the multi-angle radar data. In this case, the images are taken (from different positions) using different illumination angles (see Fig. 12.6).

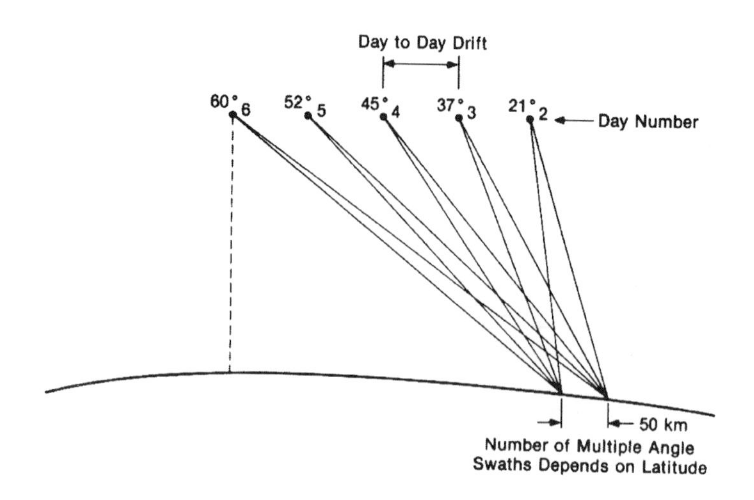

Fig. 12.6 Geometry for acquisition of multiple look angle imaging ©IEEE (Reproduced with permission of IEEE, C. Elachi, "Spaceborne Radar Remote Sensing", Fig. 5.30, Ch. 5, 1987)

Once all the polarization matrix elements have been determined on a pixel-by-pixel basis, images can be generated of various characteristic parameters, such as the Stokes parameters, ellipticity angle, etc. These characteristics can be used to identify surface units or to find polarizations which enhance certain target features relative to others (e.g., volume scattering component versus interface (surface) scattering components), or to enhance certain types of targets immersed in a set of distributed targets of different types. Examples of object classification by the scattering and Mueller matrix are given in Fig. 12.7.

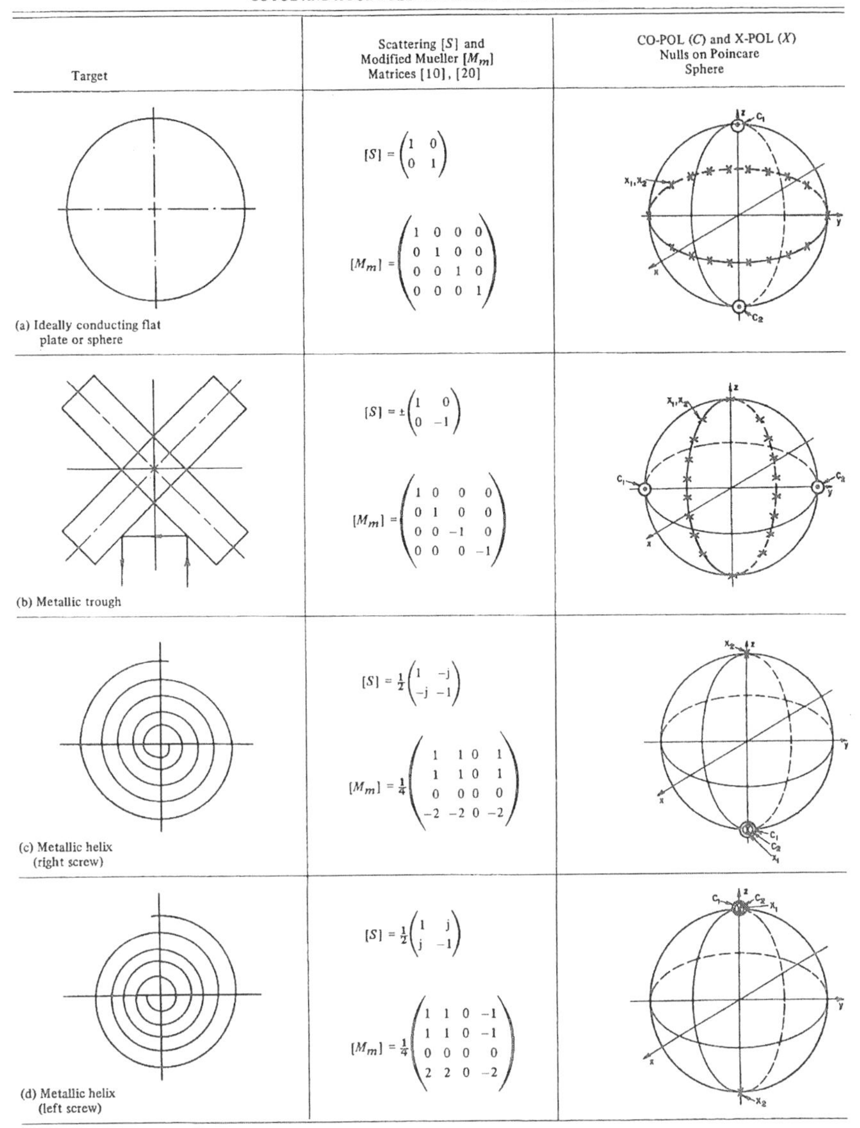

CO-POL AND X-POL NULL FOR SIMPLE TARGET SHAPES

Target	Scattering [S] and Modified Mueller $[M_m]$ Matrices [10], [20]	CO-POL (C) and X-POL (X) Nulls on Poincare Sphere
(a) Ideally conducting flat plate or sphere	$[S] = \begin{pmatrix} 1 & 0 \\ 0 & 1 \end{pmatrix}$ $[M_m] = \begin{pmatrix} 1 & 0 & 0 & 0 \\ 0 & 1 & 0 & 0 \\ 0 & 0 & 1 & 0 \\ 0 & 0 & 0 & 1 \end{pmatrix}$	
(b) Metallic trough	$[S] = \pm\begin{pmatrix} 1 & 0 \\ 0 & -1 \end{pmatrix}$ $[M_m] = \begin{pmatrix} 1 & 0 & 0 & 0 \\ 0 & 1 & 0 & 0 \\ 0 & 0 & -1 & 0 \\ 0 & 0 & 0 & -1 \end{pmatrix}$	
(c) Metallic helix (right screw)	$[S] = \frac{1}{2}\begin{pmatrix} 1 & -j \\ -j & -1 \end{pmatrix}$ $[M_m] = \frac{1}{4}\begin{pmatrix} 1 & 1 & 0 & 1 \\ 1 & 1 & 0 & 1 \\ 0 & 0 & 0 & 0 \\ -2 & -2 & 0 & -2 \end{pmatrix}$	
(d) Metallic helix (left screw)	$[S] = \frac{1}{2}\begin{pmatrix} 1 & j \\ j & -1 \end{pmatrix}$ $[M_m] = \frac{1}{4}\begin{pmatrix} 1 & 1 & 0 & -1 \\ 1 & 1 & 0 & -1 \\ 0 & 0 & 0 & 0 \\ 2 & 2 & 0 & -2 \end{pmatrix}$	

Fig. 12.7 Co-Pol and Cross-Pol Null for simple target shapes ©IEEE (Reproduced with permission of IEEE, W.M. Boerner, "Polarization Dependence in Electromagnetic Inverse Problems", IEEE Trans., Vol. AP-29, No. 2, Table 1, p. 266, March 1981)

12.3 Synthetic Aperture Radar (SAR)

The principle of operation of SAR is similar to the method of holograms. In this method, a field scattered by an interrogated object interferes with a reference coherent wave. The received interference pattern, based on the phase relationship of the scattered and the reference wave, is recorded as an intensity distribution of the sum of the reference and scattered waves. By illuminating this reference pattern (hologram) with the same coherent reference wave, it is possible to reconstruct the image of the object (scatterer). With the SAR technique, we can synthesize an interference pattern by transmitting a sequence of signal pulses from a moving "physical array" and by signal processing, i.e., by coherently summing the echoes from the range of interest, received at successive positions of the physical array. By means of this processing, we generate an array of effective aperture L (see Fig.12.8), given by

$$L = UT_s \quad (U : \text{speed of the moving "synthesizing" array}, T_s : \text{integration time}) \qquad (12.5)$$

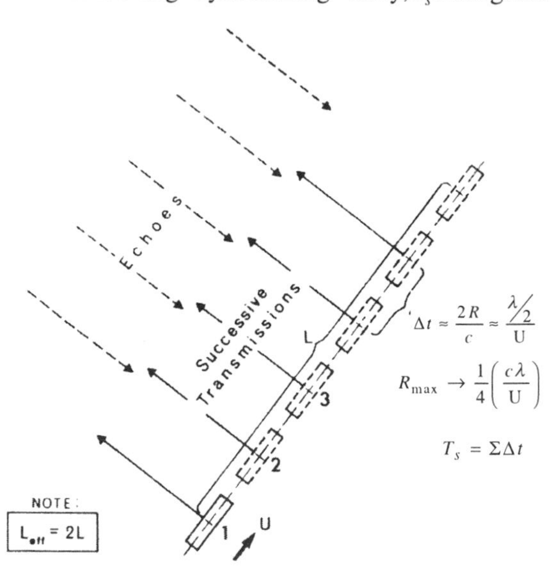

Fig. 12.8 Geometry of active synthetic aperture
(Reproduced with permission of NATO SACLANTCEN, E. Pusone, "Synthetic Aperture Performance Analysis of Beamforming and System Design", Report SR-91, Fig. 8, Ch. 2, November 1985)

U : Speed of radar platform (on which the synthesizing array is mounted)
R : Range of scatterer; λ : radio wavelength; c : speed of light

Δt : Time interval between two successive positions of the radar platform $= 2R/c$
T_s: Total integration time required to generate the array aperture
$R_{max} = \dfrac{c\lambda}{4U}$: Condition for avoiding grating lobes
The radiation pattern of the synthesized array is the "interference pattern", or hologram, generated with SAR. The signal processing used to generate the pattern is indicated in Fig. 12.9.

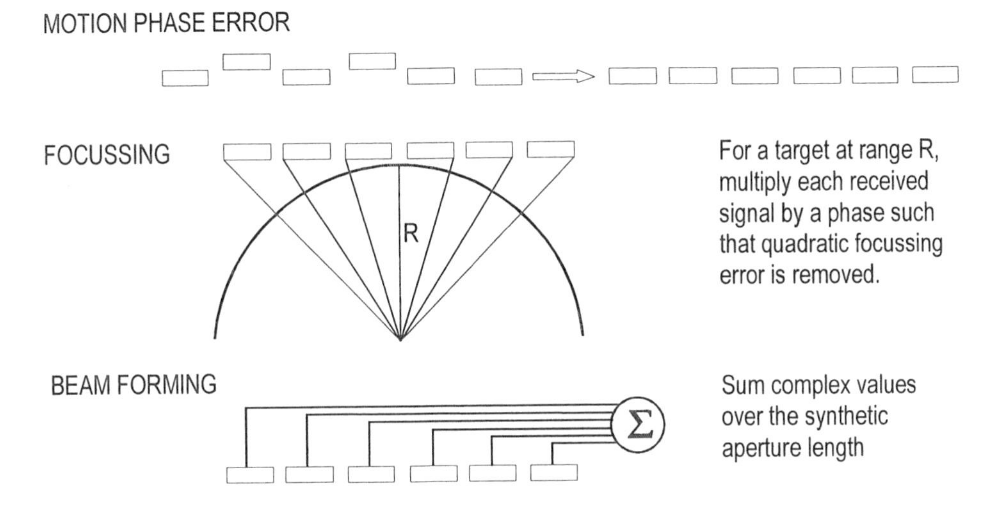

Fig. 12.9 Synthetic Aperture Processing

An example of a SAR pattern compared with the pattern of a physical array is given in Fig. 12.10.

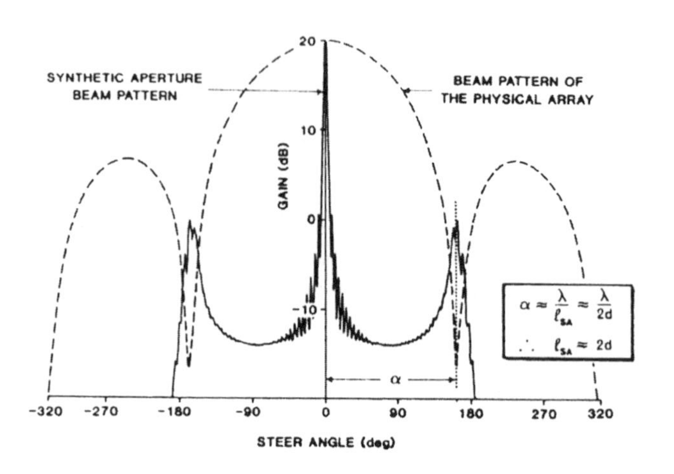

Fig. 12.10 Synthetic pattern and physical pattern
(Reproduced with permission of NATO SACLANTCEN, E. Pusone, "Synthetic Aperture Performance Analysis of Beamforming and System Design", Report SR-91, Fig. 9, Ch. 2, November 1985)

A very narrow beam can be generated by considering a long integration time T_s. The 3-dB beamwidth of the SAR pattern is given by

$$\theta_{SAR} = \lambda / UT_s \tag{12.6}$$

Fig. 12.10 shows the case in which the effect of diffraction lobes can be reduced by using a physical array with a beam pattern having its nulls at angles corresponding with the maxima of the lobes of the synthesized pattern. This can be achieved with a physical array of size D given by

$$D = 2d \tag{12.7}$$

d being the spacing between consecutive samples in the synthetic aperture process.

With the SAR technique we can produce maps of remote-sensed areas with a very fine (azimuth) resolution by using a physical array of limited size. If the signals are fully coherent, the maximum aperture L that can be synthesized is derived from the beamwidth of the physical antenna at the range of interest (see Fig. 12.11).

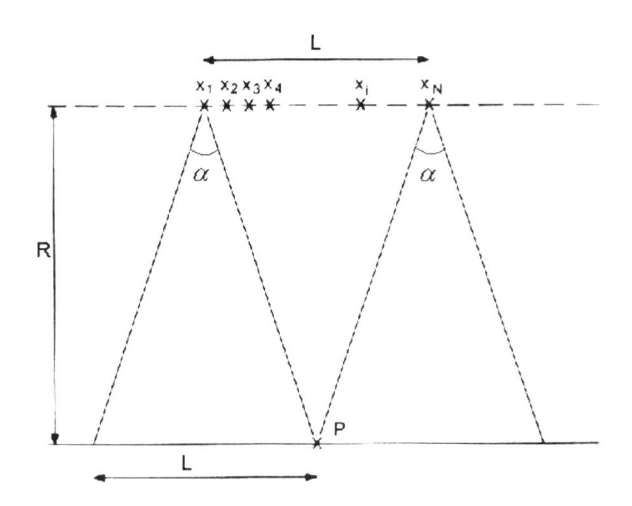

Fig. 12.11 Synthetic aperture length L, x_1, x_2,........x_N, successive positions of physical array, α the 3-dB beamwidth (λ / D) of the physical array ©IEEE (reproduced with permission of IEEE, C. Elachi, "Spaceborne Radar Remote Sensing", Fig. 3.20, Ch. 3, 1987)

The length L of the synthetic aperture is given by

$$L = \lambda R / D \tag{12.8}$$

where λ is the wavelength, D is the size of the physical array and R is the range to the scatterer. The synthesized beam is given by

$$\theta_{SAR} = \lambda / L \tag{12.9}$$

Combining Eqs (12.8) and (12.9), we have

$$\theta_{SAR} = D / R \tag{12.10}$$

The linear resolution in azimuth is given by

$$\delta_{SAR} = \theta_{SAR} R \tag{12.11}$$

Combining equations (12.10) and (12.11), we obtain

$$\delta_{SAR} = D \tag{12.12}$$

This result leads to the conclusion that by coherent signal processing of the received signals, we can achieve a maximum resolution that improves with the reduction of the physical antenna dimensions. There are, however, practical limitations in reducing D; we cite two of them below:

- the size of the physical antenna must be sufficient to provide a good signal-to-noise ratio at the input of the SAR processor;
- if we want to avoid ambiguities by reducing the grating lobes, we must have a size given by Eq. (12.7). This size is proportional to the interval between samples and, therefore, to the range of the remote-sensed area (object) under investigation (see Fig. 12.8). This means long arrays for long distances.

The synthetic aperture processing relies on the coherence of the successive pulses. The processor must compensate for changes in the signal phase in order to generate an effective aperture length. If the phase changes are known (see Fig. 12.9), the phase compensation can be done. If the phase varies randomly due to random variations of propagation path length or multi-path, the signals cannot be added in phase and the synthesized beam pattern is degraded. To overcome the effects of propagation on SAR beamforming, a good knowledge or modelling of phase fluctuations is required. This can be seen by the analysis of synthetic aperture beam-forming. The beam output voltage is given by

$$V = \sum_{k=1}^{N} w_k e^{-j\vartheta_k} \tag{12.13}$$

where w_k is the amplitude response of the physical array at the k-th position from the scatterer, ϑ_k is the phase shift of the echo at the k-th position of the physical array from the scatterer and N is the total number of synthetic array elements. The phase ϑ_k is given by

$$\vartheta_k = (\omega_0 + \omega) t_k \tag{12.14}$$

where t_k is the time delay at the k-th position of the synthetic array, ω_0 is the transmit (angular) frequency and ω (Doppler shift) is given by

$$\omega = \frac{2U}{c}\omega_0 \sin\alpha_k \tag{12.15}$$

In this expression, U is the speed of the array platform, c is the speed of light *in vacuo* and α_k is the angle of the direction of the scatterer with the normal to the direction of movement of the array (at the k-th position).

Combining Eqs (12.13), (12.14), (12.15) and filtering the Doppler components for the SAR processing we find

$$V = \sum_{k=1}^{N} w_k e^{-j\frac{2U}{c}\omega_0 t_k \sin\alpha_k} \tag{12.16}$$

For a constant interval ΔT, the sampling during the movement of the array is defined by $t_k = k\Delta T$, k=1,2,....N and the spacing between two consecutive positions is given by $d = U\Delta T$. The synthesized beam pattern is given then by

$$VV^* = \left[\frac{\sin\left(\frac{2N\cdot\pi d}{\lambda}\sin\alpha\right)}{\sin\left(2\frac{\pi d}{\lambda}\sin\alpha\right)}\right]^2 \tag{12.17}$$

Equation (12.17) represents the beam pattern synthesized by coherent summation of N echoes received at N successive positions of the array. It should be noted from Eq. (12.17) that the "active" synthesized beam pattern has a 3-dB beamwidth that is half of that for a conventional physical array of the same length.

If we have a random phase error $\Delta\varphi$ due to propagation, the beam voltage can be expressed (cf. Eq. 12.16) by

$$V = \sum_{k=1}^{N} e^{-j\left[\frac{2\omega_0 Uk\Delta T\sin\alpha_k}{c} + \Delta\varphi_k\right]} \tag{12.18}$$

The expected value of the beam voltage is given as follows:

$$E\{V\} = \sum_{k=1}^{N} e^{-j\frac{2\omega_0 Uk\Delta T \sin\alpha_k}{c}} \cdot E\{e^{-j\Delta\varphi_k}\} \tag{12.19}$$

Here, $\Delta\varphi_k = \omega\Delta\tau = \omega\tau_k$, τ_k being the differential delay around k.

The first term gives the beam pattern in the ideal case of no errors (full coherent summation). The second term modulates the first one. We expect that the effect of this random modulation is to smooth-out the interference pattern generated by the coherent summation of the echoes. This smoothing-out can result in broadening of the beam, or increasing the side lobes level (fluctuating around a constant value) [*Pusone*, 1985]. As a result of phase fluctuations, the resolution which is achievable in the (ideal) coherent case is degraded. This degradation depends on the time constant "τ" of the phase fluctuations compared to the total synthesizing time "T_s". For $\tau >> T_s$, we approach the ideal case of full coherent summation given by Eq. (12.17). (Maximum synthetic aperture length $L = 2Nd$ and maximum angular resolution $\theta_{3\text{-}dB} = \lambda / 2Nd$). For $\tau << T_s$, the random phase fluctuations will not result into the interference pattern and no beam will then be formed.

In Eq. (12.19), we write [*Papoulis,* 1962]

$$E\left\{e^{-j\Delta\varphi_k}\right\} = E\left\{e^{-j\omega\tau_k}\right\} = \int_{-\infty}^{+\infty} e^{-j\omega\tau_k}\, p(\tau_k)\, d\tau_k \tag{12.20}$$

As a specific example, we consider an exponential model for the probability density of delay τ_k, viz,

$$p(\tau_k) = e^{-\beta\tau_k},\ \tau_k \geq 0 \tag{12.21}$$

with

$$\sigma_{\tau_k} = \frac{1}{\beta} \tag{12.22}$$

the so called multi-path dispersion. Eq. (12.20) yields in this case

$$E\{e^{-j\Delta\varphi_k}\} = \int_0^{\infty} e^{-j\omega\tau_k} \cdot e^{-\beta\tau_k} d\tau_k \tag{12.23}$$

Carrying out the integration we obtain

$$E\{e^{-j\Delta\varphi_k}\} = \frac{1}{\beta + j\omega} \tag{12.24}$$

Introducing this result into Eq. (12.19), the expected value of the beam voltage becomes

$$E\{V\} = \sum_{k=1}^{N} e^{-j\theta_k} \frac{1}{\sqrt{\beta^2 + \omega^2}} e^{-j\tan^{-1}\omega\sigma_{\tau_k}} \tag{12.25}$$

where

$$\theta_k = \frac{2U\omega_0 k\Delta T \sin\alpha_k}{c} \tag{12.26}$$

is the phase term in the ideal case of no phase fluctuations and σ_{τ_k} is the multi-path dispersion as defined by the exponential model. Based on this model, we can compensate for the phase fluctuations in Eq. (12.25) from knowledge of the multi-path delay dispersion σ_{τ_k}.

The SAR radar can be realized with the Doppler-polarimetry radar described in detail in Sec. 2.4.4, where the motion "radar-scatterer" is given in terms of the movement of the scatterer relative to the stationary radar. A synthetic aperture is formed by coherently integrating the pulses corresponding to each different polarization combination. As the measurements (V or H polarization) are not simultaneous, the echoes are partially correlated. Decorrelation results from the coherence properties (in space and time) of the moving scatterer. When the radar platform moves with respect to a stationary remote-sensed area, the SAR processor must compensate for the deviation from the straight line trajectory. For air-borne radar this can be done with INS (Inertial Navigation System). For moving scatterers, the SAR processor must compensate for the phase changes due to the motion of the scatterer and the moving radar platform.

The polarimetric SAR is widely used for monitoring and imaging the earth's surface. Looking at the phase variations of the returned SAR signals it is possible to reconstruct very accurately images of terrain, vegetation, etc. *Ulaby* [1987] has derived a relationship between the polarized phase difference (PPD) $\varphi_{HH} - \varphi_{VV}$ and the reflection coefficients of the scattering area. PPD distributions have been measured for different types of surfaces. The results indicate that these distributions may lead to discrimination between various surfaces, i.e., soil and vegetation covered surfaces.

12.4 Radar altimeter

Radar altimeters provide a measure of the distance between a sensor and a scattering point (e.g. on a surface). A short pulse is transmitted towards the surface at the time t_1. The round trip distance from the reflecting object to the radar is given by [*Elachi*, 1987]

$$z_d = \frac{ct_d}{2} \tag{12.27}$$

where c is the speed of light and t_d is the round-trip delay. The accuracy with which the distance z_d is measured is given by

$$\Delta z_d = \frac{c(\Delta t_d)}{2} \tag{12.28}$$

The time difference accuracy Δt_d depends on the sharpness of the pulse which is equal to $1/B$, B being the signal bandwidth. We obtain, therefore, from Eq. (12.28)

$$\Delta z_d = \frac{c}{2B} \tag{12.29}$$

In oceanography, radar altimeters are used to measure wave heights. If the accuracy is of the order of $1\ cm$, they can discriminate between capillary waves and gravitational waves. This would require a signal bandwidth of $15\ GHz$.

From echo-shape analysis we can obtain information about surface roughness. When a pulse scatters from a (single) surface, the returned echo has a shape reflecting the (statistical) properties of the (rough) surface. In the case of the ocean, where the surface is homogeneous, the height statistics are the main factors in determining the pulse shape. In the case of terrain, the surface composition varies across the antenna footprint and its statistical properties need to be taken into account. For a perfectly smooth surface, the echo is a mirror image of the incident pulse (see Fig.12.12b). If the surface has some roughness, some return occurs in the backscatter direction at slight off-vertical angles as the pulse footprint spreads on the surface (see Fig. 12.12a). This results in a slight spread in time of the echo (see Fig. 12.12c). If the surface is very rough, some of the energy is scattered when the radio pulse intercepts the peaks of the waves (sea) and more energy is scattered as the pulse intercepts areas at various heights of the surface. This leads to a larger multi-path spread of energy which results in noticeable rise in the echo leading edge (see Fig. 12.12d). The rise time t_r depends on the *rms* height of the waves. Therefore, the time t_r can be used to measure (sea) surface roughness.

Fig. 12.12 shows examples of normalized echo shapes calculated under the assumption of a rectangular transmitted pulse and Gaussian-distributed surface heights, for different values of $2\sigma_h / c\tau$, i.e., the ratio between the surface *rms* height and the pulse length.

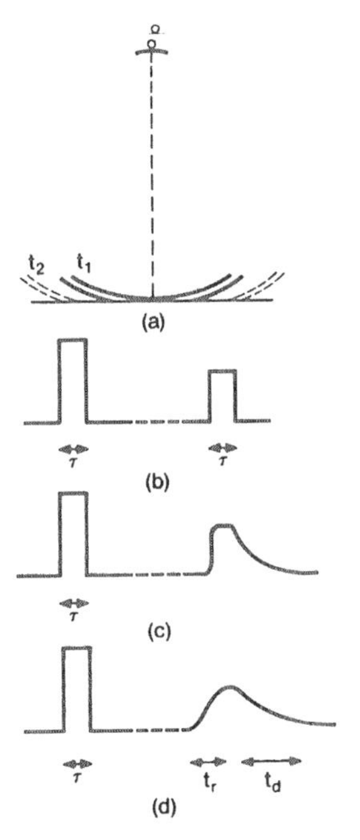

Fig. 12.12 (a) Illustration of spread of incident pulse on surface, (b), (c), (d) Echo shape for increasing rougher surface ©IEEE (Reproduced with permission of IEEE, C. Elachi, "Spaceborne Radar Remote Sensing: Applications and Techniques", Ch. 6, Fig. 6-4, 1987)

It can be seen in Fig. 12.13 that the rise time increases for larger ratios and the echo shape has a more gradual rise. The rate of decay of the echo is a function of the incident angle and the slope of the waves.

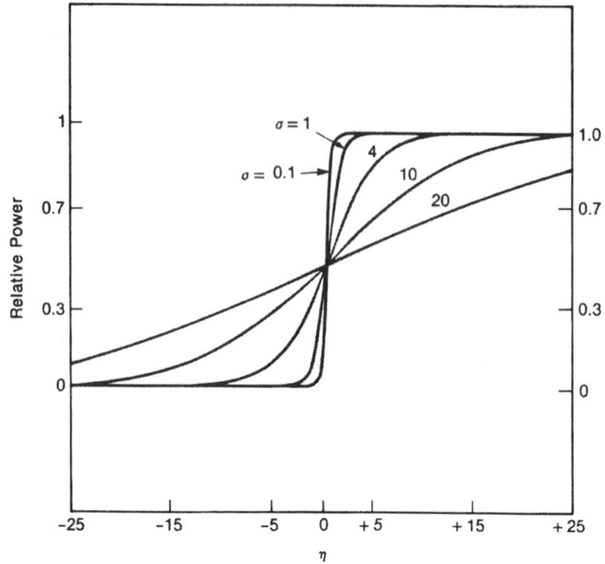

Fig. 12.13 Diagram of echo shapes for different values of $\sigma = 2\sigma_h / c\tau$ ($\eta = t/\tau$; τ = pulse length)

©IEEE (Reproduced with permission of IEEE, C. Elachi, "Spaceborne Radar Remote Sensing: Applications and Techniques", Ch. 7, Fig. 7-12, 1987)

As the surface wind increases, the ocean surface becomes more rough. We have fewer areas where there are specular returns. This leads to a decrease in the amplitude of the echo. Thus, the altimeter echo amplitude can be used as a measure for surface wind.

12.5 Tropospheric-scatter radar

Radar (usually in the UHF- or C band-frequencies) can be used to monitor the physical properties of the troposphere that may change with meteorological conditions. The main properties of the troposphere are the refractive index (related to permittivity and scale of turbulence due to fluctuations of permittivity and wind force). These properties are important for various remote sensing applications, e.g., monitoring of the air contents, weather broadcast, etc. The inverse problem consists of predicting (or measuring) these properties of the troposphere from radio-field measurements, e.g., scattered power (scattering cross section).

The scattered power is a function of permittivity and turbulence fluctuations [*Booker and Gordon,* 1950], [*Gjessing,* 1969, 1973] which depend on the meteorological changes in the atmosphere. Various models of the air properties (refractivity and turbulence) have been derived from theory [*Tatarski,* 1971], [*Chernov*, 1960], [*Obukhov,* 1953], [*Gjessing*, 1969], [*Ishimaru*, 1977]. In some cases, however, there have been discrepancies between these models and available meteorological measurements. For example, an effort was made [*Pusone,* 1980, 1981] to model the scale L of atmospheric turbulence in terms of "eddies" of sizes dependent on the thermodynamic conditions in the air. Statistics of eddy sizes were predicted by the model as a function of meteorological data measured in the upper atmosphere (from 1 to 10 km). Values were obtained in the range between 30cm to 1m. That theoretical model allowed the solution of the inverse problem using available radiosonde data and Doppler radar data. The model has been verified experimentally and has compared well with radio-field measurements. The scattered power was computed using a Hertzian potential produced by oscillating dipoles of moments proportional to permittivity fluctuations and "blobs" of turbulent air. The results showed that the scattered power depends on the size L of the scatterers (blobs) compared to wavelength. The maximum power was found around the value $L/\lambda \sim 1$.

12.6 Atmospheric monitoring with polarimetry

A polarimetric radar can be used for atmospheric monitoring. For example, by measuring the effects of rain on the microwave signal polarization, we can obtain information on the shape and the canting (angle) distribution of raindrops. By meteorological measurements, i.e., variations in atmospheric pressure, temperature and humidity, we can predict the atmospheric turbulence and its effects on polarization.

12.6.1 Precipitation

The sizes and shapes of raindrops can be estimated from measurements of the scattered field in amplitude and phase and by comparing the results with the scattered field computed for known geometries of raindrops. Various raindrop models are compared with measured values [*Oguchi*, 1977]. The shape of a raindrop can be represented by

$$r = a_0 \left(1 + \sum_{n=0}^{\infty} c_n \cos n\vartheta \right) \tag{12.30}$$

Where a_0 is the effective raindrop radius, defined as the radius of a sphere whose volume is the same as that of the deformed raindrop and c_n are deformation coefficients of the raindrop.

The characteristic parameters of a raindrop are the radius r and the deformation (coefficient ratio a/b), as illustrated in Fig. 12.14.

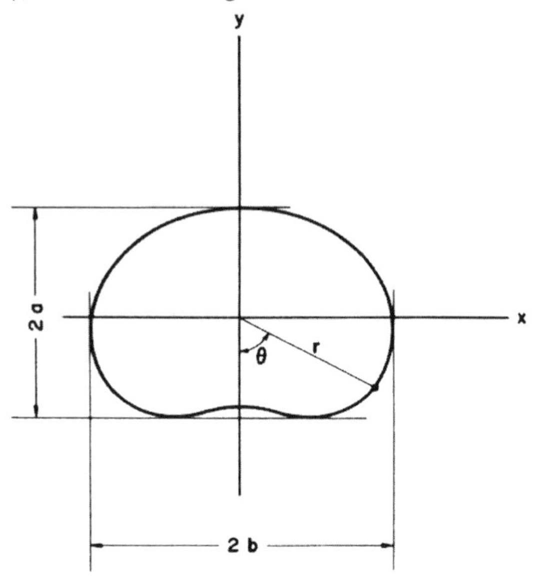

Fig. 12.14 Cross-sectional view of a deformed raindrop and the coordinate system ©AGU (American Geophysical Union), Tomohiro Oguchi, "Scattering Properties of Pruppacher-and-Pitter form raindrops and cross polarization due to rain: Calculations at 11, 13, 19.3 and 34.8 GHz", Radio Science, Vol. 12, No. 1, p. 41-51, Jan.-Feb. 1977.

From the measurement of size and fall velocity of raindrops, the degree of turbulent flow (i.e., the Reynolds number) due to rain can be estimated for given atmospheric conditions of temperature and pressure [*Unal et al.*, 1998]. The measured scattered field can be used to estimate the effects of cross-polarization due to rain. The geometry for the calculation of cross-polarization is illustrated in Fig. 12.15 as a function of two canting angles, γ in the vertical plane and θ in the horizontal plane.

Results show [*Oguchi*, 1977] that the effect of the canting angle γ on the cross-polarization is less significant than that of the canting angle θ in the $10\,GHz - 30\,GHz$ frequency band. The results also show that for a Gaussian distribution of the raindrop canting angle θ, the cross-polarization components decrease with an increase of the standard deviation of θ. This indicates that the effect of cross-polarization is more significant for rain composed of "equi-oriented" raindrops falling in a single direction.

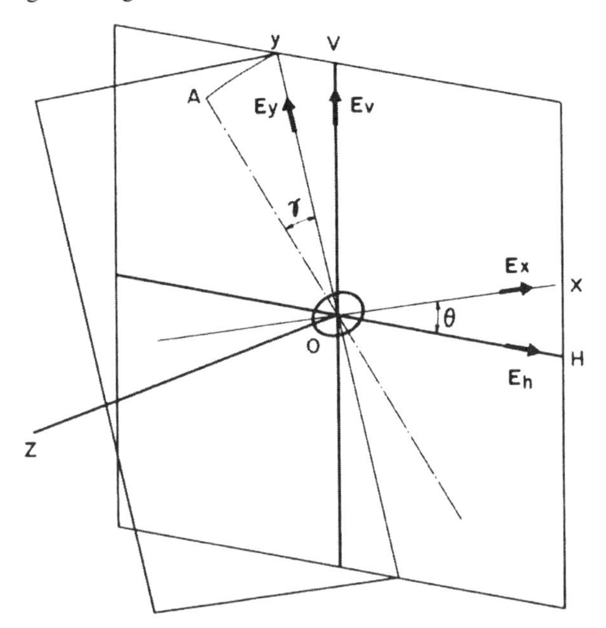

Fig. 12.15 Geometry for the calculation of cross-polarization factors ©AGU (American Geophysical Union), Tomohiro Oguchi, "Scattering Properties of Pruppacher-and-Pitter form raindrops and cross polarization due to rain: Calculations at 11, 13, 19.3 and 34.8 GHz", Radio Science, Vol. 12, No. 1, p. 41-51, Jan.-Feb. 1977.

12.6.2 Turbulence

From fundamental theory on radio scattering in the troposphere [*Booker and Gordon*, 1950; *Ishimaru*, 1977] it is found that for scattering in a direction making an angle θ_s with the direction of incidence and an angle χ with the direction of the incident electric field (Fig. 12.16), the scattered power (per unit volume) is proportional to

$$\sigma(\theta_s, \chi) \propto K_0 \frac{\sin^2 \chi}{\left[1 + \left(K_1 \sin \frac{\theta_s}{2}\right)^2\right]^2} \tag{12.31}$$

where K_0 and K_1 are functions of permittivity fluctuations and the sizes of the turbulent eddies (with respect to wavelength) in the scattering area [*Booker and Gordon*, 1950; *Pusone*, 1978].

The scattering from turbulent air (eddies) can cause a "rotation" of the polarization of an electromagnetic wave. This may result in a change of the angle χ between the electric field and the scattering direction. This effect can be significant in the case of anisotropic turbulence (anisotropic eddies).

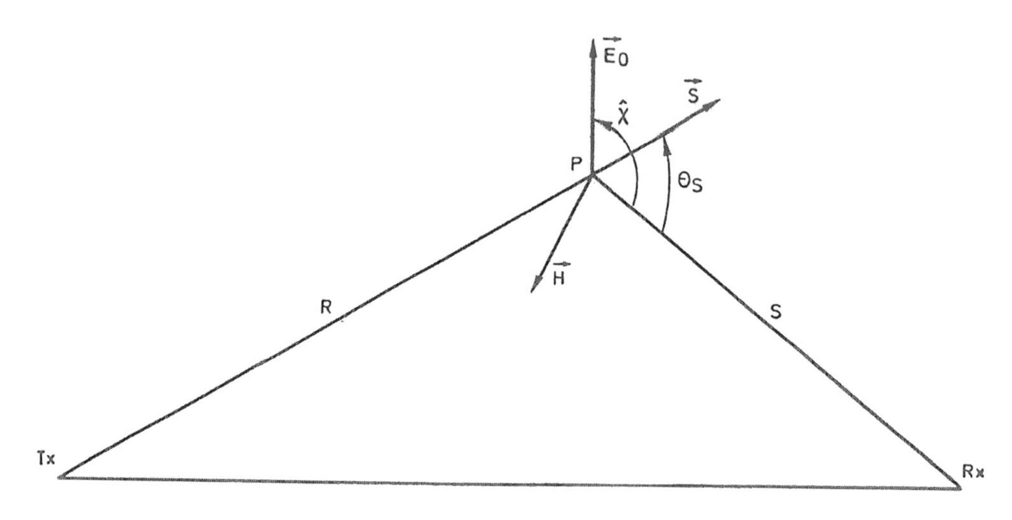

Fig. 12.16 Polarization angle χ of a scattered electromagnetic wave (From E. Pusone, "A Predictive Model Based on Physical Considerations for Troposcatter Communications Links", SHAPE TECHNICAL CENTRE, Report TM-589, Ch. 3, Fig. 4, 1978)

From Eq. (12.31), we obtain a maximum scattered power for $\chi = \pi/2$ and a minimum (zero) for $\chi = 0, \pi$. We can, therefore, measure the "depolarization" effect of air turbulence.

PART IV

CONCLUDING REMARKS

CHAPTER 13

Review of Potential Applications of Radar Polarimetry

13.1 Introduction

The problem of polarization diagnostics and ecological monitoring of the environment is complex. It has been shown that any remote sensing system contributes to the solution of inverse problems leading to the determination of the physical, geophysical, and mechanical characteristics of surface and atmosphere.

It has been emphasized that an important aspect in radar remote sensing is the solution of the inverse scattering problem. Polarization properties of scattering of electromagnetic waves at a surface and the atmosphere largely increase the information to be extracted, especially if measurements are executed at various wavelengths. Possible explanations for loss in information have been considered. Several methods required to reduce these losses are the following: special algorithms of signal processing, information analysis and the complete polarization analysis of scattering. Methods of modelling the process of scattering from surfaces in inverse problems have been discussed.

Two main kinds of modelling have been considered: mathematical and physical. Mathematical modelling deals with the analysis (and image recognition techniques) of the statistical properties of scattering from various surfaces. Physical modelling deals with the analysis of the interaction process of electromagnetic waves with rough surfaces and random scattering volumes [*Ligthart*, 1998].

A description of known methods concerning the analysis of the interaction of electromagnetic waves with surfaces has been summarized. Small (updated) perturbation methods, the Kirchhoff method, and the model of volume scattering (radiative transfer theory).

Procedures of image processing have been discussed for implementing solutions to inverse problems and some applications of polarimetric radar in remote sensing and detection problems have been given.

In this monograph, we have described extensive applications of specific radar polarimetry inverse problems for remote sensing of the environment, e.g., vegetation models, reflection of electromagnetic waves from inhomogeneous layered structures, scattering from rough boundaries, internal ruptures, etc. [*Ligthart*, 1998].

13.2 Results of polarimetric remote sensing

The derived results enable the simulation of the process of interaction of electromagnetic waves with a surface covered by vegetation. It is possible to take into account the first order approximation to analyze single scattering, and the second order approximation to analyze double scattering.

The results are intended for solving the direct problem, which is the first step for setting up the inverse problem. The results are determined by using a statistical approach.

The available models of scattered electromagnetic fields from vegetation permit us to identify agricultural fields, in which the earth's surface is covered by vegetation such as grass.

The results are summarized as follows:

- Physical processes have been specified for analyzing a model of a layer of grass above the earth's surface. On one hand, these processes lead to distinctive features of the signal: the appearance of additional phase differences of the echo signal of the components with horizontal and vertical polarizations. On the other hand, the processes are connected to parameters of the vegetation model: permittivity and volume fraction of the biomass. These effects are derived from the effects of multiple coherent scattering.

- A mathematical method for the analysis of the effects of coherent interaction between radiation and the medium characterized by extensive perturbations with spatial orientations has been developed. This method has been proposed previously [*Landau,* 1982]. We claim that this method is adequate for this remote sensing problem.

- Specific calculations of the proposed method at the level of the first order approximation have been carried out. Relatively simple relationships for the permittivity tensor in a reference frame with respect to a transmitter-receiver system have been derived:

$$\left.\begin{aligned} \varepsilon_h &= 1 + 4\pi^2 n(\varepsilon - 1) \\ \varepsilon_v &= 1 + 4\pi^2 n(\varepsilon - 1)\left[\cos^2\theta + \frac{\sin^2\theta}{\varepsilon}\right] \end{aligned}\right\} \tag{13.1}$$

Spatial dispersion effects have been analyzed on the basis of scattering in the second approximation.

Various electrodynamic models of vegetation have been analyzed . With the help of these models, it is possible to solve problems of vegetation classification, and in particular to define agricultural and biometrical characteristics of vegetation (e.g., biomass, foliage density, vegetation humidity, etc.) It should be taken into consideration that the efficiency of classification and identification of vegetation grows when various kinds of polarization (vertical-vertical, horizontal-horizontal and horizontal-vertical) for different frequencies (e.g., $\lambda = 2$ cm, 6 cm, 19 cm) are used. The definition of biometrical characteristics in direct connection with scattering is one form of coming to grips with the solutions of inverse problems in remote sensing.

The main results are summarized in the following:

1. A review of electrodynamic models of vegetation based on publications from 1983 to 1994 has been presented.

2. A simple vegetation model has been introduced as a layer of particles of a given form (cylinders, disks). The field reflected from individual particles has been determined. After this, the reflected field has been averaged over the sizes of the particles and their orientations. This averaged field has been used for calculations of the reflected field of a vegetation layer, as well as for the calculation of the backscatter coefficients of vegetation.

3. A three-dimensional vegetation model has been presented in the form of a volume occupied by scatterers with imprecisely defined geometrics. Backscattering diagrams based on the first approximation of the multiple scattering theory have been given and the field intensity has been calculated using transport theory.

4. It has been pointed out that it may be possible to connect radar remote sensing data with biometrical characteristics.

5. By means of radar remote sensing of vegetation, it is possible to classify various types of vegetation (type of crop, level of crop, etc.)

In this monograph, a literature survey has been presented which describes the connection between electrodynamic and physical characteristics of different types of earth surface materials of interest in radar remote sensing applications. These results include both theoretical and empirical models derived by different authors and classified according to the surface material types. The results are usually valid for super-high frequency radio waves. Theoretical conditions and models have been summarized. Since most of the models of different surfaces are based on available empirical data, we expect a sufficiently high degree of adequacy. Some of the theoretical surface models were proposed relatively a long time ago and were experimentally proved to be correct. For example, the four models of sea ice are in good agreement with experiments [*Vant,* 1978; *Hallikainen* 1977; *Hoekstra*, 1971].

A similar result was derived for snow in the form of a two-component mixture consisting of ice particles (inclusions) and air (the main medium) [*Sihlova*, 1988]. When snow is represented by a three-component mixture (water, air, ice), the design relations do not agree with available experimental data [*Sihlova*, 1988], meaning that this model is inadequate. Practically all models of earth covers reported in the form of diagrams, formulas, nomograms, etc., have sufficient validity to describe real covers also when experimental errors are taken into consideration. Therefore, these models may be used for conducting the required calculations in remote sensing systems, though, as usual, any model that describes a real process may demand further corrections in future.

The interrelation between empirical characteristics of remotely sensed surfaces and physical characteristics of these surfaces has been considered. The material contained in this monograph represents a survey of the available literature in this area.

Let us consider the applicability of the results in this monograph from the point of view of their practical usefulness for solving remote sensing problems.

Earth surface electrodynamic characteristics are determined (above all) by the surface permittivity, though these characteristics depend on other factors, especially geometric factors. As the final objective of remote sensing is the determination of the surface characteristics (physical, chemical, mechanical, etc.) it is important to find the relations between earth surface characteristics and permittivity, and then determine the relations between the surface permittivity and polarization characteristics of the radio waves reflected from the surface.

Relations between permittivity and soil humidity in the microwave frequency bands have been presented. For example, the linear dependence of peat permittivity upon humidity allows us to determine peat resources using radar remote sensing.

A similar linear dependence is observed for vegetation covers, which is very important for the determination of the degree of ripening of agricultural crops. A linear dependence of trees on humidity allows us to estimate the degree of forest dryness. This information may be uses to prevent forest fires, which cause vast ecological damages.

The connection between snow permittivity, snow density and humidity allows us to determine (according to special nomograms) snow and water reserves. The estimation of these reserves is important during the spring period from the point of view of snow melting, possible stream rises and floods of rivers. In addition to having data on the snow cover thickness, it is possible to make estimations on the state of winter crops and predict the future yield.

The empirical results on the relation between the surface soil permittivity and its salinity allow us to determine non-fertile (from the point of view of ecology) areas where the salt concentration compels to stop using terrain for agricultural farming, or to conduct necessary re-soiling works.

We have listed the main fields of practical usage of the derived results. Those are quite comprehensive but, of course, do not cover all other opportunities that will be discussed in the process of further investigations.

In this monograph the reflection of electromagnetic waves from non-uniform layered structures has been considered for two cases: deterministic and stochastic. The deterministic case concerns layered structures in which the permittivity varies as an exponential or polynomial. The analysis includes cases of vertical and horizontal polarization with various incidence angles. The reflection coefficient can be determined for different values of complex permittivity, layer thickness and polarization. The analysis is important for solving direct problems, since the (two- and three-layer) models are typical of many practical situations (vegetation, continent and marine ice, forest, etc.).

The more generalized case uses a stochastic approach. Here, the permittivity of one layer varies as a random function. The solution technique uses the correlation function of the permittivity fluctuations and enables computation of the average reflection coefficient and reflected power. Within the framework of various accepted approximations and assumptions, which are justified in practice, this solution technique allows us to investigate the reflection phenomena as functions of layer thickness and correlation radius.

The results can be used for surfaces that can be represented by two- and three-layer structures. One such structure is vegetation and, in particular, agricultural land. For such models, the major interest is the determination of the reflection coefficient. The models discussed in this part of the monograph enables us to solve direct problems of remote sensing. The most original results are those related to the statistical approach. In the most general case, this method also covers the deterministic approach.

In this monograph, the reflection of radiowaves from surfaces containing internal breaks with various geometrical configurations (rectangular and wedge-type pits) have been studied. Reflections from such surfaces are determined for vertical and horizontal polarizations and for different "view angles". The results are derived by using geometrical-optics methods.

For symmetric and asymmetric breaks, reflections from a spherical pit and rectangular pit with final depth have been computed and techniques for solving more complex geometries (four-media structures) have been demonstrated.

These results can be applied to radar remote sensing of agricultural fields in conjunction with irrigational systems. With the help of polarimetry, potentials for classification and identification of ruptures seem feasible. New areas of application of these models can be cartography of channels, dams and other hydraulic engineering structures.

Reflection of electromagnetic waves from a layer with a rough interface is also considered in this monograph. The chosen approach enables us to approximate the required reflection and transmission coefficients. First- and second-order approximations have been carried out. Expressions have been derived as function of surface roughness correlation.

Graphical results are shown for a vegetation layer simulated by an ensemble of cylinders. The requirements for selecting the polarization basis are formulated in the inverse problem of determining the biomass and vegetation density. The results obtained enable us to solve some inverse problems of remote sensing of the vegetation. In particular, it has been shown that it is possible to determine the density and biomass of vegetation. The information on these parameters is contained in the phase shift between various components of the scattering matrix. The problem is significant in agriculture in order to quantify parameters of vegetation.

13.3 Comparison-review of the inverse scattering models analyzed

At the end of each chapter of Part III of the monograph, the conclusions concerning the respective chapter are presented and possible fields of application of the derived results are shown. Here, we summarize the general results of the scattering models analyzed in Part III as these models are connected with different aspects of interaction of electromagnetic waves and the surface with certain particularities. When studying this interaction, we always take account of polarization effects occurring during reflection and scattering of electromagnetic waves from the surfaces with certain particularities, for example, internal fractures, etc.

At first, we considered the problem when scattering of a wave takes place at a surface covered with grass without taking account of surface roughness. This is quite a well-known problem for the solution of which various models of description of the surface with vegetation may be used. We chose the model of a grass cover in the form of cylinders strictly perpendicular to the wave incidence plane. In that case, we assumed that on the analyzed area of the surface there was a finite number of uniformly distributed cylinders. That simplified models was chosen for the purpose of derivation of concrete results in the form of closed analytic expressions. When analyzing the data, firstly we took account of single scattering of the wave on the surface, and then double scattering.

In some cases the system of differential equations describing the behavior of the scattered fields turned out to be non-closed. Consequently, we had to introduce additional transformations for the derivation of a closed system of differential equations. In that case, the derived results allowed us to continue the investigations for studying multiple scattering of radio waves on surfaces of the aforementioned type.

It was a very significant result that we managed to derive analytically the dependence of the characteristics of the scattered field on the surface permittivity. The derived results included information about the stochastic scattered electromagnetic field when the values of the surface permittivity or the values characterizing the vegetation cover state fluctuated randomly.

Our results generalize previous well-known particular cases and, naturally, those which are known for an isotropic medium. Their application allows us in some approximation (for the assumed models) to solve the inverse problem, i.e., on the basis of the available characteristics of the received scattered electromagnetic field to determine the characteristics of the sensed surface, which (in the case of a vegetation cover) include the biomass of vegetation, its humidity, degree of ripening, etc.

Further, we have more thoroughly considered particular models of vegetation covers, e.g., surfaces with agricultural crops for determining their biometric characteristics. This modelling has as purpose to understand more clearly the opportunities of application of radio polarimetry analysis for solving a specified class of inverse problems. The results presented are first of all based on experimental data and, therefore, the main relations are empirical in character. So, they require an application of correction factors and normalizing coefficients. For this reason many relations are linear in character; for example, the yield of winter wheat as a function of biomass, etc.

In terms of the process of radio wave scattering, the vegetation is considered as a multi-component complex structure, which contains an aqueous medium, air component, and vegetation. As a result, we have to consider the permittivity of vegetation as the permittivity of a mixture.

In addition to agricultural crops, we have considered the interaction of electromagnetic waves with forestlands. Here, the methods of radio polarimetry are especially efficient, as reflections of horizontally and vertically polarized waves differ substantially from each other.

When analyzing the process of scattering, a model of vegetation in the form of cylinders is often used. It is possible also to use other models, such as models of a vegetation cover in the form of volume scatterers, when the reflection of radio waves from the underlying surface is not taken into consideration.

Application of various models allows us to solve a number of inverse problems. In particular, on the basis of the variation of the reflection coefficient, we may determine the humidity, the volume of biomass (both are directly proportional) and the height of vegetation (which is inversely proportional).

Further, we have considered a more general case of interaction between the electrodynamic characteristics of layered media and their electrical and physical properties. These properties are determined by the complex permittivity of the surface. Therefore, the most interesting interaction is between the electrodynamic characteristics of layered media and the permittivity of these media. It is shown that derivation of respective analytic dependencies is possible only for some particular cases, and that in a general case only empirical relations may be used. Various surfaces are analyzed. The main conclusion is that for a particular surface it is necessary to develop appropriate models using experimental data and to choose analytic dependencies in the framework of the proposed models. It is unrealistic, for example, to try to find a unified model for snow, peat, sea ice, etc. Even within one class of surface (e.g., sea ice), it is impossible to use one universal model; sea ice, for example, differs according to aging categories, the state during different year seasons, etc.

The structure of snow cover is not less complicated. As said before, a three-dimensional model showed substantial disagreement with experimental data. This may be connected with drawbacks of the model, because in our opinion snow cover is evidently a three-dimensional mixture from a basic physics point of view.

When analyzing wet soil and soil covered with vegetation, we found out that a two-dimensional model gave analytic dependencies that were in good agreement with experimental data. The reflection from layered structures with different laws of permittivity variation with depth of the layer has been considered. Two situations have been analyzed: permittivity variation is according to a deterministic law or according to a stochastic law. The modelling we have presented is original in character and has not been considered in the literature previously. A four-layer model of the surface has been proposed. This model covers quite a large range of real surfaces. In the deterministic case, we applied exponential, polynomial and other laws of permittivity variation in one of the layers of the surface. The choice of these laws was phenomenological in character. However, the exponential version of permittivity variation with layer depth has several experimental confirmations.

Interesting results have been derived in the case of a stochastic model for the three-layer media with flat boundaries. The average power of the reflected wave changes to a great extent with comparatively small values of the mean-square deviation of the layer permittivity. The most significant result of this analysis is the fact that with certain ratios of layer depth relative to wavelength, the reflected signal power reaches significant values. A certain limitation of the derived results is the fact that the surface roughness was not taken into account when considering the effects of reflection and scattering of electromagnetic waves. However, this problem has been addressed separately.

The problems of reflection of electromagnetic waves from surfaces having internal non-uniformities such as fractures, cracks, etc. has practically not been considered in literature. Our contributions on this subject are original in character.

As a first model we have considered a symmetric wedge-like crack with an incident a flat electromagnetic wave at a given angle. In this case, we take account of multiple reflections of the wave from the walls of the crack. The wave reflection coefficient depends on the value of the complex permittivity of the reflecting wall. The results of the analysis show that the dependence of the reflection coefficients on the angle of incidence differ markedly for vertical and horizontal wave polarizations of the wave. If the wedge angles are 40° - 45° with horizontal polarization of the incident wave and up to 75° with vertical polarization, the crack on the surface may be considered as a black body for vertical sensing. If a wedge-like crack is asymmetrical in character, then the main relations hold true but the number of reflections inside the wedge changes. As a second model we have considered a hollow region of spherical form. Taking into account multiple reflections, we have observed a substantial difference for the values of the reflection coefficients for horizontal and vertical polarizations of the incident wave.

As a third model, we have chosen a finite depth crack having rectangular form. For a shallow and wide hollow crack, the reflection coefficient with horizontal polarization of the incident wave differs from zero even for small angles of incidence. A narrow and deep hollow crack may be considered as a black body for incident wave angles up to 70° - 80°. However, for a vertically polarized incident wave, a hollow in the form of a rectangular pit with finite depth is characterized by an absence of reflections.

We have also analyzed reflections of electromagnetic waves for a combined model of surface, i.e., we have considered a four-layer medium; specifically, air, a homogeneous finite depth dielectric, a homogeneous semi-infinite space with dielectric constant different from that of the previous layer, in which a rectangular hollow with finite width and different permittivity is located.

For this model, we have derived design relations. These relations allow us to determine the reflection coefficients of electromagnetic waves for various modes of polarization of the sensing wave as a function of the incidence angles and the electrical-and-physical properties of sensed media.

The above-mentioned difference of the wave reflection coefficients with horizontal and vertical polarizations of the incident wave gives us many opportunities for analysis of surface structures with irregularities in the form of hollows, fractures, pits, cracks, etc.

All results we have derived are concerned with models of surfaces in which different factors have been taken taken into consideration; in particular, multiple scattering of electromagnetic waves. However, in these models we have neglected the presence of the surface roughness. Therefore, a whole section in Part III of this monograph has been devoted to the investigation of problems of radio wave reflection from rough surfaces. The main direction in this consideration has been the fact that the roughness is described by a certain random function in the spatial coordinates. This is distinct from the traditional approach, according to which the roughness is described by a deterministic function, e.g., sinusoidal, triangular, trapezoidal, and other functions.

In this case, as a surface model we have assumed a three-layer medium, in which the first layer is air, a rough boundary interface to a second layer with homogenous characteristics located above the third layer, also with homogeneous characteristics. We have considered small-scale roughness (i.e., the mean-square value of the roughness height is much less than the incident wavelength). In this case, the influence on reflection of a vertically polarized wave is absent. The influence of roughness is investigated only for the horizontally polarized component of the incident wave.

Our results include diagrams of inverse scattering under the condition that the correlation function of the spatial roughness is exponential in character. These diagrams show a strong dependence of the wave reflection coefficients on the ratio of the correlation radius of the surface relative to incident wavelength.

Using the aforementioned approach to taking surface roughness into account, it is possible to introduce necessary changes to the pertinent equations presented in first sections of Part III of the monograph. The resulting equations are invariable more complicated. The solution of these equations has been the subject of further investigations carried out by teams of scientists headed by the authors of this monograph. The most significant aspect in these solutions is the separation of the influence of the random variation of the surface permittivity and the random character of the spatial roughness on the characteristics of reflection and scattering of radio waves from the remotely sensed surfaces.

At the end of Part III of the monograph, we have presented material connected with the description of the KLL sphere and its application for determining the complex permittivity of a surface based on measurements of the voltages and phases of the received signals corresponding to orthogonally polarized radio incident waves. We hope that the application of the KLL sphere for determining the permittivity variation as a function of the variation of electromagnetic properties of the surface will be widely used in practice comparatively with the use of the Poincare sphere for the solution of various radar-polarimetry problems.

As a whole, Part III of the monograph gives a number of examples of the solution of inverse problems of radar remote sensing for various types of surfaces and may serve as a basis for further investigations and applications.

CHAPTER 14

Historical Development of Radar Polarimetry in Russia

14.1 Introduction

Research on the polarization properties of electromagnetic waves (EMW) began in the USSR during the 1940-50 decade. The first stage of that work was completed in the mid 60's and was summarized in two monographs [*Kanarejkin et al.*,1966; *Kanarejkin et al.*, 1968]. The monographs generalized results obtained by Soviet scientists until that period. Especially it is worth to mention contributions from Central Aerological Observatory (CAO) and Voejkov's Main Geophysical Observatory (GGO) in the field of meteorology. Their publications stimulated the interest of Soviet scientists to work on theoretical and practical problems connected with the use of the polarization properties of EMWs for extending the information-seeking capability of various types of radar systems.

In the aforementioned two monographs, various aspects concerning the use of polarization properties of reflected and scattered EMWs were considered. Furthermore, practical methods for measuring the polarization parameters of EMWs were discussed. Special attention was directed toward practical problems of polarization selection, i.e., to problems of elimination of interference reflections by use of polarization distinction in signals reflected from intended interference and interfering targets. The possibility of using the polarization properties of reflected radar signals for studying the environment, in particular problems associated with meteorology and oceanography was discussed. In this sense, the work contained in these two monographs constitutes the foundation of modern theoretical and practical polarimetry.

Below is a listing of on-going Russian research in the field of radar-polarimetry:
- Development of a general theory of polarization of radiowaves;
- Polarization theory connected to complex radar targets;
- Polarization selection with appropriate radar devices;
- Development of special algorithms of polarization signal reception on the basis of classical theories of detection, discrimination, filtering and estimation of signals parameters;
- Polarization modulation;
- Polarization phenomena in passive radiolocation (radiometry);
- Polarization analysis of scattered and reflected radiowaves for environmental study, e.g., meteorology, hydrology, oceanography and geophysics;

- Radar-polarimetric methods for recognition in radar images with the help of polarization signatures, i.e., construction of "polarization portraits" of radar objects in remote sensing systems;
- Radar-polarimetry principles in navigation and communication systems.

In the following, we shall address several of the above topics in some detail.

14.2 General theory of polarization of radiowaves

Significant early contributions toward the creation of a general theory of polarization of EMWs are contained in the monographs [*Potechin et al.,* 1978; *Bogorodsky et al.,* 1981]. In [*Potechin et al.,* 1978], an in-depth discussion is provided concerning the fundamentals of EMW polarization that can be used as a methodological basis for further detailed researches in this direction. In [*Bogorodsky et al.,* 1981], a broad range of problems related to polarization of radio waves and polarization properties of deterministic and fluctuating targets, antenna systems and terrestrial covers are considered. Also, problems associated with polarization selection are discussed. As a whole, [*Bogorodsky et al.,* 1981] provides a logical development of basic polarimetric ideas stated in [*Kanarejkin et al.,* 1966; *Kanarejkin et al.*, 1968] originally, and, in some respect, generalizes research on EMW polarimetry completed in the USSR by the late 70's. In [*Bogorodsky et al.,* 1981], many new concepts regarding the theory and practice of radar-polarimetry are introduced. It is important to note, that in this monograph, polarization of both scattered and reflected radio waves, as well as polarization of thermal radio-emission of various types of terrestrial covers is considered. Additionally, the concepts of fully polarized waves in matched polarization bases of quasi-polarized waves of full polarization scanning, radar contrast of targets, statistical and covariance scattering matrices are introduced. Finally, problems of dynamic radar-polarimetry, methods of polarization scanning and the problem of detection and selection of targets, weakly contrasted from a background of a terrestrial surface, are discussed for the first time. These studies have been performed by a group of the researchers under the management of A.I. Kozlov. The ensuing results have been developed further subsequently and have found applications in actual radar systems.

Significant incentive for development of research in the field of statistical radar-polarimetry was provided by the publication of the monographs [*Pozdniak et al.,* 1974], dealing with the statistical performance of polarized signals, and [*Bass and Fuks,* 1972], dealing with scattering of radiowaves from a rough surface. In these two monographs different statistical properties of polarized waves have been studied in details. Particularly [*Bass and Fuks,* 1972] aided in development of methods of environmental remote sensing. In [*Pozdniak et al.,* 1974], probability models of partially polarized radiowaves, statistical performances of polarization parameters of such radiowaves and, in more details, statistical performances of polarization factors of reception are described. The monographs [*Pozdniak et al.,* 1974] and [*Bass and Fuks,* 1972] have motivated further developments among which we cite [*Kozlov et al.,* 1979; *Meletitsky et al.,* 1987; *Pozdniak et al.,* 1987]. During the last decade, the statistical performance of polarized signals accounting for non-Gaussian interference effects and non-Gaussian behaviour of polarization parameters has been studied e.g. in [*Agaev et al.,* 1991; *Kozlov et al.,* 1990].

The next step in the development of the theory of polarization of radiowaves was the issuing of the monograph [*Kozlov et al.,* 1994], in which a quaternion statement of a polarization condition, new methods of analysis of transformation of a polarization structure, elements of a nonlinear radar theory and methods of field simulations with a complicated polarization structure, were introduced for the first time.

14.3 The polarization theory of the radar targets

The polarization properties of deterministic and fluctuating targets were considered in [*Kanarejkin et al.,* 1966], where the scattering matrix in various polarization bases and the eigen polarization states of deterministic radar targets were described. Further development can be found in [*Bogorodsky et al.,* 1981], where the statistical description of polarization properties of radar targets is essentially extended. In particular, problems connected to the statistical scattering matrix, the power matrix of random scattering together with statistical scattering matrices and the average contrast of the fluctuating targets, the covariance scattering matrix with a probability density function of the statistical scattering matrix, are investigated.

A significant contribution to the development of the theory of the radar target (RT) polarization was made by [*Ostrovitjanov et al.,* 1982; *Varganov et al.,* 1985]. In [*Ostrovitjanov et al.,* 1982], the valuable concept of a distributed RT (RT, which can not be considered as a point-like target) is introduced and systematized. Furthermore, the statistical properties of noise signals associated with distributed targets for various polarizations of radar's antenna are determined.

In [*Varganov et al.,* 1985] general analysis of methods of the description, measurement and simulation of RT signatures, in particular of flight vehicles, is given. Within the framework of this analysis, the polarization signatures (both mono-static and bi-static) of RT's on the basis of scattering matrices are described.

Among many studies dedicated to scattering matrices, we cite specifically [*Poliansky et al.*, 1974; *Kozlov et al.*, 1976; *Krasnov et al.*, 1966]. In [*Krasnov et al.*, 1966] some ways to increase the radar receiver efficiency, to determine polarimetric properties of the least depolarized scattered wave and to built up polarimetrically adequate receiver channels are proposed. In [*Kozlov et al.*, 1976] it is suggested to describe properties of fluctuating targets by means of a 4D complex vector, which leads to a 4• 4 covariance matrix with complex-valued correlation coefficients as non-diagonal elements and rout-mean squared values as diagonal ones. In [*Poliansky et al.*, 1974] the correlation between statistical scattering and Mueller matrix is investigated.

14.4 Polarization selection

The problems of polarization selection are closely related to problems of polarization properties of radar targets. However, polarization selection is a concept, which is much wider than the polarization discrimination of two radar targets, because polarization selection can be applied to communication systems, radio-navigation systems, etc.

In [*Kanareikin et al.*, 1966], the polarization properties of some radar targets are considered from the point of view of a realization of polarization selection for point-like, surface-distributed and volume-distributed targets.

In [*Bogorodsky et al.*, 1981], in the framework of the problem of polarization selection following particular issues have been studied: polarimetric clutter suppression; discrimination of two radiowaves in matched polarization bases; selections of the radar targets by a method of full polarization scanning; synthesis of radar targets and principles of synthesis of targets with specific polarization properties.

A mode of full polarization scanning is described in [*Bogorodsky et al.*, 1981]. It is shown that this mode allows to radiate an incident wave with all kinds of polarization (spiral scanning along the Poincare sphere) within a certain time interval. Simultaneously, the technical realization of the full polarization scanning mode is considered by means of two quasi-coherent high frequency oscillations with a small, but controlled frequency shift. The application of such oscillations allows (in addition to the full polarization scanning mode) realization of wideband frequency scanning. These interesting problems are stated in more detail in [*Kozlov*, 1976; *Demidov et al.*, 1975, 1978].

Emphasis has been and is being given by Russian scientists to problems involving the synthesis and analysis of algorithms of polarization selection of various classes of reflecting objects. Representative accounts of this work can be found in the following references: [*Potechin et al.,* 1978; *Pozdniak et al.,* 1974; *Varganov et al.,* 1985; *Kozlov et al.,* 1979; *Rodimov et al.,* 1984; *Gusev et al.,* 1974; *Bogorodsky et al.,* 1985; *Maximov et al.,* 1976; *Kostrukov et al.,* 1973].

We remark, especially, on the reference [*Kozlov et al.,* 1979] in which the concept of polarization radar contrast appears for the first time. This concept constitutes the beginning of one of the new directions in radar-polarimetry. The determination of polarization contrast between two radar objects, especially between weakly-contrasted, small, moving objects is a technique frequently quoted in the literature.

Problems associated with the polarization selection are also considered in [*Rodimov et al.,* 1984; *Gusev et al.,* 1974; *Bogorodsky,* 1985]. In these references, emphasis is placed on the power contrast enhancement and the recognition of reflecting objects (problem of pattern recognition).

In [*Pozdniak et al.,* 1974; *Maximov et al.,* 1976; *Kostrukov et al.,* 1973] the polarization factor at reception is selected as a parameter for determining the polarization selection. In [*Pozdniak et al.,* 1974], the probability density of the polarization factor at reception and its statistical performance in circular and linear polarization bases has been studied.

In [*Pozdniak et al.,* 1989], the concept of polarization selection factor for a signal with interfering background is considered and the probability distribution of that factor has been obtained.

14.5 Development of algorithms for the reception of polarized signals

In the area of development of algorithms for optimal reception of polarized signals, Russian scientists have made the most significant contributions to a general radar-polarimetry theory. Problems of detection, discrimination, filtering and parameter estimation of polarized signals are considered. One of the earliest contributions has been made by [*Kiselev,* 1969], who formulated an integral equation, which determines an optimum receiver structure. In this work, as well as in [*Pozdniak et al.,* 1972], the problem of detection of an elliptically polarized signal is considered. Systematic consideration of optimal reception of polarization-modulated waves took place in [*Gusev, et al.,* 1974], where the main concepts of optimal reception are introduced. Optimal reception of polarized signals in the presence of background noise and noise immunity of (polarized signals) receiving systems are investigated as well. Further research on problems related to the detection and discrimination of polarized signals in the presence of a normal distributed partially polarized interference are discussed in [*Pozdniak et al.,* 1974]. In this work, application of the polarization matrix to the detection of a deterministic signal in the presence of a normal distributed (partially polarized) correlated interference, of a deterministic polarized signal; of a polarized signal with an arbitrary initial phase and, finally, of a polarized signal with random amplitude and phase is considered. Post-detector detection of a polarized signal in the presence of normal distributed partially polarized interference (amplitude and phase methods of detection) is discussed as well. For discrimination of polarized signals a polarization factor, i.e. the module of the ratio between envelopes of orthogonally polarized components of a signal, has been used.

The most comprehensive consideration of problems associated with estimation of parameters and filtering of polarized signals can be found in [*Maximov et al.*, 1976], where polarimetric properties and processing of partly polarized signals and interference are studied. In this work, the estimation of parameters of a coherence matrix of a partially polarized signal is conducted, linear and nonlinear algorithms of estimation of polarization parameters of signals and interferences are developed, and also adaptive algorithms of estimation of polarization parameters of signals and interferences are obtained. Algorithms of antenna system polarization parameters control and algorithms of partially polarized signals and interference control are considered.

The indicated algorithms have a rather universal character; however, they are mostly related to communications systems and, consequently, do not take into account specifically the properties of polarized radar signals. Usually, they deal with detection criteria, the choice of information parameters, the character of transmitted signals, etc.

The problem of filtering of polarized radar signals was solved by methods of nonlinear Markov optimal filtering [*Lavin,* 1983; *Lavin*, 1985; *Logvin*, 1985]. In [*Lavin et al.*, 1983], a filtering algorithm for a polarized signal with a random polarization parameter (the angle of declination of the polarization plane in relation to a selected coordinate base) is obtained. The optimal receiver contains a system of information message extraction, a phase-locked loop (PLL) system and an angle-of-EMW polarization tracking circuit. In [*Lavin,* 1985] an optimal filtering algorithm is synthesized when the random parameters of a polarized radar signal are the geometric parameters of the polarization ellipse. Finally, in [*Logvin*, 1985], the structure of an optimal receiver of a polarized signal is obtained when the amplitudes and phases of the orthogonal components of an elliptically polarized received signal are considered as the polarization parameters. In this case, the optimal receiver contains a system of information message extraction, two PLL systems and two automatic gain control (AGC) circuits with appropriate cross-connections stipulated by the availability of correlation between the orthogonal components of interferences. Such a receiver continuously matches polarimetric characteristics of antenna system to properties of the EMW. This allows us to use more completely the radar dynamic range.

In [*Kozlov et al.,* 1994], a method of polarized signals processing is described, whereby the polarization factor of the anisotropies, representing the relation of the difference of the eigenvalues of the polarization base of a remotely sensed object to their sum, is used as the main information. Based on this polarization factor of anisotropy, it is possible to discriminate objects or to detect a radar object in clutter. The merits of using this polarization factor are that its estimation is obtained from a comparison of amplitudes of received signals in the co- and cross-polar radar channels and that by realizing a measurement in a circular polarization base, the result does not depend on target distance, does not require propagation loss compensation and is therefore independent on meteorological conditions. In addition, the estimation of the polarization factor does not depend on target orientation with respect to the observation direction, which is important for air-borne radar.

The problems of optimal processing of polarized signals continue to remain at the center of the scientific attention. On-going research includes applications of digital processing, the development of robust algorithms of reception, accounting for non-Gaussian distributions of polarimetric parameters, etc.

14.6 Polarization modulation

A complete account of problems of polarization modulation are stated in [*Gusev et al.,* 1974]. In this work, elliptically polarized EMWs are represented on a double complex plane. Signals with both continuous and discrete polarization modulations are considered. The modulation of the ellipticity angle, the orientation angle of the polarization plane and their combination is described. The spectral representations of polarization-modulated signals are analyzed. The principles of construction of polarization converters and modulators have been given. Reception of polarization-modulated signals is described in details. Influence of additive and multiplicative interference on polarization-modulated signals is evaluated. Optimization of polarization parameters for detection and discrimination polarization-modulated signals is performed.

Further development of the theory of polarization modulation has been undertaken in [*Badulin et al.*, 1988; *Tatarinov et al.*, 1989; *Stepanenko et al.*, 1987]. In [*Badulin et al.,* 1988], the spectral structure of a linearly polarized signal with a rotated plane of polarization is used to identify and classify radar targets. As an example, the second harmonics of the frequency of polarization rotation in the received signal spectrum is used. For various polarization parameters of RT, the amplitudes and phases of their spectral components on these frequency harmonics are calculated. For polarization parameters of RT, the orientation of a target coordinate system, the residual of the eigen-values of the target scattering matrix and the anisotropy factor were used. It has been shown that if one or two polarization performances of the target are known *a-priori*, it is possible to obtain an unambigious relationship between the spectral component parameters and the polarization performances of the target, e.g., for meteorological targets. In [*Tatarinov,* 1989], the polarization manipulation of a type "linear polarization – circular polarization" is considered. With the help of such "polarization-manipulated" signal, it is possible to estimate the polarization anisotropy of a dispersing object. Particular examples of such estimations are indicated in this reference.

The information cited above indicates that polarization-modulated or polarization-manipulated signals can be effectively used in radar remote sensing. This problem is considered below in more detail.

14.7 The polarization analysis of scattered and reflected radiowaves for studying the environment

In [*Kanarejkin et al.*, 1966], the first indication appears of the applicability of polarization analysis for deriving information about hydrometers; specifically, a study on rain intensity, and rain drop sizes and orientation was formulated. Subsequently, [*Kanarejkin et al.,* 1968] applied polarization analysis of scattered EMWs to the investigation of a marine surface. This period coincides with intense activities in the USSR in connection with the development of remote sensing systems and their application to the study of the environment. Here, there is no possibility to account even for a small part the work of Russian scientists, who used principles of polarization analysis in their studies on marine surfaces, marine and continental ice, the atmosphere and hydrometeors, surfaces of natural space objects (moon, Mars, Venus, etc.), agricultural production, and many others. All we can do is to point out a few references that have played an important role in the introduction of radar-polarimetric methods in modern remote sensing systems.

For studying the properties of ice surfaces, the modification of the polarization characteristics of scattered EMWs allows one to solve various problems of glaciology, for example, definitions of salinity of ice and its volumetric moisture content. The development of radar-polarimetry methods for studying marine and continental ice began with the work of [*Bogorodsky*, 1976], where it is shown experimentally how the polarization of radar signals for vertical sensing of glaciers is changed. Similar results are obtained for shelf ice. Interesting results based on radar-polarimetric investigations of ice are contained in [*Nikitin et al.*, 1985]. For glaciers, measurements of the Stokes parameters have been conducted and the correlation processing of functions of these parameters have been carried out; the latter justifies the model for a glacier as a double refracting plate with linear eigen polarization. The radar-polarimetry analysis of a glacier allows us to determine its state of stress, to find directions of principal stresses and module of principal stresses difference. The definition of stress state of ice is a very important practical problem, and only with the help of radar-polarimetry methods it gives solutions.

A large range of radar-polarimetric investigations of marine surfaces has been carried out by Russian scientists. The relations between scattered field characteristics with experimentally observed parameters and characteristics of the sea surface for various polarizations of the incident field have been studied. The representation of a statistical polarization matrix for the scattered field, with elements defined by the geometric characteristics and electrophysical parameters of the sea surface, has been constructed. Anomaly high levels of horizontally polarized wave backscattering have been investigated and explained using an appropriate theoretical model. All these radar-polarimetry problems are of general theoretical importance, and are of interest not only for concrete marine surfaces, but also for a much wider choice of appropriate applications. Along this vein we cite the following references: [*Eshenko et al.*, 1972; *Zujkov et al.*, 1981; *Melnichuk et al.*, 1975].

A number of studies involving meteorological applications of radar-polarimetric analyses has been performed by scientists from the Central Aerological Observatory (CAO) and Voejkov's Main Geophysical Observatory (GGO). Similar investigations have been carried out by various organizations for studies of vegetation covers, woods, agricultural land and other Earth covers. More detail applications of radar-polarimetry in remote sensing systems is considered below.

14.8 Applications of radar-polarimetry in remote sensing systems

The wide application of remote sensing to environmental studies and ecological monitoring during the two last decades has demanded the introduction of new approaches and methods of radar remote sensing. Among these approaches radar-polarimetric methods are of primary importance because many essential results in remote sensing can be obtained only by using different polarization properties of scattered or reflected radio waves.

Below we provide a partial list [cf. e.g., *Melnichuk et al.,* 1975] of specific applications of radar-polarimetry in remote sensing systems:

- Estimation of the characteristics of the environment and ecosystems (from regional up to global);
- Estimation of the physical condition of objects, such as oil pipelines, cables, gas conduits, etc;
- Description of woods, agricultural and fishing ecosystems;
- Classification and evaluation of conditions of terrestrial cover, soils, bogs, lakes, etc.;
- Realization of hydrological and glaciological observations, evaluation of hydrology and humidity of soils, performances of snow covers, icebergs, glaciers, permafrost condition of ground;
- Realization of cartography of marine ice and estimation of blocking ice areas;
- Estimation of the bio-mass of agricultural crops;
- Estimation of the growth and condition of forests;
- Determination of the extent and consequences of wood fires; also of drought and floods;
- Cartography of marine surfaces;
- Evaluation and consequences of volcanic activity, including lava currents and dirt streams;
- Determination of sea state and wind parameters;
- Determination of salinity zones and corrosion of soil;
- Observation and assessment of petroleum contamination on a water surface.

Many of the problems listed above cannot be solved without the application of radar-polarimetric methods. A large number of such examples is indicated in [*Kozlov et al.,* 1992; *Kozlov et al.*, 1993].

One of the most popular radar-polarimetric activity has become the technique of determining polarization signatures. It consists of special functions, which are created in three-dimensional space; the ellipticity angle of the EMW polarization ellipse and the orientation angle of the polarization plane are placed on two axes and the backscattering factor is placed on the third axis. For each remotely sensed object, we may obtain an unique polarization signature.

In the case of reception of scattered radiation with background noise characterized by a Gaussian distribution model, the polarization signature of noise is constant, as noise is an unpolarized signal. This unpolarized component introduces a "pedestal" in the construction of the polarization signature.

Polarization signatures can be used to distinguish among different surfaces. The accepted models can take into account variants of a smoothed surface, weak and strong surface roughness, or combinations. Such models can describe many kinds of terrestrial surfaces; therefore, with the help of known outcomes of remote sensing, the construction of polarization signatures for different kinds of surfaces seems to be possible, i.e., with the help of polarization images ("portraits"), the creation of a data bank for classification and identification of surfaces seems to be feasible. The most productive outcomes can be reached by a combination of polarization images of surfaces and other indicators, for example spectral information, etc.

An important aspect of remote sensing is the solution of inverse problems, e.g., when geometric, physical, chemical, mechanical and other properties of remotely sensed objects or surfaces are determined from characteristic properties of reflected or scattered radar signals. The solution of inverse problems, which are usually very difficult, can be aided significantly by radar-polarimetric means.

Generally, multi-dimensional inverse problems as arising in remote sensing systems are mathematically ill-posed. This results into ambiguities and instabilities under perturbation of a set of measurements. Mathematically, ill-posed problems are handled by some sort of regularization. A description of a special regularization method applied to the solution of an inverse problem of radar positioning using radar-polarimetry is discussed in [*Stepanenko et al.*, 1987].

APPENDIX A (Ref. Chapter 2)

Object Motion Analysis
(Translation + Rotation)

Computation of Doppler spread and correlation time

The calculation refers to the geometry of Fig. 2.15 (chapter 2) that we report below for convenience as Fig. A.1.

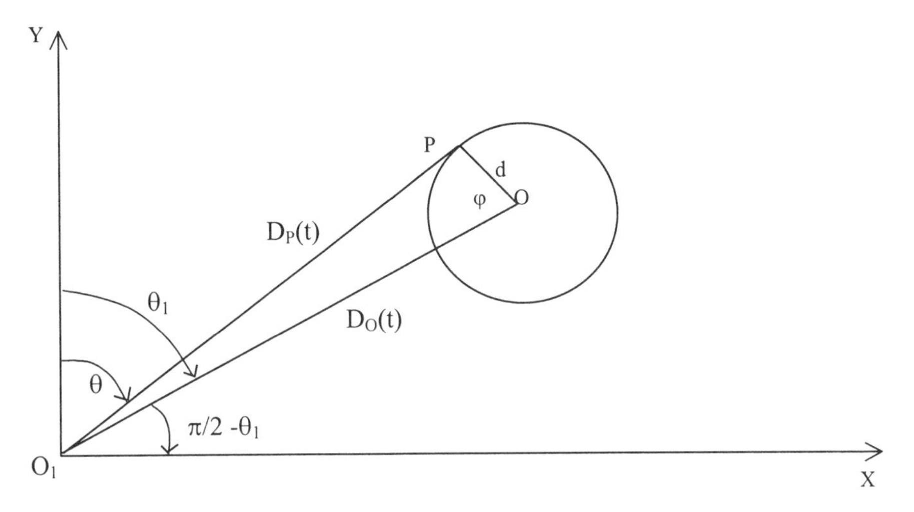

Fig. A.1 Geometry of motion (translation + rotation) of a two-scatterer object model

The distance P-O_1 of point P of the object from the radar position O_1 is calculated from the triangle POO_1 as follows:

$$P - O_1 = \sqrt{(O - O_1)^2 + (P - O)^2 - 2(O - O_1)(P - O)\cos\varphi} \tag{A.1}$$

Here:

O-O_1 = $D_0(t)$ is the distance of object's center of gravity O from the radar
P-O_1 = $D_P(t)$ is the distance of point P from the radar
P-O = d is the distance of point P from the center of gravity

φ : rotation angle of point P around the center of gravity O
t : time

We derive from Eq. (A.1) the motion of point P from the radar given under the approximation of d<< $D_0(t)$:

$$D_P(t) \cong D_0(t) - d\cos\varphi \qquad (A.2)$$

For the center of gravity O we consider the motion given by:

$$D_0(t) = v_0 t \qquad (A.3)$$

The variation of the bearing angle θ due to the rotation of point P around O can be computed from the triangle POO_1 :

$$\frac{\sin(\theta_1 - \theta)}{d} = \frac{\sin\varphi}{P - O_1} \qquad (A.4)$$

Here, θ_1 is the bearing angle of the center of gravity O. In the approximation d/(P-O) <<1, we derive from Eq. (A.3) the bearing angle θ:

$$\theta \approx \theta_1 - \frac{d}{D_P}\sin\varphi \qquad (A.5)$$

The total signal received from the two scatterer points P and O is given by the sum

$$S_T = A_0 e^{\left[j\omega\left(t - \frac{2D_0(t)}{c}\right)\right]} + A_P e^{\left[j\omega\left(t - \frac{2D_P(t)}{c}\right)\right]} \qquad (A.6)$$

where ω is the angular frequency of the electromagnetic field, c is speed of light, and the distances $D_P(t)$ and $D_0(t)$ are defined by the equations of motion A.2 and A.3.

Eq. (A.6) can also be written (as constant-envelope wave after limiter) as

$$S_T(t) = e^{j\omega\left(1 - \frac{2v_0}{c}\right)t}\left\{1 + e^{j\frac{2\omega}{c}d\cos\varphi(t)}\right\} \qquad (A.7)$$

After mixing at the receiver, we obtain the Doppler signal given by:

$$S_T(t) = e^{-j\frac{2\omega}{c}v_0 t}\left\{1 + e^{j\frac{2\omega}{c}d\cos\varphi}\right\} \tag{A.8}$$

By differentiating Eq. (A.8), we have the Doppler information in the envelope of the wave:

$$\dot{S}_{T_D} = -2\frac{\omega}{c}\left\{v_0\left[1 + e^{j2\frac{\omega}{c}d\cos\varphi}\right] + d\sin\varphi\cdot\dot{\varphi}\right\}e^{j2\frac{\omega}{c}v_0 t} \tag{A.9}$$

This expression can be re-written as

$$\dot{S}_{T_D} = \mathrm{Re} + j\,\mathrm{Im} \tag{A.10}$$

where

$$\mathrm{Re} = -2\frac{\omega}{c}v_0\left[1 + \cos\left(2\frac{\omega}{c}d\cos\varphi\right)\right] - 2\frac{\omega}{c}d\sin\varphi\dot{\varphi} \tag{A.11}$$

$$\mathrm{Im} = -2\frac{\omega}{c}v_0\sin\left(2\frac{\omega}{c}d\cos\varphi\right) \tag{A.12}$$

The amplitude is given by

$$\sqrt{\dot{S}_{T_D}} = \{\ \mathrm{Re}^2 + \mathrm{Im}^2\ \}^{1/2} \tag{A.13}$$

From Eqs (A.11)-(A.13), taking into account that $\dot{\varphi} = \frac{v_n}{d}$, where v_n is the velocity of rotation of point P, d the distance P-O and $\alpha = \frac{2\omega d}{c}\cos\varphi$, we have

$$\sqrt{\dot{S}^2_{T_{Doppler}}} = \frac{4\omega v_0}{c}\left[2(1+\cos\alpha)\left(1 + \frac{v_n}{v_0}\sin\varphi\right) + \left(\frac{v_n}{v_0}\sin\varphi\right)^2\right]^{1/2} \tag{A.14}$$

If we have an object of length 3m and moving at a speed v_n of 5m/s (~ 20km/hour), the angular velocity is computed for d=3L as

$$\dot{\varphi} = \frac{v_n}{d} \tag{A.15}$$

$$\dot{\varphi} = \frac{v_n}{3L} \sim 30 \text{ degrees/second} \tag{A.16}$$

In this example, the time of covering a circular path of 18m diameter is

$$\tau_m = \frac{2\pi d}{v_n} = \frac{6\pi L}{v_n} \quad \sim 12\text{s} \tag{A.17}$$

APPENDIX B (Ref. Chapter 6)

The Stokes Matrix

Information about polarization properties of a reflected signal is specified by the Stokes matrix [*Potapov*, 1992; *Durden*, 1989; *Born,* 1970; *Ishimaru*, 1981].

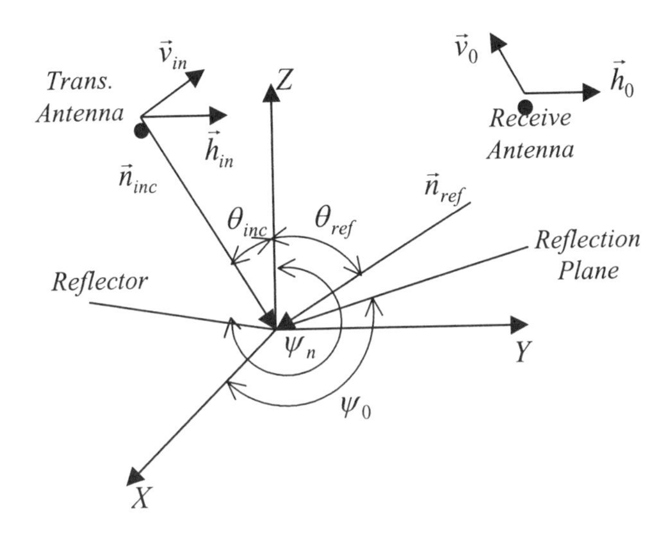

Fig. B.1 Geometry of reflection

The incident wave electric field (see Fig. B.1) can be described by

$$\vec{E}_{in} = E_{in}^{h}\vec{h}_{in} + E_{in}^{v}\vec{v}_{in} \tag{B.1}$$

where $\vec{h}_{in}, \vec{v}_{in}$ are unit vectors along the horizontal and vertical directions respectively. Then, the reflected (scattered) field becomes

$$\vec{E}_{ref} = E_{ref}^{h}\vec{h}_{ref} + E_{ref}^{v}\vec{v}_{ref} \tag{B.2}$$

We introduce, also, the notation

$$\vec{n}_{ref} = \vec{h}_{ref} \times \vec{v}_{ref}\,; \vec{n}_{in} = \vec{h}_{in} \times \vec{v}_{in} \qquad \text{(B.3)}$$

The relationship between the amplitudes of the incident and reflected waves is given by

$$\begin{pmatrix} E_h \\ E_v \end{pmatrix}_{ref} = \begin{pmatrix} s_{hh} & s_{hv} \\ s_{vh} & s_{vv} \end{pmatrix} \begin{pmatrix} E_h \\ E_v \end{pmatrix}_{in} \qquad \text{(B.4)}$$

where r is the distance from a scatterer to the point of observation.

A wave can be described by means of Stokes vector $\vec{S}$, defined as

$$\vec{S} = \begin{pmatrix} S_0 \\ S_1 \\ S_2 \\ S_3 \end{pmatrix} = \begin{pmatrix} E_h E_h^* + E_v E_v^* \\ E_h E_h^* - E_v E_v^* \\ 2\,\mathrm{Re}\left(E_h E_v^*\right) \\ 2\,\mathrm{Im}\left(E_h E_v^*\right) \end{pmatrix} \qquad \text{(B.5)}$$

The relationship between the Stokes vectors of an incident and a reflected wave assumes the form

$$\vec{S}_{ref} = RR^T M \vec{S}_{in} \qquad \text{(B.6)}$$

where

$$R = \begin{pmatrix} 1 & 1 & 0 & 0 \\ 1 & -1 & 0 & 0 \\ 0 & 0 & 1 & 1 \\ 0 & 0 & -j & -j \end{pmatrix} \qquad \text{(B.7)}$$

and M is the Stokes matrix, defined as

$$M = \left(R^T\right)^{-1} W R^{-1} \tag{B.8}$$

with

$$W = \begin{pmatrix} S_{hh}S_{hh}^* & S_{hv}S_{hv}^* & S_{hh}S_{hv}^* & S_{hv}S_{hh}^* \\ S_{vh}S_{vh}^* & S_{vv}S_{vv}^* & S_{vh}S_{vv}^* & S_{vv}S_{vh}^* \\ S_{hh}S_{vh}^* & S_{hv}S_{vv}^* & S_{hh}S_{vv}^* & S_{hv}S_{vh}^* \\ S_{vh}S_{hh}^* & S_{vv}S_{hv}^* & S_{vh}S_{hv}^* & S_{vv}S_{hh}^* \end{pmatrix} \tag{B.9}$$

the Mueller matrix.

The superscript T in Eqs (B.6) and (B.8) means transposition of a matrix, and R^{-1} denotes the inverse of matrix R. If we have a number of scatterers, the average value of the Stokes matrix is the sum of corresponding matrices of the individual scatterers.

The reflected power P is given by

$$P = k\vec{S}^r M \vec{S}^t \tag{B.10}$$

where k is a system constant and $\vec{S}^r, \vec{S}^t$ are vectors describing receiving and transmitting antenna matrices, respectively.

APPENDIX C (Ref. Chapter 8)

$$\alpha = e^{ik_0 d\sqrt{\varepsilon_2 - \sin^2\beta}}$$

$$\delta_1 = J_\nu(X)$$

$$\delta_2 = N_\nu(X)$$

$$\delta_3 = \frac{i}{k_0}\cdot\frac{dJ_\nu(X)}{dz}\bigg|_{z=-d_2} = \frac{i}{2}\sqrt{\varepsilon_2}\left[J_{\nu+1}(X) - J_{\nu-1}(X)\right]$$

$$\delta_4 = \frac{i}{k_0}\cdot\frac{dN_\nu(X)}{dz}\bigg|_{z=-d_2} = \frac{i}{2}\sqrt{\varepsilon_2}\left[N_{\nu+1}(X) - N_{\nu-1}(X)\right]$$

$$\delta_5 = J_\nu(Y)$$

$$\delta_6 = N_\nu(Y)$$

$$\delta_7 = \frac{i}{k_0}\cdot\frac{dJ_\nu(Y)}{dz}\bigg|_{z=-d_2-d_3} = \frac{i}{2}\sqrt{\varepsilon_4'}\left[J_{\nu+1}(Y) - J_{\nu-1}(Y)\right]$$

$$\delta_8 = \frac{i}{k_0}\cdot\frac{dN_\nu(Y)}{dz}\bigg|_{z=-d_2-d_3} = \frac{i}{2}\sqrt{\varepsilon_4'}\left[N_{\nu+1}(Y) - N_{\nu-1}(Y)\right]$$

$$X = \frac{k_0\sqrt{N}}{a}e^{ad_2} = \frac{k_0\sqrt{\varepsilon_2}}{a}$$

$$Y = \frac{k_0\sqrt{N}}{a}e^{a(d_2+d_3)} = \frac{k_0\sqrt{\varepsilon_4'}}{a}$$

$$g = e^{-ik_0\sqrt{\varepsilon_4' - \sin^2\beta}(d_2+d_3)}$$

APPENDIX D (Ref. Chapter 8)

$$\delta_1 = e^{ad_2} J_\nu\left(\tilde{X}\right)$$

$$\delta_2 = e^{ad_2} N_\nu\left(\tilde{X}\right)$$

$$\delta_3 = ie^{ad_2} \cdot \left\{-\frac{a}{k_0} J_\nu\left(\tilde{X}\right) + \frac{\sqrt{\varepsilon_2}}{2}\left[J_{\nu+1}\left(\tilde{X}\right) - J_{\nu-1}\left(\tilde{X}\right)\right]\right\}$$

$$\delta_4 = -ie^{ad_2} \cdot \left\{\frac{a}{k_0} N_\nu\left(\tilde{X}\right) - \frac{\sqrt{\varepsilon_2}}{2}\left[N_{\nu+1}\left(\tilde{X}\right) - N_{\nu-1}\left(\tilde{X}\right)\right]\right\}$$

$$\delta_5 = e^{a(d_2+d_3)} J_\nu\left(\tilde{Y}\right)$$

$$\delta_6 = e^{a(d_2+d_3)} N_\nu\left(\tilde{Y}\right)$$

$$\delta_7 = ie^{a(d_2+d_3)} \cdot \left\{-\frac{a}{k_0} J_\nu\left(\tilde{Y}\right) + \frac{\sqrt{\varepsilon_4}}{2}\left[J_{\nu+1}\left(\tilde{Y}\right) - J_{\nu-1}\left(\tilde{Y}\right)\right]\right\}$$

$$\delta_8 = -ie^{a(d_2+d_3)} \cdot \left\{\frac{a}{k_0} N_\nu\left(\tilde{Y}\right) - \frac{\sqrt{\varepsilon_4}}{2}\left[N_{\nu+1}\left(\tilde{Y}\right) - N_{\nu-1}\left(\tilde{Y}\right)\right]\right\}$$

APPENDIX E (Ref. Chapter 8)

$$\alpha_{12} = \sqrt{b} J_{\frac{1}{n+2}}(\beta)$$

$$\alpha_{13} = \sqrt{b} J_{-\frac{1}{n+2}}(\beta)$$

$$\alpha_{22} = \frac{a}{2k_0 b}\alpha_{12} - \frac{b^{\frac{n+1}{2}}}{2} J_{\frac{1}{n+2}-1}(\beta) + \frac{b^{\frac{n+1}{2}}}{2} J_{\frac{1}{n+2}+1}(\beta)$$

$$\alpha_{23} = \frac{a}{2k_0 b}\alpha_{13} - \frac{b^{\frac{n+1}{2}}}{2} J_{-\frac{1}{n+2}-1}(\beta) + \frac{b^{\frac{n+1}{2}}}{2} J_{-\frac{1}{n+2}+1}(\beta)$$

$$\alpha_{32} = \sqrt{b-ad}\, J_{\frac{1}{n+2}}(\gamma)$$

$$\alpha_{33} = \sqrt{b-ad}\, J_{-\frac{1}{n+2}}(\gamma)$$

$$\alpha_{42} = \frac{a}{2k_0(b-ad)}\alpha_{32} - \frac{(b-ad)^{\frac{n+1}{2}}}{2} J_{\frac{1}{n+2}-1}(\gamma) - \frac{(b-ad)^{\frac{n+1}{2}}}{2} J_{\frac{1}{n+2}+1}(\gamma)$$

$$\alpha_{43} = \frac{a}{2k_0(b-ad)}\alpha_{33} - \frac{(b-ad)^{\frac{n+1}{2}}}{2} J_{-\frac{1}{n+2}-1}(\gamma) - \frac{(b-ad)^{\frac{n+1}{2}}}{2} J_{-\frac{1}{n+2}+1}(\gamma)$$

$$a = \frac{1}{d}\left[\varepsilon_2^{\frac{1}{n}}(0) - \varepsilon_3^{\frac{1}{n}}(0)\right]$$

$$b = \varepsilon_2^{\frac{1}{n}}(0)$$

$$\frac{a}{k_0\sqrt{b}} = \frac{\varepsilon_2^{\frac{1}{n}}(0) - \varepsilon_3^{\frac{1}{n}}(0)}{2\pi\varepsilon_2^{\frac{1}{2n}}(0)}\frac{\lambda}{d}$$

$$\frac{a}{k_0\sqrt{b-ad}}=\frac{\varepsilon_2^{\frac{1}{n}}(0)-\varepsilon_3^{\frac{1}{n}}(0)}{2\pi\varepsilon_3^{\frac{1}{2n}}(0)}\frac{\lambda}{d}$$

$$\beta=4\pi\frac{\varepsilon_2^{\frac{1}{n}+\frac{1}{2}}(0)}{(n+2)\left(\varepsilon_2^{\frac{1}{n}}(0)-\varepsilon_3^{\frac{1}{n}}(0)\right)}\frac{d}{\lambda}$$

$$\gamma=4\pi\frac{\varepsilon_3^{\frac{1}{n}+\frac{1}{2}}(0)}{(n+2)\left(\varepsilon_2^{\frac{1}{n}}(0)-\varepsilon_3^{\frac{1}{n}}(0)\right)}\frac{d}{\lambda}$$

APPENDIX F (Ref. Chapter 8)

For evaluation of expression (8.85) it is necessary to compute some integrals:

$$B_1 = \int_{-\infty}^{\infty}\int_{-\infty}^{\infty} \frac{\cos pz \cos p'z'}{\left(p^2 - k_0^2\varepsilon_{02}\right)\left(p'^2 - k_0^2\varepsilon_{02}^*\right)} dpdp'$$

$$B_2 = \int_{-\infty}^{\infty}\int_{-\infty}^{\infty} \frac{\cos pd \cos p'd \cos pz \cos p'z'}{\left(p^2 - k_0^2\varepsilon_{02}\right)\left(p'^2 - k_0^2\varepsilon_{02}^*\right)} dpdp'$$

$$B_3 = \int_{-\infty}^{\infty}\int_{-\infty}^{\infty} \frac{\cos p'd \cos pz \cos p'z'}{\left(p^2 - k_0^2\varepsilon_{02}\right)\left(p'^2 - k_0^2\varepsilon_{02}^*\right)} dpdp'$$

$$B_4 = \int_{-\infty}^{\infty}\int_{-\infty}^{\infty} \frac{\cos pd \cos pz \cos p'z'}{\left(p^2 - k_0^2\varepsilon_{02}\right)\left(p'^2 - k_0^2\varepsilon_{02}^*\right)} dpdp'$$

$$B_5 = \int_{-\infty}^{\infty} \frac{\cos pd \cos pz}{\left(p^2 - k_0^2\varepsilon_{02}\right)} dp$$

$$B_6 = \int_{-\infty}^{\infty} \frac{\cos p'd \cos p'z'}{\left(p'^2 - k_0^2\varepsilon_{02}^*\right)} dp'$$

The integrands in B_5 and B_6 have the following poles:

$$p_{1,2} = \pm k_0\sqrt{\varepsilon_{02}} = k_0\left(\pm\alpha_1 \mp i\alpha_2\right)$$

$$p_{1,2} = \pm k_0\sqrt{\varepsilon_{02}^*} = k_0\left(\pm\alpha_1 \pm i\alpha_2\right)$$

Of four available poles p_1 and p_2', two poles are located in the top half plane, and two other poles p_1' and p_2 are located in the bottom half plane of the complex variable $p(p')$.

$$B_5 = \frac{1}{4}\int_{-\infty}^{\infty} \frac{dp}{p^2 - k_0^2\varepsilon_{02}}\left[e^{ip(d+z)} + e^{-ip(d+z)} + e^{ip(d-z)} + e^{-ip(d-z)}\right] =$$

$$= \frac{1}{2}\int_{-\infty}^{\infty} \frac{dp}{p^2 - k_0^2\varepsilon_{02}}\left[e^{ip(d+z)} + e^{ip(d-z)}\right] = \pi i \cdot res(p_2) =$$

$$= -\frac{\pi i}{2k_0\sqrt{\varepsilon_{02}}} e^{-ik_0 d\sqrt{\varepsilon_{02}}}\left[e^{-ik_0 z\sqrt{\varepsilon_{02}}} + e^{ik_0 z\sqrt{\varepsilon_{02}}}\right] =$$

$$= -\frac{\pi i}{k_0\sqrt{\varepsilon_{02}}} e^{-ik_0 d\sqrt{\varepsilon_{02}}} \cos\left(k_0 z\sqrt{\varepsilon_{02}}\right)$$

Evaluating other integrals similarly, the following expressions are found:

$$B_1 = \frac{\pi^2}{k_0^2\left|\varepsilon_{02}\right|} e^{ik_0 z\sqrt{\varepsilon_{02}}} e^{-ik_0 z'\sqrt{\varepsilon_{02}^*}}$$

$$B_2 = \frac{\pi^2}{k_0^2\left|\varepsilon_{02}\right|} e^{-2k_0\alpha_2 d} \cos\left(k_0 z\sqrt{\varepsilon_{02}}\right)\cos\left(k_0 z'\sqrt{\varepsilon_{02}^*}\right)$$

$$B_3 = \frac{\pi^2}{k_0^2\left|\varepsilon_{02}\right|} e^{-ik_0 d\sqrt{\varepsilon_{02}}} e^{-ik_0 z'\sqrt{\varepsilon_{02}^*}} \cos\left(k_0 z\sqrt{\varepsilon_{02}}\right)$$

$$B_4 = \frac{\pi^2}{k_0^2\left|\varepsilon_{02}\right|} e^{ik_0 d\sqrt{\varepsilon_{02}^*}} e^{-ik_0 z'\sqrt{\varepsilon_{02}^*}} \cos\left(k_0 z'\sqrt{\varepsilon_{02}^*}\right)$$

$$B_6 = \frac{\pi i}{k_0\sqrt{\varepsilon_{02}^*}} e^{ik_0 d\sqrt{\varepsilon_{02}^*}} \cos\left(k_0 z'\sqrt{\varepsilon_{02}^*}\right)$$

$$B_7 = \int_{-\infty}^{\infty} \frac{\cos pz}{p^2 - k_0^2\varepsilon_{02}} dp = -\frac{\pi i}{k_0\sqrt{\varepsilon_{02}}} e^{ik_0 z\sqrt{\varepsilon_{02}}}$$

$$B_8 = \int_{-\infty}^{\infty} \frac{\cos p'z'}{p'^2 - k_0^2\varepsilon_{02}^*} dp' = \frac{\pi i}{k_0\sqrt{\varepsilon_{02}^*}} e^{-ik_0 z\sqrt{\varepsilon_{02}^*}}$$

REFERENCES

Achmanov C.A., Diakov U.E. and Chirkin A.L., "Introduction in Statistical Radiophysics and Optics," Nauka, 1981.

Agaev S.K., Kozlov A.I., Rusinov V.R., "Statistical characteristics bending around of a non-Gaussian signal for availability of a non-Gaussian interferences," Izvesty VUZ, Radio electronics, 1991, No. 4, pp. 93-96.

Ament W.S., "Toward a Theory of Reflection by a Rough Surface," Proc. IRE, Vol. 41, pp. 142-146, 1953.

Andreev A.V., Ponomarev U.V. and Smolin A.A., "X-Ray Diffraction from an Acoustic Wave Surface," Letters to ZTF, Vol. 14, pp. 1260-1264, 1988.

Andreev, A.V., "X-rays Optics of Surface," UFN Vol. 145, pp. 113-136, 1985.

Andreev, A.V., Beliaev, D.V., "X-Ray Reflection from a Diffraction Lattice," Optics and Spectroscopy, Vol. 67, pp. 714-720, 1989

Armand, N.A. et al., "Exploration of the Natural Environment by Radiophysical Methods, Review." Radiophysics, Vol. 20, pp. 809-841, 1977.

Assur A., "Composition of sea ice and its tensile strength," US Army Snow and Ice and Permafrost Research Establishment, Wilmettle Illinois, 1960.

Badulin N.N., Gulko V.L. Spectral characteristics of Echo-signal for polarization modulation of a radar radiation. Izvesty VUZ, Radio electronics, 1988, V. 31, No. 4, pp. 74-76.

Barabanenkov U.N., "Multiple Wave Scattering on an Ensemble of Particles and Radiation Transition Theories," UFN, Vol. 117, p. 49, 1975.

Basharinov A.E. et al., "Scattering of waves by damp soil," Radiotechnica, Moscow, 14, pp. 76-78, 1989.

Bass F.G. and Fuks I.M., "Wave Scattering from Statistically Rough Surfaces," Oxford: Pergamon, 1979.

Bass F.G., Fuks I.M, "Scattering of waves on a statistically rough surface," M.: Nauka, 1972.

Baum Carl E., "Discrimination of Buried Targets via the Singularity Expansion," Inverse Problems, Vol. 13, pp. 557-570, 1997., IOP, Bristol, UK.

Baum Carl E., Rothwell E.J., Chen K.M. and Nyquist D.P., "The Singularity Expansion Method and its Application to Target Identification," Proc. IEEE, Vol. 79, No.10, pp. 1481-1491, 1991.

Beard C.I., "Coherent and Incoherent scattering of Microwaves from the Ocean," IRE Trans. Antennas Propag. Vol. AP-9, pp. 470-483, 1961.

Beckmann P. and Spizzichino A., The Scattering of Electromagnetic Waves from Rough Surfaces, Oxford: Pergamon. Reprinted by Artech House, USA, 1987.

Beckmann P., "Shadowing of Random Rough Surfaces," IEEE Trans. Antennas Propag. Vol. AP-13, pp. 384-388, 1965.

Berkowitz R.S., Modern Radar. Analysis, Evaluation and System Design, John Wiley & Sons, Inc. 1965.

Boerner W.M., El-Arini M.B., Chan C.Y. and Mastoris P.M., "Polarization Dependence in Electromagnetic Inverse Problems," IEEE Trans. Antennas Propag. Vol. AP-29, No.2, pp. 262-271, 1981.

Boerner W.M., Mott H., Luneburg E., Livingstone C., Paterson J.S. Polarimetry in remote sensing, 3-d edition, ASPRS Publishing Bethesda, MD, 1997.

Boerner W.M., W.L. Lang, An-Qing Xi and Yamaguchi Y., "On the Basic Principles of Radar Polarimetry: The Target Characteristics Polarization State Theory of Kennaugh, Huynen's Poalrization Fork Concept and its Extension to the Partially Polarized Case," Proc. IEEE, Vol. 79, No. 10, pp. 1538-1539, 1991.

Bogorodsky V.V. "Remote sensing of glaciers," Leningrad, Gidrometeoizdat, pp. 64, 1975.

Bogorodsky V.V., Kanarejkin D.B., Kozlov A.I. Polarization of scattered and own radio emission of terrestrial covers. L.: Gidrometeoizdat, 1981.

Bogorodsky V.V., Kozlov A.I. and Logvin A.I., "Microwave Radar of Earth Covers," Leningrad. Gidrometeoizat, p. 272, 1985.

Bogorodsky V.V., Trepov G.V., Fedorov B.A. Modification of polarization of radar signals for vertical sensing of glaciers. ZTF, 1976, V. 46, v. 2.

Bojarsky D.A. et al., "Frequency-dependent model of effective complex dielectric permittivity of a damp snow," Radiotechnica and electronica, Moscow, Vol. 39, ^{1}10, pp. 1479-1485, 1995.

Bojarsky D.A. et al., "Modeling of electrical properties of a dry snow cover," Izvestija VUZ, ser. Radiophysica, Vol. 34, ^{1}8, pp. 859-862, 1991.

Booker H.G. and Gordon "A Theory of Radio Scattering in the Troposphere," Proc. IRE Vol. 38, pp. 401-412, 1950.

Born, M., Fundamentals of Optics, M., Pergamon Press, p. 856, 1970.

Brancaccio A., Leone G., Liseno A., Pierri R., Soldovieri F. " Research Activities at the Second University of Naples towards GPR Applications," Proceedings of International Symposium on Antennas for Radar Earth Observation, Delft University of Technology, the Netherlands, 8-9 June 2000.

Brancaccio A., Leone G., Pierri R., 'Information content of Born scattered fields: results in the circular cylindrical case," Journal of Optical Society of America Part A, Vol. 15, n.7, pp.1909-1917, July 1998.

Brehovskih, L.M., Waves in Layered Media, Nauka, Moscow, p 502, 1957.

Brooks J.W. and Maier M.W., "Detection of Abandoned Land Mines," IEE Conference Publication 7-9 October 1996.

Brown G.S. "Backscattering from a Gaussian-distributed Perfectly Conducting Rough Surface." IEEE Trans. Antennas Propag. Vol. AP-26, pp. 472-482, 1978.; corrections in Vol. AP-28, pp. 943-946, 1980.

Bruchovecky A.C. and Fuks I.M., "The Effective Impedance Tensor of a Statistically Rough Impedance Surface," Radiophysics, Vol. 28, pp. 1400-1407, 1985.

Brussaard G., Rogers, D.V., "Propagation Considerations in Satellite Communications Systems," Proc. IEEE, Vol. 78, pp. 1275-1282, 1990.

Cerniakov M. and Donskoi L., "Optimisation of the Radar for Buried Object Detection," Radar 97, 14-16 October 1997, Publication No. 449 IEE 1997.

Chain H.T. and Tan H.S.A., "High Order of Renormalization Method for Radar Backscatter from a Random Medium," IEE Trans. on Geoscience and Remote Sensing, Vol. 27, pp. 79-85, 1989.

Chandrasekhar S., Radiative Transfer, Clarendon, Oxford, 1950; Dover, New York, 1960.

Chauhan, N. et al., "Polarization Utilization in the Microwave Inversion of Leaf Angle Distributions," IEEE Transactions on Geoscience and Remote Sensing, Vol. 27, pp. 395-402, 1989.

Chernov L.A., "Wave Propagation in a Random Medium," McGraw Hill, N.Y., 1960.

Chuhlantzev, A.A., "Effective Dielectric Permittivity of Vegetation," Radio Engineering and Electronics, Vol. 33, pp. 2310-2319, 1988.

Chuhlantzev, A.A., "Microwave Emission from the Ground Surface in the Presence of Vegetation Cover," Radio Engineering and Electronics, Vol. 24, pp. 256-264, 1979.

Chuhlantzev, A.A., "Remote Sensing of Vegetation in the Microwave Range," All-Union Conference on Application of Radio-Physical Methods in Rural Environment Explorations, Erevan, pp. 71-76, 1980.

Chuhlantzev, A.A., "Scattering and Absorption of Microwave Emission by Vegetation Elements," Radio Engineering and Electronics, Vol 31, pp. 1044-1095, 1986.

Chuhlantzev, A.A., "Vegetation Modeling by a Set of Scatterers," Radio Engineering and Electronics, Vol. 34, pp. 240-244, 1989.

Chuhlantzev, A.A. et al., "Attenuation of Microwave Emission in Vegetation," Radio Engineering and Electronics, Vol. 34, pp. 2269-2278, 1989.

Chuhlantzev, A.A. et al., "Radar Characteristics of Vegetation at Microwave Frequencies," Radio Engineering, Vol. 34, pp. 16-24, 1979.

Collin, R.E., "Antennas and Radiowave Propagation," Mc Graw Hill Book Company, 1985.

Daniels D.J., "Surface – Penetrating Radar," I.E.E., 1996

Demidov J.M. Kozlov A.I., Krasnitsky J.A. the Antenna system with improvement of a signal on polarization. Izvesty VUZ, Radio electronics, 1978, V. 21, No. 8, pp. 122-124.

Demidov Ju.M., Kozlov A.I., Ustinovich V.V., "About polarization selection of reflected signals," A radio engineering and electronics engineering, 1975, V. 20, No. 5, pp. 1099-1100.

Dmitriev V.V. et al., "Radiating ability of a snow cover," Izvestija VUZ, ser. Radiophysica, Moscow, Vol. 33, [1]9, pp. 1020-1026, 1990.

Dobson M.G., Ulaby F.T., "Active microwave soil moisture research," IEEE Trans. on Geosc. and remote sensing, Vol. GE-24, [1]1, pp. 23-36, 1986.

Doviak R. and Zrnich D., "Doppler Radars and Weather Observations," Academic Press, 1988.

Durden, S. et al., "Modeling and Observation of the Radar Polarization Signature of Forested Areas," IEEE Transactions on Geoscience and Remote Sensing, Vol. 27, pp. 290-301, 1989.

Durden, S. et al., "The Unpolarized Component in Polarimetric Radar Observation of Forested Areas," IEEE Transactions on Geoscience and Remote Sensing, Vol. 28, pp. 268-271, 1990.

Elachi, C., "Spaceborne Radar Remote Sensing: Applications and Techniques," IEEE Press, 1987.

Eshenko S.D., Lande V.Sh., "To a problem on a radar image of a marine surface", Radio engineering and electronics engineering, 1972, V. 17, No. 8.

Ferrazzolli, P. et al., "Comparison between the Microwave Emissivity and Backscatter Coefficients of Crops," IEEE Transactions on Geoscience and Remote Sensing, Vol. 27, pp. 772-777, 1989.

Fieschi R., Enciclopedia della Fisica Vol. 2, Chapter 10.1.5: Dielectric Properties, ISEDI, Milano, Italy, April 1976.

Finkelshtein M.I. et al., "Underground radiolocation," Moscow, Radio and Sviaz, pp. 216, 1994.

Finkelshtein M.I., Mendelson V.L. Undersurface radar. Moscow, Radio and communication, 1984.

Finkelshtein, M.I. et al., "Subsurface Radar Location System," Radio and Communication, p. 216, 1994.

Finkelshtein, M.I., Mendelson, V.L. and Kutiev, V.A., "Radiolocation of Layered Ground Covers," Sov. Radio, p. 174, 1977.

Frankenstein G., Garner R., "Equations for determining the brine volume of sea ice from -0,5° C.-22,9° C," Journal of Glaciology, Vol. 6, [1]48, pp. 943-947, 1965.

Fuks I, "Contribution to the Theory of Radio Wave Scattering on the Perturbed Sea Surface," Izv. vuzov, Radiofizika Vol. 9, pp. 876-887, 1966.

Fuks I., "Shadowing by a Statistically Rough Surface," Izv. vuzov, Radiofizika Vol 12 pp. 552-561, 1969.

Fung A.K., "Mechanisms of Polarized and Depolarized Scattering from a Rough Dielectric Surface," J. Franklin Institute, 285, pp. 125-133, 1968.

Fung A.K. and Eom H.J., "A Theory of Wave Scattering from an Inhomogeneous Layer with an Irregular Interface," IEEE Trans. Antennas Propag. Vol. AP-29, pp. 899-910, 1981.

Fung A.K. and Eom H.J., "Multiple Scattering and Depolarization by a Randomly Rough Kirchhoff Surface," IEEE Trans. Antennas Propag. Vol. AP-29, pp. 463-471, 1981.

Fung Adrian K., Microwave Scattering and Emission Models and their Applications, Artech House, Inc, 1994.

Furutsu, K., "On Statistical Theory of Electromagnetic Waves in a Fluctuating Medium," J. Res. NBS, Vol. 67, p. 303, 1963.

Gantmacher, F.R., "Matrix Theory," Nauka, 1988.

Gjessing D.T, "Recent Advances in Radio and Optics Propagation for Modern Communications, Navigation and Detection Systems," AGARD-LS-93, pp. 12-1, 12-24: Target Detection and Identification Methods based on Radio-and Optical Waves, April 1978.

Gjessing D.T., "Atmospheric Structure Deduced from Forward-Scatter Wave Propagation Experiments," Radio Science Vol. 4, No.12 pp.1195-1201, 1969.

Gjessing D.T., "On the Use of Forward Scatter techniques in the Study of Turbulent Stratified Layers in the Troposphere," Boundary-Layer Meteorology Vol. 4, pp. 377-396, 1973.

Gjessing D.T., H.Jeske and N.K.Hansen, "An Investigation of the Tropospheric Fine Scale Properties Using Radio, Radar and Direct Methods," Journal of Atmospheric and Terrestrial Physics Vol. 31, pp. 1157-1182, 1969.

Gusev K.G., Filatov A.D., Sopolev A.L. Polarization modulation. M.: Radio and Svjaz, 1974.

Hallikainen M. et al., "Dielectric properties of snow in the 3 to 37 GHz range," IEEE Trans. on Ant. and Prop., Vol. AP-34, ¹11, pp. 1329-1340, 1986.

Hallikainen M., "Dielectric properties of sea ice at microwave frequencies," Helsinki Univ. of Techn, Radio Lab. Reports, Report S-94, pp. 1-53, 1977.

Hanai T., "Dielectric theory on the interfacial polarization for two-phase mixtures," Bull Inst. Chem Res. Kyoto Univ., Vol. 39, pp. 341-368, 1961.

Hanna F.F. et al., "Permittivity in the microwave region of some natural minerals," PA Geoph., Vol. 120, pp. 772-777, 1982.

Hitney H.V., "Propagation Modelling and decision Aids for Communications, radar and navigation Systems," AGARD North Atlantic Treaty Organization, Chapter 4B, Sept. 1994.

Hoekstra P., Capplino P., "Dielectric properties of sea ice and sodium chloride ice at UHF and microwave frequencies," Journal of Geophys. Res., Vol. 76, 120, pp. 4922-4932, 1971.

Huynen J.R., "Knowledge Aspects in Radar Target Polarimetry," PIERS, Proc. of the Fourth International Workshop on Radar Polarimetry, Nantes, July 1998.

Huynen J.R., "Measurements of the Target Scattering Matrix," IEEE Proceedings pp.936-946, August 1965

Iljin V.A. et al., "Laboratory researches of complex dielectric permittivity of a frozen sand," Radiotechnica and Electronica, Moscow, Russia, Vol. 38, ¹6, pp. 1036-1041,1993.

Iljin V.A., Slobodchikova S.V., "Laboratory researches of the radiating characteristics of frozen sandy ground," Radiotechnica and electronica, Moscow, v.39, ¹5, pp. 800-806, 1994.

Iljin V.A., Sosnovsky J.M., "Laboratory researches of influence of a index of salinity on the dielectric properties of a sand," Radiotechnica and electronica, Moscow, Vol. 40, [1]1, pp. 48-54, 1995.

Ishimaru A., "Theory and Applications of Wave Propagation and Scattering in Random Media," Proc IEEE Vol. 65 No.7, 1977.

Ishimaru, A., "Propagation and Scattering of Waves in Random Inhomogeneous Media," Mir, Vols. 1 and 2, 1981.

Jackson J.D., "Classical Electrodynamics," John Wiley & Sons, Inc., 1975.

Kanarejkin D.B., Pavlov N.F., Potechin V.A. Radar signals polarization. M.: Radio and Svjaz, 1966.

Kanarejkin D.B., Potechin V.A., Shishkin M.F. Marine radiopolarimetry. L.: Sudostroenie, 1968.

Karam M. et al., "Electromagnetic Wave Scattering from Some Vegetation Samples," IEEE Transactions on Geoscience and Remote Sensing, Vol. 26, pp. 799-808, 1988.

Karam M. et al., "Leaf Shape Effects in Electromagnetic Wave Scattering from Vegetation," IEEE Transactions on Geoscience and Remote Sensing, Vol. 27, pp. 687-697, 1989.

Karam M.A. Fung A.K., Lang R.H., Chanhan N.S., "A Microwave Scattering Model for Layered Vegetation, IEEE Trans. on Geoscience and Remote Sensing, Vol. 30, pp. 767-784, 1992.

Karam M.A., Fung A.K., "Propagation and Scattering in Multi-layered Random Media with Rough Interfaces," Electromagnetics Vol. 2, pp. 239-256, 1982.

Karpuhin, V.I., Peshkov, A.N. and Finkelshtein, M.I., "Sensing of Agricultural Crops by Radioimpulse of Nanosecond Width," Radio Engineering and Electronics, Vol. 33, pp. 550-556, 1989.

Kireev V., "Cours de Chimie Physique," MIR, Moscow, 1968.

Kiselev A.Z. An optimum reception of a polarized signal for availability accidentally of polarized noise. A radio engineering and electronics engineering, 1969, V. 14, No. 2, pp. 219-229.

Klyatskin, V. I., Statistical Description of Dynamic Systems with Fluctuating Characteristics, Nauka, Moscow, 1975.

Kochin A.V., Kuznetsov V.L., "The Method of Measuring Distribution Parameters of Volume Scatterers," Claim No. 4862886/09, 1990.

Kochin A.V.Kuznetsov V.L., "On the Definition of the Reflection Structure of a Meteo Target within the Radar Pulse Volume" Conference: Remote Sensing Application in Natural Environments, Moscow, 1992.

Kondratiev, K.Y. and Fedtchenko, P.P., "Spectral Reflectance and Recognition of Vegetation," Gidrometeoizdat, p 216, Leningrad, 1986.

Kondratiev, K.Y. et al., "Airborne Explorations of Soils and Vegetation," Gidrometeoizdat, p. 231, Leningrad, 1984.

Kostinski A.B., Boerner W.M., "On Foundations of Radar Polarimetry," IEE Trans. Antenas Propag. Vol. AP-34, No.12, pp. 1395-1404; 1470-1473, Dec. 1986.

Kostrukov A.M., Gusev K.G Estimation of efficiency of suppression of fluctuation polarized interferences by a polarization selection method. Izvesty VUZ, Radio electronics, 1973, V. 16, No. 1, pp. 73-78.

Kozlov A.I. Property of statistical parameters of a scattering matrix elements of the radar targets. Izvesty VUZ, Radio electronics, 1979, V. 22, No. 1, pp. 14-18.

Kozlov A.I. Radar contrast of two objects. Izvesty VUZ, Radio electronics, 1979, V. 22, No. 7, pp. 63-67.

Kozlov A.I., Demidov J.M. Some properties of a covariance scattering matrix. Radio engineering and electronics engineering, 1976, V. 21, No. 11.

Kozlov A.I., Ligthart L.P., Logvin A.I. Deterministic and stochastic modeling of objects. The Report about cooperative works between The Moscow State Technical University of Civil Aviation and The Delft University of Technology. 1998.

Kozlov A.I., Ligthart L.P., Logvin A.I. Relation between the electrodynamic characteristics and the radar polarization state. The Report about cooperative works between The Moscow State Technical University of Civil Aviation and The Delft University of Technology. 1997.

Kozlov A.I., Logvin A.I Remote sensing of marine ice. St.Petersburg, Gidrometeoizdat, 1993, pp. 290.

Kozlov A.I., Rusinov V.R., Mosionzhic A.I. Statistical characteristics of polarization parameters of non-Gaussian periodically non-stationary radio signals. Radio engineering and electronics engineering, 1990, V. 35, No. 4, pp. 883-888.

Kozlov A.I., Sarichev V.A. etc. Polarization of signals in complicated transport radio electronic complexes. Saint Pietersburg, Chronograf, 1994.

Kozlov, A.I., Logvin, A.I., Lutin, E.A., "Methods and Instruments of Underlying Surface Radar Remote Sensing for National Needs," V. 24, VINITI, 1992.

Krasnov O.A. Polarization structure of an electromagnetic wave, scattered stables radar target. In book.: Scattering of electromagnetic waves. Taganrog, 1966, V. 6, pp. 68-71.

Kuusk, "Application of Remote Methods for Evaluation of Agricultural Crops States," Obninsk, p 52, 1982.

Kuznetsov V.L., "On to Mechanisms of Signal Scintillations by Wave Propagation in Precipitations," Preprints Conference "Climate Parameters in Radiowave Propagation Prediction," Moscow, 1994.

Kuznetsov V.L., "On Transformation of the Electromagnetic Angular Spectrum in a Transition Layer of a Regular Surface with Random Roughness," Conference on Remote Sensing in Natural Environments, Murom, 1992.

Kuznetsov V.L., "The Model of a Transition Layer in Problems of Interaction between Electromagnetic Radiation and Substantially Rough Surface," MSTUCA, 1993.

Kuznetsov, V.L., "Electromagnetic Wave Scattering from a Periodic Surface with Random Roughness," M.: MSTUCA, 1991

Kuznetsov, V. L., Budanov, V.G., "Polarization Characteristics of Electromagnetic Radiation Multiply Scattered in a Cloud of Small Particles," Trans. In. Radiophysics, Vol. 31, pp. 493-495, 1988.

Landau L.D., Lifschitz E.M. and Pitaevskii L.P., Electrodynamics of Continuous Media: Course of Theoretical Physics Vol. 8, Pergamon Press, edition 1998.

Landau L.D., Lifshitz E.M., Fluid Mechanics: Course of Theoretical Physics Vol. 6, Pergamon Press, edition 1979.

Landau, L. D., Lifshitz, E. M., Electrodynamics of Continuous Media, Nauka, Moscow, 1982.

Lane I., Saxton J., "Dielectric dispersion in pure polar liquids at very high radio frequencies, II. The Effect of Electrolytes in Solution," Proc. of the Royal Sjc., Vol. 2144, pp. 531-545, 1953.

Lang R.H. et al., "Electromagnetic Backscattering from a Layer of Vegetation: Discrete Approach," IEEE Transactions on Geoscience and Remote Sensing, Vol. GE-21, pp. 62-71, 1983.

Lang, R.H. et al., "Microwave Inversion of Leaf Area and Inclination Angle Distributions from Backscattered Data," IEEE Transactions on Geoscience and Remote Sensing, Vol. GE-23, pp. 685-693, 1985.

Lasinski, M. et al., "Estimation of Subpixel Vegetation Cover Using Red-Infrared Scatterograms," IEEE Transactions on Geoscience and Remote Sensing, Vol. 28, pp. 258-263, 1990.

Lasinski, M. et al., "The Structure of Red-Infrared Scatterograms of Semi-vegetated Landscapes," IEEE Transactions on Geoscience and Remote Sensing, Vol. 27, pp. 441-451, 1989.

Lavin A.I. Nonlinear filtration of a polarized impulse signal. Izvesty VUZ, Radio electronics, 1985, V. 28, No. 3, pp. 72-74.

Lavin A.I. Nonlinear filtration of polarized radar signals. Radio engineering, 1983, No. 12, pp. 32-34.

Lax, M., "Multiple Scattering of Waves," Rev.Mod.Phys. 23(4., pp. 287-310, 1951

Levanon N., "Radar Principles," John Wiley & Sons, 1988.

Lestari A.A., Yarovoy A.G., Ligthart L.P., "Capacitatively-Tapered Bowtie Antenna," Proc. on CD-ROM of the Millennium Conference on Antennas & Propagation, Davos, Switzerland, April 2000.

Le Vine, D.M. et al., "Scattering from Arbitrary Oriented Dielectric Discs in the Physical Optics Regime," J. Opt. Soc. Am., Vol. 73, pp. 1255-1262, 1983.

Levy M.F. and Craig K.G., "Case Studies of Transhorizon Propagation: Reliability of Predictions Using Radiosonde Data," IEE, URSI Sixth International Conference on

Antennas and Propagation ICAP 1989, Part 2: Propagation, Conference Publication No.301.

Li Bao-Wen, “Sound Propagation in Turbulent Media,” Ph. D. dissertation, Oldenburg, 1992.

Liang S.H. and Strachler A.H., “Calculation of the Angular Radiance Distribution for a Coupled Atmosphere and Canopy,” IEEE Trans. on Geoscience and Remote Sensing, Vol. 31, pp. 491-501, 1993.

Lifshitz E.M. and L.P. Pitaevskii, Physical Kinetics, Vol. 10 of Course of Theoretical Physics by Landau L.D. & Lifschitz E.M., Butterworth–Heinenann, 1997, ISBN: 0 7506 2635 6; Fizicheskaya kinetika, Nauka, Moscow 1979.

Ligthart L.P., Kozlov A.I. and Logvin A.I., “Theoretical Modelling of Microwave Scattering,” PIERS, Proc. of the Fourth International Workshop on Radar Polarimetry, Nantes, July 1998.

Ligthart L.P., Tatarinov V and Tatarinov S., “Polarization Properties of Distributed Radar Targets,” PIERS, Proc. of the Fourth International Workshop on Radar Polarimetry, Nantes, July 1998.

Logvin A.I. Nonlinear filtration of radar signals with random polarization parameters of an electromagnetic wave. Radio engineering, 1985, No. 6, pp. 56

Logvin A.I., Kozlov A.I. and Ligthart L.P., “Polarimetric Method for Measuring and Visualizing Permittivity Characteristics of the Earth Surface,” PIERS, Proc. of the Fourth International Workshop on Radar Polarimetry, Nantes, July 1998.

Lysanov Yu P, “Mean Coefficient of Reflection from an Uneven Surface Bounding an Inhomogeneous Medium,” Soviet Physics-Acoustics Vol. 15, pp. 340-344, 1970.

Malmberg C., Maryott A., “Dielectric constant of water from 0°C to 100°C,” J. Res. Nat. Bur. Stand, pp. 1-8, 1956.

Marin L., “Natural Mode Representation of Transient Scattered Field,” IEEE Trans. Antennas Propag. Vol. AP-24, pp. 809-810, 1973.; Vol. AP-22, pp. 266-274, 1974.

Massey B.S., Mechanics of Fluid, D. Van Nostrand Co., London 1968.

Matzler C., Wegmuller U., “Dielectric properties of fresh-water ice at microwave frequencies,” J. Appl. Phys., Vol. 20, pp. 1623-1630, 1987.

Maximov M.V. Editor, "Guard from radio interference- M: Radio and Svjaz," 1976.

Meletitsky V.A., Mosionzhic A.I. Probability model of periodically non-Gaussian non-stationary radio signals. Radio engineering and electronics engineering, 1987, V. 32, No. 4, pp. 747-754.

Melnichuk J.V., Chernikov A.A. About a matrix of return scattering of centimetric radiowaves of the excited surface of the sea. Proc. Central Aerological Observatory, 1975, v. 121, pp. 58-70.

Mikhaylovskiy A.I. and Fuks I.M., "Statistical Characteristics of the Number of Specular Points on a Random Surface for Small Grazing Angles," Journal of Communications Technology and Electronics, 38 (7), 1993.

Miller A.R., Brown R.M. and Vegh E., "New Derivation for Rough-Surface Reflection Coefficient and for the Distribution of Sea-Wave Elevations," IEE Proc. Vol. 131, No.2, pp. 114-116, 1984.

Mo, T. et al., "Calculation of the Microwave Brightness Temperature of Rough Soil Surface: Bare Field," IEEE Transactions on Geoscience and Remote Sensing, Vol. GE-25, pp. 47-55, 1987.

Mosetti, F., "Fondamenti di Oceanologia e Idrologia," Vol. 6, UTET, Torino, 1979.

Nathanson F.E., "Radar Design Principles," Scitech Publishing, Inc., N.J., 1999.

Newton R., Scattering Theory of Waves and Particles, Moscow, Mir, 1969.

Niemeijer R.J., "Doppler-Polarimetric Radar Signal Processing," Ph.D. Thesis, Delft University of Technology, The Netherlands, May 1996.

Nikitin S.A., Menshikov V.A., Vesnin A.V. etc. Study of glaciers of Altai by methods of an impulse radar-location and microwave-radiopolarimetry. Proc. Artic Antartic Scientific Research Institute, 1985, No. 395, pp. 68-80.

Novikov, E. A., "Functionals and Method of Random Forces in the Theory of Turbulence," JETP, Vol. 47, p. 1919, 1964.

Obukhov A.M., "Amplitude and Phase Fluctuations in an Extended Random Medium," Izv. Akad. Nauk USSR, No.2, p. 155, 1953.

Ogilvy J.A., Theory of Wave Scattering from Random Rough Surfaces, IOP Publishing Ltd, 1991.

Oguchi T., "Pruppacher-and-Pitter Form Raindrops and Cross-Polarization Due to Rain: Calculations at 11, 13, 19.3 and 34.8 GHz., "Radio Science Vol. 12 No. 1, pp., 45-51, 1977.

Olhoeft G.R., Strangway D.W., "Earth and Planetary," Sc.Left., Vol. 24, pp. 394, 1975.

Ostrovitjanov R.V., Basalov F.A. Statistical theory of the radar extended targets. M.: Radio and Svjaz, 1982.

Papoulis A. "Probability, Random Variables and Stochastic Processes," Mc Graw – Hill Book Company, 1965

Peake,W.N., "Reflection of Radiowaves from Rough Surfaces," Transactions IRE, Vol. AP-7, p. 324, 1959.

Pierri R., Brancaccio A., De Blasio F., "Multifrequency dielectric profile inversion for a cylindrical stratified medium," IEEE Trans. Geoscience and Remote Sensing, Vol. 38, n.4, July 2000.

Pierri R., De Blasio F., Brancaccio A. "Multifrequency apprroach to inverse scattering: the linear and quadratic models," IGARSS'99, Proc. Vol. V, page 2522-2524, Hamburg, BRD.

Pitts D. et al., "The Use of a Helicopter-Mounted Ranging Scatterometer for Estimation of Extinction and Scattering Properties of Forest Canopies," IEEE Transactions on Geoscience and Remote Sensing, Vol. 26, pp. 144-151, 1988.

Podkovkov N.F., "The model of complex complex dielectric permittivity of ground in a microwaves range," Voprosi radioelectronici, Moscow, pp. 73-80, 1990.

Poelman A.J., "Virtual Polarization Adaptation. A method of Increasing the Detection Capability of a Radar System Through Polarization Vector Processing," IEE Proc. Vol. 128, Pt.F, No.5, October 1981.

Poliansky V.A., Kanarejkin D.B., About connection between a statistical scattering matrix and Muller matrix. Radio engineering and electronics engineering, 1974. V. 19, No. 11, pp. 2407-2410.

Potapov, A.A., "Radio-Physical Effects Connected with the Interaction of Millimetre-Wave Band Electromagnetic Emission and an Ambient Medium," Foreign Radio Electronics, N 11, pp. 23-48,1992.

Potechin V.A., Tatarinov V.N. "Theory of a coherence of an electromagnetic field" M.: Radio and Svjaz, 1978.

Pozdniak S.I.. "Distribution of factor of polarization selection of a signal on a background of interferences". Radio engineering and electronics engineering, 1989, V. 34, No. 4, pp. 880-882.

Pozdniak S.I., Meletitsky V.A., "Introduction in the statistical theory of polarization of radiowaves". M.: Radio and Svjaz, 1974.

Pozdniak S.I., Mitz J.K., "Coherence matrix and parameters of Stokes is partial polarized waves in three-dimensional space". Radio engineering, 1987, No. 4, pp. 80-82.

Pozdnjak S.I., Radzievsky V.G., Trifonov A.L., "Analysis of an optimum reception of a polarized signal". Radio engineering, 1972, V. 27, pp. 6-10.

Pusone E., "A Channel Model for Prediction of Delay and Doppler Power Spectra and Frequency Correlation Function for Troposcatter Communications Links," Proc. IEEE No. 80CH 1521-4-COM, Zurich Seminar, March 1980.

Pusone E., "A Predictive Model Based on Physical Considerations for Troposcatter Communications Links" SHAPE Technical Centre Technical Memorandum TM-589, Den Haag, November 1978

Pusone E. and Lloyd L., "Synthetic–Aperture Sonar: Performance Analysis of Beamforming and System Design," NATO Saclantcen ASW Research Centre, SR-91, November 1985.

Pusone E., "A Troposcatter Prediction Model of Long-Term Statistics of Multi-Path Dispersion and Doppler Spread Based on Atmospheric Parameters," IEE Colloquium on Troposcatter, 28 October 1981, Savoy Place London.

Pusone E., J.S. van Sinttruijen and P. van Genderen," A Mathematical Model for Analyzing Pulse Shape Distortion Due to Propagation Over a Sea Surface," Proc. of the 5th International Conference on Radar Systems, May 17-21, 1999, Brest, France.

Pusone E., P. van Genderen, "Effects of propagation over sea at low grazing angles on the shape of backscattered wide X-band signals," NATO-RTO-SET Sensor & Electronic Technology Panel Symposium on 'Low Grazing Angle Clutter: its Characterization, Measurement and Application', Paper no. 19 of the Proceedings, Laurel Maryland USA, April 2000.

Pusone E., P. van Genderen, J.S. van Sinttruijen, "Multipath Effects on Polarimetric Radar Response form Sea Backscatter for Low Grazing Angles, at X-Band Frequencies," International Conference on Electromagnetics in Advanced Applications (ICEAA 01), Torino, Italy, 12 Sept. 2001.

Pusone E., P. van Genderen, J.S. van Sinttruijen, "A Predictive Model of Sea Backscatter Doppler Spectrum at X-band Frequencies, at Low Grazing Angles," International Conference on Electromagnetics in Advanced Applications (ICEAA 01), Torino, Italy, 12 Sept. 2001.

Radio Engineering: Results of Science and Technology, VINITI, p. 175, 1977.

Ratchkulik, V.I. and Sitnikova, M.V., "Reflective Properties of Vegetation," Gidrometeoizdat, p. 287, 1981.

Redkin, B.A. et al., "Calculation of the Dielectric Capacitivity Tensor of Vegetation Media," Radio Engineering and Electronics, Vol. 22, pp. 1596-1599, 1977.

Redkin, B.A. et al., "Theoretical and Experimental Research on Reflections at Small Grazing Angles," Radiophysics, Vol. 16, pp. 1172-1175, 1973.

Richardson, I., "L-band Radar Backscatter Modeling of Forest Stands," IEEE Transactions on Geoscience and Remote Sensing, Vo. GE-25, pp. 487-498, 1987.

Richter J.H., Editor. "Radio Waves Propagation Modelling, Prediction and Assessment" AGARD AG-326 Chapter 2, pp. 21-22, Dec. 90.

Rino, C. L., "A Spectral Domain Method for Multiply Scattering in Continuous Randomly Irregular Media". IEEE Trans. on Antennas and Propagation, Vol. 36, pp. 114-128, 1988.

Rodimov A.L., Popovsky V.V., "Statistical theory is polarizable-temporary processings signals and interferences," M.: Radio and Svjaz, 1984.

Ross U.K. "Radiation to and from Vegetation," Gidrometeoizdat, p. 270, Leningrad, 1975.

Ruck, G.T. et al., Radar Cross Section Handbook, New York, John Wiley & Sons, 1970.

Rytov, S.M., Kravtsov, U.A., Tatarskii, V.I., "Introduction to Statistical Radiophysics: Random Fields," Nauka, Moscow, part 2, 1978.

Sarabandi K., Ulaby F.T., Tassoudij M.A., "Calibration of Polarimetric Radar Systems with Good Polarization Isolation," IEEE Transactions on Geoscience and Remote Sensing, Vol. 28, No. 1, Jan. 1990.

Schiffer, R., Thicheim, K.O., "Light Scattering by Dielectric Needles and Disks," J. App.l. Phys., Vol. 50, pp. 2476-2483, 1979.

Schmulevitch S.A., Troitsky V.S., "About dependence of dielectric properties of mountain rocks on their volumetric weight," DAN USSR, Vol. 201, ¹3, pp. 593-594, 1971.

Schwartz M, Bennet W.R. and Stein S., Communication Systems and Techniques, Mc Graw Hill Book Company, 1966.; Part III, Para 2, "Phenomenological Description of Multi-path and Fading," pp. 347-374.

Sherman P. ed., "Electrical Properties of Emulsions," N.-Y., Academic Press, 1968.

Shishov V.I., "Limiting Form of Pulse Shape After Multiple Scatter" Astron. Zh, Vol. 50, p. 941, 1973.

Shishov V.I., Izv.Vysh., Ucheb. Zaved. Radiofizika Vol. 2, No.6, p. 866, 1968.

Shuji F. et al., "Measurement of the dielectric properties of acid-doped ice at 9,7 GHz," IEEE Geosc. and Remote Sensing, Vol. GE-30, ¹4, pp. 799-803, 1992.

Shutko, A.M., Microwave Radiometry of Open Water and Lands, Moscow, Nauka, p 189, 1986.

Shwering, F. et al., "A Transport Theory of Millimeter Wave Propagation in Woods and Forests," Journal of Wave Material Interaction, Vol. 1, pp. 205-235, 1986.

Shwering, F. et al., "Millimeter Wave Propagation in Vegetation: Experiment and Theory," IEEE Transactions on Geoscience and Remote Sensing, Vol. 26, pp. 355-367, 1988.

Sihlova A.H. et al., "Permittivity of dielectric mixtures," IEEE Trans. On Geoscience and Remote Sensing, Vol. 26, ¹4, pp. 420-429, 1988.

Silver S., "Microwave Antenna Theory and Design," IEE Electromagnetic Series 19, Edition 1984, Peter Peregrinus Ltd, London, UK

Skolnik, M.I., Radar Handbook, Mc Graw Hill, 1970., Chapter 27.4, "Polarization Scattering Matrix."

Slutsky A.G., Iakushkin I.G., "On the Combined Description of Surface and Volume Random Roughness Effects on Electromagnetic Wave Reflection from the Interface of two Media," Radiophysics Vol. 32, pp.183-192, 1989.

Sommerfeld A., Thermodynamics of Statistical Mechanics, Academic Press, NY, 1956.

Stepanenko B.D., Shukin G.G., Bobilev L.P., Matrosov S.J. Radiometry in a meteorology. L.: Gidrometeoizdat, 1987.

Stogryn A. "Equations for calculating dielectric constant of saline ice," IEEE Transactions on Microwave Theory and Techniques, Vol. MTTT-19, [1]8, pp. 733-736, 1971.

Stratton J.A., Electromagnetic Theory, McGraw Hill, 1941.

Tatarinov V., Ligthart L. and Tatarinov S., "Polarization Properties of Complex Radar Objects Having Random Distribution of the Scattering Centers," PIERS, Proc. of the Fourth International Workshop on Radar Polarimetry, Nantes, July 1998.

Tatarinov V.N., Lukjanov S.P, Masalov E.V., "Rejector Comb Filtration for Polarization – Modulated Radar Signals," Izvesty VUZ, Radio electronics, 1989, V. 32, No. 5, pp. 3-7.

Tatarski V.I., "Wave Propagation in a Turbulent Media," McGraw Hill, 1971.

Taylor L., "Dielectric properties of mixtures," IEEE Trans. on Antennas and Propagation, Vol. AP-13, [1]6, pp. 943-947, 1965.

Teocharov, A.N., "Wave Scattering from a Surface with High Roughness," 9-th All-Union Symposium on Diffraction and Wave propagation, Tbilisi, pp. 195-198, 1985.

Thurai M. and Goddard. I.W.F., "Precipitation Scatter Measurement from a Transhorizon Experiment at 11.2 GH.," IEE Proc. – H – Vol. 139, No. 1, pp. 53-58, Feb. 1992.

Tinga H.R., Voss W.A.G., "Generalized approach to multiphase dielectric mixture theory," J. App.l. Phys. Vol. 44, [1]9, pp. 3897-3902, 1973,

Tiuri M. et al. "The complex dielectric constant of snow at microwave frequencies," IEEE Journal of oceanic Eng., Vol. OE-9, [1]5, pp. 377-382, 1984.

Tiuri M. et al., "The use radiowave probe and subsurface interface radar in peat resource inventory," Proc. of the Symp. of IPS Commission I, Aberden, Scotland, 1983.

Tiuri M., Toikka M., "Radiowave probe in situ water content measurement of peat," SUO, Vol. 33, [1]3, pp. 65-70, 1982.

Toan, T. et al., "Multitemporal and Dual-Polarization Observations of Agricultural Vegetation Covers by X-band SAR Images," IEEE Transactions on Geoscience and Remote Sensing, Vol. 27, pp. 709-718, 1989.

Tsang L., Kong J.A., Ding K.H., "Scattering of Electromagnetic Waves; Theory and Applications," John Wiley 2000

Tuchkov L.T., Editor., "Radar Characteristics of Aircrafts," Radio and Communications, 1985.

Twersky V., "On the Scattering and Reflection of Electromagnetic Waves by Rough Surfaces," IRE Trans. Antennas Propag. Vol. AP-5, pp. 81-90, 1957.

Ulaby F.T. and Dobson M.C., Handbook of Radar Scattering Statistics for Terrain, Artech House, Inc., 1989.

Ulaby F.T., Elachi C., Editors., Radar Polarimetry for Geoscience Applications, Artech House, Inc. MA, USA., ISBN: 0-89006-406-7., 1990.

Ulaby F.T. et al. "Relating Polarization Phase Difference of SAR Signals to Scene Properties," IEEE Transactions on Geoscience and Remote Sensing Vol. GE-25, pp. 83-92, No.1, January 1987.

Ulaby F.T., Moore R.K., Fung A.K. Microwave Remote Sensing, Vol. 2: "Radar remote Sensing and Surface Scattering and Emission Theory," Addison Wesley, USA, 1982.

Ulaby, F.T. et al., "Effect of Vegetation on the Microwave Radiometric Sensitivity to Soil Moisture," IEEE Transactions on Geoscience and Remote Sensing, Vol. GE-21, pp. 51-61, 1983.

Ulaby, F.T. et al., "Michigan Microwave Canopy Scattering Model," International Journal of Remote Sensing, Vol. 11, pp. 1223-1253, 1990.

Ulaby, F.T. et al., "Microwave Backscatter Dependence on Surface Roughness Soil Moisture and Soil Texture," Part I – Bare Soil, IEEE Transactions on Geoscience and Remote Sensing, Vol GE-16, pp. 286-295, 1978.

Ulaby, F.T. et al., "Microwave Dielectric Properties of Plant Materials," IEEE Transactions on Geoscience and Remote Sensing, Vol. GE-22, pp. 406-414, 1984.

Ulaby, F.T. et al., "Microwave Dielectric Spectrum of Vegetation. Part 1. Experimental Observations, Part 2. Dual Dispersion Model," IEEE Transactions on Geoscience and Remote Sensing, Vol. GE-25, pp. 541-557, 1987.

Ulaby, F.T. et al., "Microwave Propagation Constant for a Vegetation Canopy with Vertical Stalks," IEEE Transactions on Geoscience and Remote Sensing, Vol. GE-25, pp. 714-725, 1987.

Ulaby, F.T. et al., "Millimeter-Wave Bistatic Scattering from Ground and Vegetation Targets," IEEE Transactions on Geoscience and Remote Sensing, Vol. 26, pp. 229-243, 1988.

Unal C.M.H., Moisseev D.N. and L.P. Ligthart, "Doppler-Polarimetric Radar Measurements of Precipitation," PIERS, Proc. of the Fourth International Workshop on Radar Polarimetry, Nantes, July, 1998.

Uscinski B.J., "The Elements of Wave Propagation in Random Media," McGraw Hill, 1977.

Valenzuela G.R., "Depolarization of EM waves by Slightly Rough Surface," IEEE Trans. Ant. and Prop., Vol. 15, pp. 552-557, 1967.

Vant M.R. et al., "The complex dielectric constant of sea ice at frequencies in the range 0,1-40GHz," J. Appl. Phys., Vol. 49, ¹3á, pp. 1264-1280, 1978.

Varganov M.E, Zinovjev V.S, Astanin L.J. etc. "Radar characteristics of flight vehicles" (Under edit) Tuchkov L.T. M.: Radio and Svjaz, 1985.

Vasiliev V.I. et al, "The mathematical model of freezing of a salinity frozen ground, Pricladnaja Mechanica and Theoretichescaja Fisika," Vol. 36, ¹5, pp. 57-66, 1995.

Voronovich A.G., "Phase Operator in Problems of Wave Scattering from Rough Surfaces," 9-th All-Union Symposium on Diffraction and Wave propagation, Tbilisi, pp. 168-171, 1985.

Voronovich A.G., "One Approximate Method for Calculating Sound Scattering by a Rough Free Surface," Doklady Akademii Nauk SSSR, Vol. 272, No. 6, pp. 1351-1355, 1983.

Voronovich A.G., "On the Theory of Electromagnetic Wave Scattering form the Sea Surface at Low Grazing Angels," Radio Science, Vol. 31, No. 6, pp. 1519-1530, Nov.-Dec. 1996.

Voronovich A.G. and Zavorotny, "The Effects of Steep Sea-waves on Polarization Ratio at Low Grazing Angels," IEEE Trans. on Geoscience and Remote Sensing, Vol. 38, No. 1, January 2000.

Walsh I. And Srivastava S.K., "Rough Surface Propagation and Scatter: General Formulation and Solution for Periodic Surfaces" Radio Science Vol. 22, pp. 193-208, 1987.

Wang J.R., "The dielectric properties of soil water mixtures at microwave frequencies," Radio Sc., Vol. 15, [1]5, pp. 997-985, 1980,.

Watts S., "Optimum Radar Polarization for Target Detection in Sea Clutter," IEE Colloquium in Polarization in Radar, pp. 4/1-4/5, 22 March 1996, Savoy Place, London.

Wentworth F.L., Cohn M., "Electrical properties of ice at 0,1 to 30 MHz," Radio Sc. J. of Res. of NBS IUSNC-URSI, Vol. 68D, [1]6, pp. 681-691, 1964.

Williams W.D., "Conductivity and salinity of Australian salt lakes," Aus. J. Mar. Freshw. Res., Vol. 37, pp. 177-182, 1986.

Windle J.J., Shaw T.M., J. Chem. Phys., Vol. 22, p. 1752, 1954.

Wobxhall, "A theory of the complex complex dielectric permittivity of soil containing water. The semi disperse model," IEEE Trans. on Geoscience Electronics, Vol. GE-15, [1]1, pp. 49-58, 1977.

Yakovlev V.P., "Radar remote researches of a wood," Zarubeznaj radioelectronica, [1]9/10, pp. 23-35. 1994.

Yakovlev, V.P., "Applications of Radar Remote Sensing to Agriculture". Foreign Radio Electronics, N 718, pp. 53-62, 1994.

Yarovoy A.G., R.V. de Jongh and L.P. Ligthart, "IRCTR activities in Modeling of Electromagnetic Wave Transmission Through an Air-Ground Interface," PIERS, Proc. of the Fourth International Workshop in Radar Polarimetry, Nantes, July 1998.

Zebker H.A., J.J. van Zyl, "Imaging Radar Polarimetry: A Review," Proc. IEEE, Vol. 79, No. 11, pp. 1583-1606, 1991.

Zhuk N.P., Tret'yakov O.A., Yarovoy A.G., "Statistical perturbation theory for an electromagnetic field in a medium with a rough boundary," Sov.Phys.JETP Vol 71, 5. Nov. 1990, published in English by American Institute of Physics, 1991.

Zhukovsky A.P., Onoprienko E.I., Chizhov V.I., "Theoretical principles of altitude-finding radar". Moscow. Sov. Radio. 1979, pp.320.

Zujkov V.A., Kulekin G.P., Lutcenko V.I., "Feature of scattering of a microwave–radiation by the sea for small yaw angles". Izvesty VUZ, Radio electronics, 1981, V. 24, No. 7, pp. 831-839.